SCIENCE AND CIVILISATION
IN CHINA

The subject of discourse, briefly put, is the free travel and inward and outward movement of the divine *ch'i*; it is not skin, flesh, sinews and bones.

HUANG TI NEI CHING, LING SHU
First century BC

Chinese medicine heals in a world of unceasing transformation. This condition of constant change, this fluidity of material forms, stands in sharp
contrast to a (modern Western) commonsense world of discrete entities
characterized by fixed essences, which seem to be exhaustively describable
in structural terms . . . In the early Chinese sciences, by contrast, where
generation and transformation are intrinsic to existence, fixity and stasis
occur only as a result of concerted action and therefore demand explanation;
motion and change are a given and seldom need be explained with reference
to their causes. One consequence of this dynamic bias in Chinese medicine
is that the body and its organs (i.e., anatomical structure) appear as merely
contingent effects or by-products of physiological processes.

JUDITH FARQUHAR
1994

THE PICTURE OF THE TAOIST GENII PRINTED ON THE COVER

of this book is part of a painted temple scroll, recent but traditional, given to Mr Brian Harland in Szechuan province (1946). Concerning these four divinities, of respectable rank in the Taoist bureaucracy, the following particulars have been handed down. The title of the first of the four signifies 'Heavenly Prince', that of the other three 'Mysterious Commander'.

At the top, on the left, is Liu *Thien Chün*, Comptroller-General of Crops and Weather. Before his deification (so it was said) he was a rain-making magician and weather forecaster named Liu Chün, born in the Chin dynasty about +340. Among his attributes may be seen the sun and moon, and a measuring-rod or carpenter's square. The two great luminaries imply the making of the calendar, so important for a primarily agricultural society, the efforts, ever renewed, to reconcile celestial periodicities. The carpenter's square is no ordinary tool, but the gnomon for measuring the lengths of the sun's solstitial shadows. The Comptroller-General also carries a bell because in ancient and mediaeval times there was thought to be a close connection between calendrical calculations and the arithmetical acoustics of bells and pitch-pipes.

At the top, on the right, is Wên *Yuan Shuai*, Intendant of the Spiritual Officials of the Sacred Mountain, Thai Shan. He was taken to be an incarnation of one of the Hour-Presidents (*Chia Shen*), i.e., tutelary deities of the twelve cyclical characters (see Vol. 4, pt 2, p. 440). During his earthly pilgrimage his name was Huan Tzu-Yü and he was a scholar and astronomer in the Later Han (b. +142). He is seen holding an armillary ring.

Below, on the left, is Kou *Yuan Shuai*, Assistant Secretary of State in the Ministry of Thunder. He is therefore a late emanation of a very ancient god, Lei Kung. Before he became deified he was Hsin Hsing, a poor woodcutter, but no doubt an incarnation of the spirit of the constellation Kou-Chhen (the Angular Arranger), part of the group of stars which we know as Ursa Minor. He is equipped with hammer and chisel.

Below, on the right, is Pi *Yuan Shuai*, Commander of the Lightning, with his flashing sword, a deity with distinct alchemical and cosmological interests. According to tradition, in his early life he was a countryman whose name was Thien Hua. Together with the colleague on his right, he controlled the Spirits of the Five Directions.

Such is the legendary folklore of common men canonised by popular acclamation. An interesting scroll, of no great artistic merit, destined to decorate a temple wall, to be looked upon by humble people, it symbolises something which this book has to say. Chinese art and literature have been so profuse, Chinese mythological imagery so fertile, that the West has often missed other aspects, perhaps more important, of Chinese civilisation. Here the graduated scale of Liu Chün, at first sight unexpected in this setting, reminds us of the ever-present theme of quantitative measurement in Chinese culture; there were rain-gauges already in the Sung (+12th century) and sliding calipers in the Han (+1st). The armillary ring of Huan Tzu-Yü bears witness that Naburiannu and Hipparchus, al-Naqqash and Tycho, had worthy counterparts in China. The tools of Hsin Hsing symbolise that great empirical tradition which informed the work of Chinese artisans and technicians all through the ages.

SCIENCE AND CIVILISATION IN CHINA

Joseph Needham
(1900–1995)

'Certain it is that no people or group of peoples has had a monopoly in contributing to the development of Science. Their achievements should be mutually recognised and freely celebrated with the joined hands of universal brotherhood.'

Science and Civilisation in China VOLUME I, PREFACE.

*

Joseph Needham directly supervised the publication of seventeen books in the *Science and Civilisation in China* series, from the first volume, which appeared in 1954, through to Volume 6.3, which was in press at the time of his death in March 1995.

The planning and preparation of further volumes will continue. Responsibility for the commissioning and approval of work for publication in the series is now taken by the Publications Board of the Needham Research Institute in Cambridge, under the chairmanship of Dr Christopher Cullen, who acts as general editor of the series.

中國科學技術史

李約瑟 著

萬朝鼎

SCIENCE AND CIVILISATION IN CHINA

VOLUME 6

BIOLOGY AND BIOLOGICAL TECHNOLOGY

PART VI: MEDICINE

BY

JOSEPH NEEDHAM

with the collaboration of

LU GWEI-DJEN

edited and with an introduction by

NATHAN SIVIN

PROFESSOR OF CHINESE CULTURE AND OF THE HISTORY OF SCIENCE
UNIVERSITY OF PENNSYLVANIA

CAMBRIDGE
UNIVERSITY PRESS

PUBLISHED BY THE PRESS SYNDICATE OF THE UNIVERSITY OF CAMBRIDGE
The Pitt Building, Trumpington Street, Cambridge CB2 1RP, United Kingdom

CAMBRIDGE UNIVERSITY PRESS
The Edinburgh Building, Cambridge CB2 2RU, UK http://www.cup.cam.ac.uk
40 West 20th Street, New York, NY 10011–4211, USA http://www.cup.org
10 Stamford Road, Oakleigh, Melbourne 3166, Australia

First published 2000

Printed in the United Kingdom at the University Press, Cambridge

Typeset in Monotype Imprint 10/13pt, in QuarkXpress™   [GC]

A catalogue record for this book is available from the British Library

ISBN 0 521 63262 5 hardback

To the memory of

LU GWEI-DJEN
Fellow of Robinson College, Cambridge

DOROTHY NEEDHAM
Founding Fellow of Lucy Cavendish College, Cambridge

JOSEPH NEEDHAM
Sometime Master of Gonville and Caius College, Cambridge

and splendid days in many parts of the world shared in a
quest for understanding

CONTENTS

Editor's introduction, *p.*1

    The contents of this volume, *p.* 3

    Recurrent themes, *p.* 6

    Problematic foundations, *p.* 9

    Medical history and Chinese studies, *p.* 16

    Research in Asia, *p.* 18

    Current and future research issues, *p.* 21

    Terra incognita: the scientific value of therapies, *p.* 34

    Editing conventions, *p.* 36

    Acknowledgements, *p.* 37

44 MEDICINE . . . . *page* 38

(*a*) Medicine in Chinese culture, *p.* 38

    (1) The general position of medicine and medical doctors in traditional-Chinese society, *p.* 38

    (2) The principal doctrines of Chinese medicine, *p.* 42

    (3) The fathers and their history, *p.* 45

    (4) Influences of bureaucratism on Chinese medicine, *p.* 52

    (5) Influences of the Chinese religious systems on medicine, *p.* 57

    (6) Acupuncture, *p.* 60

    (7) The contrast between traditional-Chinese and modern-Western medicine, *p.* 65

    (8) The possible integration of traditional-Chinese and modern-Western medicine, *p.* 65

ILLUSTRATIONS

TABLES

SERIES EDITOR'S PREFACE

When Joseph Needham laid out the plan for the Science and Civilisation in China project half a century ago, he envisaged it as consisting of seven volumes. For the sake of readers familiar with the multi-part volumes of SCC as it is today, I must explain that each volume was originally supposed to be a single book. Contained within the seven-volume structure was a finer division of subjects into fifty sections. Sections 43 to 45 broadly covered the topics of medicine in China.

The first three volumes showed the steady tendency to growth, which was the result of Needham's immense energies of assimilation and synthesis. Volume 3 was of a size which in itself would have constituted a life's work for many less ambitious scholars. At this point considerations of what we may very appropriately call binding energy dictated that the nucleons of knowledge should be reformed into smaller and more stable units. Every volume from number 4 onwards has been split into physically separate parts, which in some cases amount to more than a dozen substantial books.

The problem was, as Needham himself was the first to point out, that an ever-lengthening SCC would eventually outstretch the span of normal human energies or even life itself. Needham's response to this was (as one might have expected) heroic. Although responsibility for some parts of the plan was passed to collaborators, he made no compromise with the length or depth of treatment that seemed appropriate for those topics on which he worked himself. His routine of daily labour continued through his eighties into his nineties, and it was only with difficulty that he was persuaded to stay at home for a rest the day before he died on 24 March 1995.

One of the principal tasks that occupied him during his final years was the mass of research and writing on medicine by himself and Lu Gwei-djen, his lifetime collaborator (1904–1991). Both of them had worked on this topic from the very beginnings of their cooperation on the SCC project. Some of their results had appeared in draft versions as journal articles or contributions to conference proceedings, and one portion had been published as a major book on acupuncture and moxibustion – *Celestial Lancets* (1980). But much remained to be done, since a large part of what had been written was in need of updating, and some of it required substantial expansion and rewriting.

At the age of ninety Joseph Needham retired as Director of the Needham Research Institute, and was succeeded by Professor Ho Peng-Yoke, who had been one of his long-term collaborators. For Professor Ho, this meant lengthy periods of exile from his home and family in Australia. In the following year I was asked by Joseph Needham to join the team working under Professor Ho by acting as Chairman of the Needham Research Institute Publications Board. One of the first issues to be dealt with was how the Board might help to ensure that Needham's work on medicine was speedily brought to press. As Joseph Needham himself was well aware, this was not likely to be a simple task.

It was already clear that the labours of updating and editing were becoming an insupportable burden for the author alone. But it was not easy to find a suitable helper. Much more than a research assistant or amanuensis was required – the job was one which could be done only by a senior scholar who could combine a sympathetic understanding of Needham and Lu's *oeuvre* with a deep and wide knowledge of the field. Moreover, such a scholar would have to be prepared to give up much of his own time to bringing to publication the work of another. So it was only after much deliberation within the NRI that Joseph Needham wrote to Professor Nathan Sivin in April 1993, making an appeal in the words of the young man in Saint Paul's dream 'Come over into Macedonia and help us.'

Nathan Sivin's response was rapid and generous. It was soon agreed that he should take on the job of editing the Needham and Lu material for publication, and of providing the book with an introduction reviewing the state of the field. And so the work began. Over the next two years there was repeated and close consultation between editor and author on many points, often through the medium of letters in giant print as Needham's eyesight began to fail. It is never an easy task to finish another scholar's work, but Professor Sivin did just that in a way that carefully preserved the spirit and substance of the writing of both authors while maintaining his own exacting standards to the full. In his introduction he has fully explained the approach he took to this task, and I shall not repeat his description here. It is enough to say that the result stands as a monument of broad enquiry, deep understanding of Chinese culture, and exact scholarship.

Finally thanks are due to those others who have helped this volume along its path to publication. First must be the two scholars who gave freely of their advice when asked to review the original material before the invitation was issued to Professor Sivin. These were Professor Francesca Bray and Professor Judith Farquhar. Also Professor Ma Boying went through the entire text during a visit to Cambridge and contributed many valuable suggestions.

Throughout the lifetime of this project many bodies have contributed funding to support the work of Needham and his collaborators. Most specifically, the Chiang Ching-kuo Foundation for Scholarly Exchange gave generous funding, which enabled Joseph Needham to have the essential services of a research assistant during the period when he was working on this book. Three holders of this post must be mentioned, Miss Jovanna Muir, Miss Corinne Richeux and Mrs Tracey Humphries (*née* Sinclair). Their contribution to the completion of this giant task took place behind the scenes, but it is one which Joseph Needham himself would certainly have acknowledged with gratitude had he lived to write a preface to this book. So I will do so in his place.

Dr Christopher Cullen

EDITOR'S INTRODUCTION

Nearly twenty years ago Joseph Needham asked me if I would like to contribute a volume on medicine to *Science and Civilisation in China*. That seemed a splendid challenge, extended too early. The problem lay not in the book but in the field of enquiry.

By 1970 Chinese and Japanese historians of medicine, and their few colleagues elsewhere, had largely reconstructed the most important achievements of the Chinese tradition as measured by the yardstick of modern biomedicine. Lu Gwei-djen and Joseph Needham built on this foundation, mostly in the 1960s, several comprehensive accounts of ancient Chinese priorities in such areas as variolation for the prevention of smallpox, acupuncture and its spread outside China, and qualifying examinations for medical practice.[1] These studies shared the ecumenical vision of steady Chinese contributions to world science that has animated the many volumes of this book since 1954. The sophisticated and cumulative character of the Chinese tradition came as a surprise to Western scholars whose histories of medicine regularly had either ignored or dismissed as 'folk practices' the therapeutic experience of non-European civilisations. Lu and Needham were also innovative in desiring to see the evolution of medicine as a social enterprise rather than as a succession of breakthroughs by individual geniuses. But their gaze remained fixed on the emergence of modern biomedical knowledge from discoveries and concepts that originated in many parts of the world.

There lay the source of my predicament. Like most people who explore the history of science today, I do not see knowledge, no matter where, as converging toward a predestined state. I see today's knowledge, not as an endpoint, but as a fleeting moment in a long sweep of creation. My experience in research has led me to view science as something that people invent and reinvent bit by bit, never completely constrained by what is already there, never pulled by some immutable goal, often mistaken, always on the edge of obsolescence. That view makes its history not a procession of destined triumphs but a meandering journey, its direction often changing, with no end but where it turns out to be on a given day. Despite the remarkable rigour and power of science, in this sense of open evolution it is like the history of everything else human beings do. Like other humanists, I find the mis-steps and failures as fascinating and instructive as the successes. The issue is not how A or B anticipated the modern Z, but how people went from A to B, and what we can learn from that about the process of historical change.

---

[1] Readers will notice that sometimes I write of Lu and Needham, and sometimes of Needham. It is impossible to separate the contributions of two people who collaborated for more than half a century. Lu Gwei-djen was particularly interested in and knowledgeable about Chinese medicine, did a great deal of the reading in the enormous literature of that field, and discussed with Joseph Needham almost everything he drafted across the breadth of *Science and Civilisation in China*. The general lines of interpretation and final formulations in this volume were almost entirely his formulation of their shared understanding.

And of course medicine and science, although public spokesmen for medical associations often confuse them, are not the same thing. Through most of history doctors have drawn on the sciences of the time to broaden their understanding, and appealed to them for prestige (see p. 16). Nevertheless medicine remains (and those whose health depends on it hope it will remain) first and foremost an art of caring for suffering people. Medicine, today dependent on physics, chemistry, biology and their combinations, consumes scientific knowledge and provides data for many kinds of research. But one hopes to be treated by a physician mindful of the Hippocratic oath, not one who approaches patients as an experimenter approaches a laboratory animal. Practitioners in China as in Europe were eloquent about their ethical obligations.[2]

For an earlier volume of *Science and Civilisation in China* I had studied the theoretical foundations of alchemy. I worked out how those foundations appeared, not to modern chemists, but to the alchemists themselves. Their aims turned out to be, not learning about the properties, composition and reactions of substances, but using known chemical processes to create small models of cosmic cycles and using them for spiritual self-cultivation, or else manufacturing elixirs of immortality to ingest themselves or to provide to others. The unexpected outcome of this investigation left me greatly doubting that Chinese alchemy (or for that matter Hellenistic alchemy) can be described accurately as a precursor of chemistry.[3] In this and other studies, like many historians of science I was finding the positivist view of early enquirers as modern scientists out of their time to be more distracting than helpful.

Medicine, even more than alchemy, has yielded unanticipated conclusions. Historians of European medicine after 1970 were exploring new issues, a matter that I will take up later (p. 22). These new issues were suggestive, but on the whole so ethnocentrically defined that it was hard to see where they pointed once one began thinking of therapy as a basic activity of every culture. Students of +18th-century France were discovering that physicians played a minor role in health care. Their social status was of course higher than that of priests, laymen, and a great assortment of practitioners that the physicians derided as quacks and mountebanks. But the doctors were so few, so largely urban in an agrarian society, and so seldom devoted to the care of the poor, that their impact on the health of the public was a great deal smaller than the chroniclers of medical progress had earlier admitted.[4]

This was also the case in imperial China. There, as we shall see, the character of the record (massive though it is) makes it more difficult to document the diversity of care, the thoughts and acts of healers, and the perceptions of patients. As more evidence has been excavated (literally, in many cases), it has not fallen into the same patterns, nor has it been investigated in the ways that were proving fruitful in the West. As the scope and quantity of scholarship have increased, attempts to construct a history of medicine on the old model of the march of scientific progress have come to look more and more antiquated, less and less credible.

---

[2] See Section (*a*).   [3] See Vol. 5, pt 4, pp. 210–305, esp. p. 244, carried a bit further in Sivin (1976).
[4] See, among many innovative studies, Brown (1982) and MacDonald (1981) for England, and Ramsey (1987) for France.

That is where I had to leave it. Much though I had learned from my earlier work on *Science and Civilisation in China*, the medical tradition seemed to me beset by enigmas that I could not ignore for the sake of a neat summary of what historians already know. My first priority, probably for some years, had to be finding out what the questions are.

I was convinced that the first step is to see each aspect of Chinese medicine in its relation to the whole, without distraction by foreign or modern assumptions. Only in the light of such a comprehensive understanding can we compare confidently that tradition with those of other cultures, or evaluate it from the viewpoint of modern biomedicine.[5] Conclusions about individual items taken out of context, without this discipline, are more likely to confirm our preconceptions than to correct them.

But this search for fruitful questions was not to be accomplished overnight. Reconstructing the alchemists' view of alchemy, with its hundred or so surviving books, was the work of a couple of years. The nearly 10,000 surviving books on medicine written before 1900 present a very different scale of historiographic effort.[6] Whether or not it is possible to encompass them in one lifetime of specialised research, as a generalist I am committed to a broader set of problems. For that reason, with great regret, I was unable to take up Needham's challenge.

THE CONTENTS OF THIS VOLUME

Lu and Needham intended to write a comprehensive survey of medicine's many aspects, to become part of *Science and Civilisation in China*. They wrote a number of important essays, mostly between 1939 and 1970, toward this overview. Despite their long collaboration, they were unable to complete it.

By now a growing band of able scholars, most of them just beginning their careers, have reconnoitred some of the questions and have blazed several promising trails. The work of Lu and Needham on medicine inspired many of these newcomers. Nevertheless, as *Science and Civilisation in China* nears completion, we have had to face the fact that in the near future no one is yet ready to survey the whole of medical history in a way that will meet the high standards of the series: based on a thorough acquaintance with the archaeological evidence and the overwhelming mass of primary sources; adducing the best modern scholarship from China, Japan and the West on the topic; bringing to bear the most powerful tools that constitute the state of the art in the history of medicine and Sinology. The state of the art is very different at our century's end than it was in the 1950s and 1960s, and mastery of it is even more daunting than it was then.

In 1992 and 1993 several scholars were asked to advise on how medicine should be represented in *Science and Civilisation in China*, and how the published essays should be

---

[5] On this matter Joseph Needham and I did not agree; he believed that 'to write the history of science we have to take modern science as our yardstick – that is the only thing we can do'. On this assumption see below, p. 7. For explicit comparisons of our views, see his comments in Vol. 5, pt 5, pp. xxxvi–xxxviii, xli–xliv, and mine in Sivin (1982).

[6] Chung-kuo Chung-i Yen-chiu-yüan Thu-shu-kuan 中國中醫研究院圖書館 (*1991*) is a union catalogue of medical books in 113 Chinese libraries. It contains over 12,000 titles, but some of these are variants.

used. On the basis of their advice, Joseph Needham asked me to edit this volume. After considering the possibilities that colleagues had suggested, it seemed to me entirely fitting and feasible to compile, as a volume of the series, a selection of the essays in which Needham and Lu presented their own insights as a volume of the series.

I have revised all of them to incorporate the results of recent research worldwide and to cite pertinent publications.[7] I have endeavoured at the same time not to obscure the authors' basic interpretations. This book will be useful to many readers not only as a record of the pioneering work of an earlier generation but as a guide to recent insights.

This fascicle contains five sections, originally published as essays and revised here:

(a) *Introduction: Medicine in Chinese Culture (1966)*: a compact introduction to the wide variety of themes originally envisaged for this volume. It first argues that medicine was shaped by China's 'feudal bureaucratism', and gives evidence that it was esteemed among professions. It then examines the several dimensions of medical doctrine bearing on vital and pathological processes, stressing the importance of prevention as a natural outgrowth of feudal bureaucratism. It asserts that magical therapies became 'fringe activities' early on, so that 'from the beginning Chinese medicine was rational through and through'. A brief historical survey identifies the earliest written sources and describes institutions that grew out of China's bureaucratic traditions, among them education and examinations to qualify physicians for practice (examined fully in Section (*a*), Subsection (3)), the national medical service, and official and private hospitals. A discussion of religion and medicine gauges the contributions of Confucianism, Taoism and Buddhism. The authors set out some broad comparisons between Chinese medicine and that of other civilisations. The conclusion looks at efforts to integrate traditional and modern medicine since 1949, with special attention to acupuncture.[8]

(b) *Hygiene and Preventive Medicine in Ancient China (1962)*: on means to longevity due to 'the philosophers who may broadly be termed Taoist', and efforts of physicians to prevent disease. Lu and Needham cite early sources to show a concern with private and public hygiene and nutritional regimen, and use rabies as an example of social organisation against disease. They conclude that the attitudes of ancient and mediaeval doctors and scholars to hygiene and preventive medicine compares favourably with those of their counterparts in Greece and Rome.[9]

(c) *China and the Origin of Qualifying Examinations in Medicine (1962)*: this essay was the first to describe in detail the medical examinations that began in the Middle Ages. It sets out representative evidence, filling in the background of the medical civil service and institutions for educating doctors. In order to prove that the effects of these innovations far transcended the Chinese cultural sphere, it

---

[7] On editing conventions see p. 36, below.

[8] A lecture presented in 1966, published in Needham & Lu (1969) and Needham (1970). For the most complete bibliography of Needham's writings to 1980 see Li Guohao *et al.* (1982), pp. 703–30. Lu & Needham (1980) is a detailed study of acupuncture.

[9] This material first appeared as Needham & Lu (1962), and was reprinted in Needham (1980), pp. 340–78.

traces to China the origins of Muslim examinations and licensing practices in the +10th century, and from Islam those of Europe, beginning via Salerno a century later.[10]

(d) *China and the Origins of Immunology (1980)*: the authors explore the vast primary literature of eruptive diseases of childhood. These sources have been little studied in China, and are mostly unknown outside it.[11] Lu and Needham show, with a wealth of quotation, that immunisation for smallpox is amply documented *ca.* +1500, and that it made its way via Turkey to England and thence elsewhere in the Occident by *ca.* +1700, to be gradually supplanted after +1800 by variolation. They go on to speculate that immunisation actually originated *ca.* +1000, but, kept secret by Taoist inoculators, was not recorded for five centuries longer. An interesting section on 'The ethnographical dimension' uses evidence from the history of immunology to disprove the old assumption that 'precursors' of inoculation such as scarification must have originated with 'primitive' peoples. It also adduces European parallels to such Chinese popular practices as wearing the clothes of smallpox victims to ward off disease.[12]

(e) *Forensic Medicine in Ancient China (1988)*: this is a broad survey of medicine as it was applied to jurisprudence. Magistrates were expected to examine corpses to settle doubts about the cause of death, and to use medical evidence to resolve conflicting accounts by the living. The history of this fascinating topic, which draws on almost every department of medical practice, is largely that of the world's oldest surviving monographic handbook on the subject, the Washing away of wrongs (*Hsi yüan chi lu* 洗冤集錄, +1247), magisterial both literally and figuratively. Officials periodically rewrote and expanded it to reflect current practice until well into the 19th century. The Washing away of wrongs was not the first book of its kind. The authors look at what is known of its predecessors. They also document a breakthrough in 1975, when an excavated manuscript showed that this use of medicine in the practice of criminal law was already firmly in place in the −3rd century. The section ends with a look at the prehistory and early history of forensic medicine in Europe.[13]

Although 'Medicine in Chinese culture' is still available in *Clerks and Craftsmen*, as the authors' only general survey of medical themes it merits inclusion here, with minor revision, as an introductory section. With that exception, the sections are set forth in order of original publication.[14]

This book omits several useful publications by Lu and Needham. *Celestial Lancets*, a book-length study of acupuncture, its history and its scientific rationale, is still in print.

---

[10] Published in Lu & Needham (1963), and reprinted in Needham (1970), pp. 340–78. I have added some materials from sources not available to the authors.

[11] Chang Chia-feng (1996) has cited an even wider range of sources.

[12] The authors revised and considerably expanded this subsection from Needham (1987), itself a longer version of *ibid.* (1980).

[13] Published as Lu & Needham (1988).

[14] Discussions of medical interest are also scattered through other volumes of this book, especially Vol. 5, pt 5, on physiological alchemy.

It is summarised at the end of Section (a). 'Proto-endocrinology in medieval China' is available in Vol. V, pt 5 (Section 33 (*k*)). 'A Contribution to the history of Chinese dietetics', one of the authors' first collaborative historical studies, has not been superseded since they wrote it in 1939, but the topic would now demand a much broader study. Section (b) reiterates some of its findings. 'Records of diseases in ancient China' is a study of archaic words for diseases in the early classics, based on, and slightly extending, a monographic study in Chinese by Yü Yen (*1953*). Yü's book stops before the earliest medical books emerge. The essay still provides the only starting point in Western languages for students of palaeopathology, but such work, to be publishable today, would have to draw on a systematic study of oracle script and bronze inscriptions as well as advances in classical studies.[15]

In what remains of this introductory essay I first review certain themes that permeate Needham's work and that are prominent in the chapters that follow. Then I take up several characteristic approaches that have turned out in the light of later research to be problematic. I then examine the changing disciplines, the history of medicine and Chinese studies, at the overlap of which Lu and Needham worked. Finally, I summarise the most interesting new questions and new results to arise from recent work. I assess where they have led and where they appear to be leading.

RECURRENT THEMES

This volume represents roughly fifty years of research on Chinese science and civilisation, and forty-five years of publication on medicine, by each of the authors. Sections (*b*) to (*e*) provide a sampling of historical themes, examined in some depth, to complement the broad survey in Section (*a*). All were written with a view toward incorporating them, after revision, into the medical part of this series.

These five chapters reflect a foundation of assumptions and themes on which the edifice of *Science and Civilisation in China* is built. Let me set out some of them. All are also found in the other volumes.

(1)   The unit of exploration is the world. China presents a particularly rich and interesting complex of contributions, but their significance becomes clear only through comparison. The comparanda are particular techniques, features of institutions, items of knowledge or conceptions, and what Needham calls 'factors'. These can be any discrete aspect of culture or value, in particular the 'inhibiting factors' that stop or slow down a given development in one civilisation.

In this respect the work of Lu and Needham differs strikingly from that of their contemporaries. The reigning assumption, today as a generation ago, among the Young Turks of Euro-American history as well as its dotards, is that from the ancient Greeks onward, science and 'scientific' medicine has been an exclusively European enterprise, which other peoples have advanced only to the

---

[15] Lu & Needham (1951, 1967, 1980). The essay on dietetics was submitted to *ISIS* before World War II, but was not published until six years after the war ended.

extent that they have accepted European learning. Those who specialise in the non-Western parts of the globe have massively refuted this parochial bias, but the conventional wisdom sturdily ignores them.

Needham's case is based on a remarkable command of European as well as non-European history. The lists of references in this work are sometimes the best available for someone who wants to study a given topic in the West as well as in China. Needham has earned much of the credit for slightly eroding the general provinciality. Historians of Europe at least now entitle their textbooks 'A history of Western science' rather than 'A history of science', even though they still feel no need to explain why a history of Western science is enough.[16] Given the lowering of language requirements for the doctorate, it is unlikely that world-wide histories will be the next step. Those sufficiently intrepid to contemplate writing one will not find better models than parts of this volume.

(2) There are two quite different kinds of comparison. One type compares the achievements of different civilisations, primarily to locate priorities. The items generally come from different times. As Needham noted in the Introduction to *The Grand Titration*, 'we are always trying to fix dates' in order 'to "titrate" the great civilisations against one another, to find out and give credit where credit is due . . .'.[17] The other, often implicit and even more common, compares items of Chinese knowledge or practice with those of today.

The second type of comparison is as important as the first, because Needham's judgements of significance depend on the criteria of modern science. Here, he is titrating against what he considered to be a fixed standard of known purity. This reflects the positivism that was normal in the technical histories of the 1950s.

(3) Many of Needham's assessments are based on a view of science not of the present but of the future. He always had in mind a future in which the transition from physical to organic models of theory-making has been completed and, not incidentally, one in which the human community and the community of scientific exploration are united, no longer culturally and politically riven. In the past we can discern differences in science in each of the great civilisations, but Needham was convinced that they must inevitably converge to form one universal science.

(4) Historians of physics tend to write of the 'connections' between physics and philosophy, as though they were inherently unrelated realms, rather than seeing physics as a subset of thought about the external world. Needham disregarded these narrow professional borders. His definition of the minimal field that one must explore to understand the emergence of science is uncompromisingly poly-mathic: 'language and logic, religion and philosophy, theology, music, humanit-arianism, attitudes to time and change'.[18] These are not abstract desiderata;

[16] Lindberg (1992, on science) and Conrad *et al.* (1995, on medicine) are typical of textbooks that assume, despite Needham's evidence to the contrary, that one can safely disregard the influence of Chinese science and medicine on that of Europe, that a 'Western' history is preferable to one that portrays the interaction of all cultures.
[17] Needham (1969), p. 21.   [18] *Ibid.*, p. 216.

he realises all of them in this volume. Religion, for example, is central to the next item.

(5) Needham traced the origin of natural knowledge in ancient China primarily to two opposed institutions and the ideologies that supported them. One is the feudal bureaucratism of the Confucian State, with no concern for abstract theory but a strong interest in utility and rationalisation. The other is Taoism, a mystical religion of Nature devoted to contemplating and observing it without preconceptions or prejudices. These are fixtures in the sorts of history of philosophy that are more interested in disembodied isms than in the activity of particular human beings, but Needham's use of them has little in common with such familiar idealisations.

His feudal Confucianism is in principle reactionary, but it wields authority over the economic circumstances in which scientists work. Taoism, although it never defined an experimental method or a scientific logic, was creative in a way strikingly analogous to certain characteristics of modern scientists. Buddhism, because it rejects the phenomenal world, is an insignificant part of the picture.[19]

(6) One of Needham's most fundamental convictions is that 'analysable differences in social and economic pattern between China and Western Europe will in the end illuminate, as far as anything can ever throw light on it, both the earlier predominance of Chinese science and technology and also the later rise of modern science in Europe alone'.[20] This does not, curiously enough, lead to an analysis that draws strongly on modern economics and sociology. On the contrary, economic data rarely appear, and discussions of social patterns largely depend on the dichotomy of Confucianism vs. Taoism, defined more as philosophical than as social categories. There is some concern with tensions between élite and non-élite over control of the means of production, but these again tend to devolve into issues of Confucianism vs. Taoism – isms, not identifiable collectivities. The major figures of this century's social theory, from Weber on, do not figure explicitly or implicitly in the arguments.

(7) Lu and Needham's studies of medicine, with hardly an exception, have been histories of medical and surgical disorders, seen as the best-known and best-educated European and American practitioners understood and treated them. Even childbirth, to the extent that it entered the picture, entered it as a disease. I consider, below, the many dimensions missing from this picture. But it is pertinent first to notice that the authors were unfashionably broad in another important respect.

(8) From their first collaboration in this area, a study of dietetics, Lu Gwei-djen's professional competence as a nutritional scientist prompted them to look not only at the stuggle against illness but at the maintenance of health. Section (b) of this volume, based on an essay of 1962, is an overview of hygiene and preventive medicine in early China. The discussion of hygiene pays as much attention

---

[19] These matters are discussed in detail in Vol. 2, Sections 9, 10 and 15.   [20] Needham (1969), p. 217.

to mental as to physical health. It is attentive to environmental and personal clean-liness, and to sanitation in cooking. It includes a mini-monograph on detergents in typical Needham style.

PROBLEMATIC FOUNDATIONS

Lu and Needham by 1960 were using some of the most innovative history of science and Sinological methodologies available. They were doing so not to write definitive history but to meet a different challenge, not for academe but for the educated public. The justification of the project 'is that a vast and scattered literature does already exist, and that it has never before been digested into the compass of a single book' concerned with a coherent set of issues. They aimed to produce amply documented hypotheses that could be straightforwardly tested and bettered.[21] Inevitably, certain of their basic methods of approach appear in hindsight to be as distracting as they were suggestive. I now examine some instances that bear on the study of medicine.

Science, magic and religion

One instance is the conviction of Lu and Needham that the borders between science, magic and religion were heavily travelled, and that before modern times this was a mat-ter of benefit to science. This idea is not trivial, but attempts to draw concrete conclu-sions from it have for generations led repeatedly to frustration and unresolved polemic. The humanist Lynn Thorndike, beginning in 1923, massively documented the overlap of the three realms for Europe, but historians of science who had grown up in the tech-nical world remained more likely to consider the study of 'superstition' pernicious than to exploit it.[22]

Robert Merton's work on the Puritan origins of modern science in 1938 was sim-ilarly important. Merton did not argue that English Calvinist theology generated scientific innovation. He merely claimed, with many qualifications, that Puritans shared certain clusters of values that encouraged worldly endeavour, of which enquiries into Nature were a part. Modest though this claim was, at first historians largely ignored it. Later, despite Merton's fame in sociology, they were little influenced by it. The same can be said for Frances Yates's several volumes devoted to her more sophisticated proposition that Renaissance magic played a crucial role in prompting and forming the Scientific Revolution.[23]

'Magic', in anthropology and history, has come to be a dubious term, decreasingly used and increasingly contentious. Its earlier vogue owed much to Bronislaw Malinowski's (1948) view of it as a kind of failed technology, 'a false technical act' but 'a true social act'. Peoples who wanted to control Nature, but had no means to do so,

---

[21] Vol. 1, p. 5. [22] See the manifestos of Neugebauer (1951) and Thorndike (1955).
[23] Yates (1964, 1979, 1982, 1983, 1984). See the critique in Westman (1977).

performed elaborate rituals to convince themselves communally that they had that power. They could always explain their lack of success in changing their environment by some flaw in performance of the ceremony. What this view overlooked, as S. J. Tambiah (1968) showed in his classic study of Malinowski's field notes, was the fact that these rituals were religious, that they depended on an appeal, explicit or implicit, to divine authority. There was no delusion that the perfect mechanical performance of rites would force Nature to do human bidding. What those present changed was themselves.[24]

Needham was more inclusive than Merton or Yates in asserting a three-way bond between magic, religion and science, rather than the binary links that those authors studied. Not surprisingly, his claims were as vague as theirs. Historians of science have not extracted from them propositions well enough defined to test and apply to further studies. The centrality of Taoism in these linkages is one of several reasons to scrutinise Needham's conception of it.

Taoism

Needham held Taoism particularly responsible for originating scientific attitudes and accomplishments. For example, in Section (*d*), 'The origins of immunology', in this volume, he argues that Taoists not only invented inoculation for smallpox but kept it utterly secret for at least five centuries. This openness to historical linkages between science and religion, as we have seen, is one of the strengths of his writings. But many of his arguments about Taoism here and elsewhere have not aged gracefully because of great changes since the 1960s in the way that students of religion understand that tradition. When most of Lu and Needham's writings were drafted, Sinologists were still thinking of Taoism as a philosophical stance that, after about −300, survived in two forms. One was a degenerate, superstitious religion that, along with Buddhism, served the needs of the uneducated masses. This Taoism was often at the root of rebellion, and some scholars affirmed that 'we always find the Taoists with the party opposing the literati'. The other was a free-floating set of attitudes toward Nature and society. As the leading textbook of its time put it, 'The man in power was usually a Confucian positivist, seeking to save society. The same man out of power became a Taoist quietist, intent on blending with nature around him.'[25]

By 1980 this understanding had changed decisively. Scholars had made considerable strides in studying the massive collections of Taoist scriptures. They had found and observed Taoism as a living religion in Chinese communities outside the mainland (and strongly reviving on the mainland, especially in the southeast coastal regions, after 1980).

---

[24] Malinowski (1948), Tambiah (1968). See the contemporary discussion in Tomlinson (1993), pp. 249–50.
[25] Fairbank & Reischauer (1958), p. 76. Weber (1922/1964), equally irrelevantly, saw organised Taoism as rebellious. Historians now connect 'Neo-Taoism', which used to be considered a third legacy, primarily with Mādhyamika Buddhism.

For Joseph Needham's main discussion of Taoism, see Vol. 2, pp. 33–164. Sivin (1995d) examines in detail the various claims over the past century that there was a special relationship between Taoism and the evolution of science. See also the summary of discussions on the relations between popular religion and Taoism in *ibid.* (1979).

They came to understand that the earlier view, which had been conventional in China before Western students picked it up, had forced on the *Lao-tzu* 老子 and the *Chuang-tzu* 莊子, two texts with some overlapping mystical content, a most unstable marriage. These two writings, and the miscellaneous books classed with them in library catalogues, did not contain a single, consistent philosophy. Rather than exerting a distinct 'Taoist' influence on later thought, by −100 they had been absorbed into a State-sponsored synthesis that as a whole was neither quietist nor absorbed in Nature.[26] In this synthesis the legacy of *Lao-tzu* and *Chuang-tzu* is not an ism in any meaningful sense, but, as the final quotation in the previous paragraph puts it, a mood into (and out of) which any conventional person in given circumstances may drift. When 'Taoism' refers to that mood, it no longer makes sense to speak of 'the Taoists' as a social entity.

The organised Taoist religious movements, with their hereditary priesthoods and elaborate liturgies, are not degraded embodiments of the early classics, nor even their direct offshoots. Their origin lies instead in the popular religion of China. This religion is perhaps the most significant discovery of recent decades – hidden by centuries of sporadic official persecution and the contempt of ideologists who saw it as a potential source of heterodoxy. Even today some Sinologists who have not followed these discoveries speak of the priests of popular religion as 'sorcerers' or 'shamans', reflecting the way the disdainful State ideologists of imperial times saw folk religious leaders. These are questionable translations of *fa-shih* 法師 (originally, a teacher of the Buddhist doctrine) and *wu* 巫 (literally, 'medium'), the former eventually the common term for popular priests, and the latter (from the Han on) a term of contempt for them.

Nor have specialists in recent years found a single case in which Taoist organisations after the +2nd century were responsible for rebellion or other political activity directed against the government. In this sense they differ from the popular Maitreya cults, which, although in no sense activist, were periodically taken over by rebellious groups.[27]

Since historians of religion have already clarified these matters,[28] let me sample a few problematic assertions about Taoism. Readers who wish to see one of Joseph Needham's more sustained arguments examined in detail may consult the Appendix to Section (*d*).

He writes in Section (*d*) of 'the Taoist conception of *yang sheng* or "nourishing the vitality"' (p. 115), of 'Ko Hung's strong belief in the efficacy of Taoist respiratory techniques', of 'Taoist alchemy' (p. 121), and of written charms and incantations as Taoist (p. 139). He asserts that allusions to an inoculum as 'elixir [*tan* 丹], numinous, of the immortals' 'clearly point to the Taoist origins of inoculation' (p. 156). Physical and meditational techniques of self-cultivation, we now know, pre-dated the Taoist movements, were widely practised by non-initiates, and were not in any sense peculiar to Taoist masters. Written charms and incantations are as characteristic of popular religion as of Taoism; only charms that bore seals of authority were reliably Taoist. The primary

---

[26] Sivin (1995f).   [27] Naquin (1976, 1981).

[28] In addition to the writings of such innovative historical scholars as Bokenkamp, Robinet, Schipper, Strickmann and Verellen, see the bibliography by Thompson & Seaman (1993) and the account of Japanese scholarship by Fukui (1995).

meaning of *tan* was cinnabar, an important ingredient in medical formulae. The sense 'elixir' is historically derivative. In medicine *tan* referred to a variety of mixed formulations that need not have alchemical associations. Immortality and numinosity were beliefs universal among Chinese; their associations were in no sense restricted to any group that can be called 'the Taoists'. In any case, Needham practically never specifies which of the many senses of 'Taoist' he means.[29]

In Section (*a*), he asserts a link between alchemy and medicine with even less qualification: 'The macrobiotic preoccupation made Chinese alchemy, as it were, iatrochemistry, almost from the first, and many of the most important physicians and medical writers in Chinese history were wholly or partly Taoist' (p. 58). It is impossible to evaluate this assertion without knowing what it means by 'Taoist'. In what sense can one call 'wholly or partly Taoist' the majority of important figures in Chinese medicine, who were perfectly conventional, or even firmly ensconced in the 'Confucian' bureaucratic hierarchy?[30] If we adopt the criteria given at the beginning of Section (*b*), which make Taoists, *in contrast to* Confucians, devoted students of Nature, religious mystics, believers in immortality and alchemists (p. 67), few of the more than 20,000 physicians known to us would qualify.[31] The notion that Taoists by definition study Nature and Confucians do not, venerable cliché though it is, has not survived into modern studies of Taoism.

Confucianism *vs.* Taoism is not only a religious interpretation but by implication a political one. By 1960, few historians of science or medicine were raising questions about the disposition of power. Biographies tended to portray discoverers as politically conformist, or to overlook their politics. But Needham's Taoists are indeed by definition the political fringe: 'they were in complete opposition to the very structure of feudal society, and their withdrawal was part of their protest'. They were 'the spokesmen of some kind of primitive agrarian collectivism, and were opposed to the feudal nobility and to the merchants alike'. Passively or actively, they stood against the Confucians, who 'were entirely on the side of the literate administrators, and lacked all sympathy with artisans and manual workers'. This Taoism was no mere mood.

Needham describes one of his Taoist exemplars, Thao Hung-ching 陶弘景 (+456 to +536), the scholar of alchemy and materia medica, as belonging to 'a long line of scholars who excluded themselves from the ranks of the Confucian bureaucratic hierarchy and earned their living by selling medicinal herbs'.[32] There is no evidence, however, that Thao was politically unconventional. He did not pursue a career of drug peddling, but, as Mugitani, Strickmann and others have shown, lived on lavish patronage from the emperors of two successive dynasties. Leaving aside the significance of such instances, Needham's willingness to raise issues of science and politics established a healthy precedent.

---

[29] On this ambiguity, almost universal among Sinologists before the 1980s, see Sivin (1978).
[30] See the survey of individuals in Sivin (1995d), pp. 51–4.
[31] Ho Shih-hsi 何時希 (*1991*) provides biographical information of varying quality on over 21,000 traditional physicians.
[32] Vol. 2, pp. 86, 100, 132.

Nevertheless, his attempt to reduce the social, political and religious tensions that bore on Chinese scientific and medical activity to a running battle between two isms was abortive. The next generation of scholars has moved in very different directions as their research has continued to reveal the complexity of technical culture.

Confluence

'Viewing the whole march of humanity in the study of Nature as one single enterprise' implies a model of confluence. Such a model is in fact central. 'Although modern science originated in Europe and in Europe only, it was built upon a foundation of medieval science and technology much of which was non-European. . . . One would rather prefer the image of the great rivers of past science and technology flowing into the ocean of modern natural knowledge, so that all peoples have been testators and all are now inheritors, each in their several ways.' This asserts that the confluence has been accomplished in the past.

One can hardly deny, in the face of Needham's evidence, that the scientific phase of the Renaissance drew on ideas and practices that flowed together in the Islamic world and transformed Europe. But he is claiming a good deal more, namely that as a result science has matured, transcending the bounds of particular cultures as it freed itself from European characteristics: 'The Scientific Revolution, the mathematisation of hypotheses about Nature, combined with relentless experimentation, have made modern science absolutely international.' When challenged, Needham has qualified this judgement, allowing that there may remain important cultural differences. That qualification did not, however, lead him to redesign his larger framework.[33]

How do what begin as separate traditions flow together? Needham uses two complex images to portray this flow. One portrays 'the ancient and medieval sciences of all the peoples and cultures as rivers flowing into the ocean of modern science'. The second metaphor identifies for a given realm of knowledge or practice the 'transcurrent point', when Europe decisively took the lead over China, and the 'fusion point' when the differences vanished as Chinese and Western knowledge coalesced. For example, in an extended reflection on this trope, Needham concluded that in mathematics, astronomy and physics, the transcurrent point happened *ca.* +1610, and the fusion point soon afterward, by +1640. In botany the first transition occurred between +1710 and +1780, but the fusion point not until 1880. In medicine the transcurrent point cannot be located more precisely than between +1800 and 1900, and the fusion point is yet to come.[34]

This more or less metallurgical metaphor does not point to a process. The hydraulic metaphor, at least, implies a spontaneous, linear flow. The discussions of both suggest something similar to a scientific conference: learned scholars compare two approaches, decide impartially which is superior, and thenceforward accept it. It becomes, to use a term absent from Needham's historiographical vocabulary, a paradigm. The sum of these accumulating multicultural paradigms is a true synthesis.

---

[33] Needham (1969), p. 56; Vol. 5, pt 2, pp. xxvii–xxix. [34] Vol. 5, pt 2, *ibid.*; Needham (1967).

One might object, however, that 'fusion' is not an apt metaphor for the worldwide spread of modern science. We can find traces of ancient China in the early modern evolution of biology or physics, but it would be difficult to specify what aspect of their modern theories is culturally Chinese in more than this genealogical sense. Various authors, attempting to produce such specifications, have pointed to traces of mysticism or subjectivity in modern scientific theory. These trophies, however, reside in interpretations for laymen, not in the equations with which scientists communicate with each other. Such claims also do not take into account that there is nothing peculiarly East Asian about either mysticism or subjectivity.[35]

The transition to modern science in 20th-century China was not exactly a 'fusion point', a matter of two bodies melting together. Chinese who had the power to determine what would be taught in educational institutions explicitly rejected the traditional form for the modern-Western one. What happened in medicine was a good deal more complicated, but that makes it the exception, not the rule. The process that determined its near-abolition and survival was intensely political.

In recent research on Chinese science, Needham's collegial model of encounters between technical traditions has given way to one more attentive to social conflict. Western astronomy triumphed in China in +1644, not because Chinese and Western astronomers had agreed that its time had come, but because Johann Adam Schall von Bell, given supreme power as Astronomer-Royal by the new Manchu overlords of China, commanded the career officials of the Bureau of Astronomy to learn Western astronomy without delay, or lose their jobs. Chemistry took over in the 19th century when, as a result of the Opium Wars, peace treaties allowed foreign missionaries to impose their own curricula on students that they were free to recruit, and so on. Between 1927 and 1929, the Nanjing government might well have outlawed the practice of traditional medicine had the practitioners not learned modern methods of interest-group organising and lobbying.[36]

Decisions that ended the distinct histories of Chinese sciences were, to sum up, only to a limited extent intellectual. Ultimately they were administrative decisions about educational or regulatory systems. Future work on questions of confluence is likely to pay a great deal more attention than has been true in the past to political process and its social circumstances.

Boundaries between science and technology

In the 1950s and 1960s, most historians of Europe and North America agreed that modern technology is applied science. Up to some unspecified point in the past, however, the two were scarcely related. Science was done by gentlemen and technology by artisans. When knowledge passed back and forth, on the whole gentlemen gained more from the exchange than craftsmen did. By 1980 the controversies over boundaries and

[35] E.g. Capra (1977), Holbrook (1981).
[36] Huang I-nung 黃一農 (*1990*); Sivin (1982), pp. 63–5; Zhao Hongjun (1991), p. 18.

transitions in the relationship were generating some heat and even a bit of light.[37] What we now understand as great overlaps in the Renaissance between new approaches in painting and the emergence of optics, or among navigators, book-keepers and mathematical theorists, were barely being explored.[38]

Needham and his collaborators, despite their abstract interest in the role of commerce in the Scientific Revolution, were not involved in the skirmishes that settled such issues. The picture they limned for China had little to do with any of the debates about Europe.[39] Because in Chinese 'feudal-bureaucratic society' sustained large-scale investment could come only from the centre, the scientists and engineers tended to group together in the civil service, and particularly in its lower reaches. Both were on the same side of the gap that separated officials and commoners, those whose careers depended on their hearts and minds and those who (at least in the eyes of their betters) had only bodily toil to offer. When Needham considered technical talent among commoners, he also discussed science and technology together, for 'the Taoists' were responsible for both.

This association encouraged a remarkable elision in Needham's writing. I have cited, above, one of many places in which he speaks of the early 'predominance of Chinese science and technology'. This is a key proposition, because on its basis Needham sought to explain 'the later rise of modern science in Europe alone' (p. 8). When we examine assertions about early predominance one by one, however, those well supported by evidence and argument tend to be matters of mechanical invention. As for civil engineering, the weighty volume on that topic (Vol. 4, pt 3) suggests that the Han Empire and the Romans were building and innovating on quite comparable scales. One cannot speak of predominance until centuries later, when Western civilisation had self-destructed and was gradually rebuilding. Chinese history can hardly explain that self-destruction. As for science, Needham has shown in case after case that the emphases and techniques of the two cultural areas in mathematics, astronomy, mathematical harmonics and so on differed decisively. Surely that leaves claims about superiority beside the point. They are equally irrelevant to the comparison of alchemy and astrology East and West.

To return to the subject of this volume, one would have to be equally non-committal about medicine. The first outcome of any comparison is the striking contrast in ways of thinking about the body: a structure of organs and tissues threatened by lesions in Europe, a congeries of vital processes responding to pathogenic *chhi* in China. Despite parallels in notions of health as a metabolic balance, there were fundamental differences in what had to remain balanced. No three historians are going to agree about which end of Eurasia had the more effective therapies, or even about when it would be appropriate to compare cures. The notion that either was predominant is simply not testable.

There is no need to pause longer over this issue. Once one sets aside the untenable assumption that China in −1000, −500, −200 or +200 was demonstrably ahead of

[37] See e.g. Layton (1976, 1988), Hughes (1981).
[38] See, among many, Bennett (1986), Biagioli (1989) and Jewson (1974).
[39] Detailed discussion in Vol. 4, pt 2, pp. 10–42.

Europe in technical activity (or vice versa), what is left of the 'Scientific Revolution Problem'?[40]

Positivism

I have mentioned above the positivism that pervades Needham's judgements about what is pertinent to the history of science and medicine (p. 7). Even with respect to medicine, scientific understandings, much more than clinical practices, set his norms. If modern knowledge is the end of all earlier striving everywhere, it becomes the only obvious criterion of value.

'Modern' is, of course, a broad word, and historical writings seldom define what period they mean by it. Some of the criteria of assessment in this book are based on practices long since abandoned. This is scarcely surprising, for example, in an era of laboratory tests. Medical schools no longer teach many fine points of signs-and-symptoms diagnosis, on which doctors depended until well into the 20th century.

Historians today are more likely than Needham and his contemporaries to aim for an integral understanding of technical phenomena in the time and place they are studying, and to define their criteria as this aim requires. This redirection has greatly limited the influence of Needham's methodology on young scholars.

Medical history and Chinese studies

Medical history

The history of medicine took new directions late in the 19th century as physicians canvassed the past to map the march of medical progress. The issue that engaged them was how their forebears had come to apply biological, chemical and physical knowledge to clinical research and practice, giving medicine the cachet of modern science, despite its undiminished responsibility to care for the sick and suffering.

Medicine's cachet did not begin recently. Researchers have traced the shared intellectual horizon of science and medicine ever further backward, beyond William Harvey and his experimental, quantitative proof that the blood circulates, beyond the mediaeval professors of medicine who argued that what they taught was a *scientia*, beyond Galen's trove of detailed observation and philosophical controversy, to the beginnings of systematic thought in −5th-century Greece, when the Hippocratic physicians were as prominent as the natural philosophers and hardly less insistent on rational argument.[41]

The history of medicine, when by 1960 it became a recognised discipline, was concerned with the accomplishments of doctors, primarily with their knowledge and secondarily with their practice. It recognised the social dimension of medicine only to the

---

[40] For a discussion of this science–technology issue in relation to many other issues that bear on the 'Scientific Revolution Problem', see Sivin (1982), pp. 46–7.

[41] See, on Greece, Lloyd (1987), esp. ch. 5; on the Middle Ages, among others, O'Boyle (1994); and, on Harvey, Frank (1980).

extent of studying physicians' statuses and professional organisations.[42] This tight focus led to considerable strides in ascertaining what physicians knew, establishing its relation to scientific knowledge, and justifying intellectually what was consistently portrayed as the medical profession's well-earned prestige and income. Not surprisingly in view of these emphases, historical studies found a secure home in medical schools.

One can imagine quite different histories of medicine emerging from the sources and methods available, devoted perhaps to therapeutic practice or to the patient's experience. Eventually some of them came into being.[43] A metamorphosis of the history of science in the 1970s greatly affected the history of medicine. The change was to a large extent due to a change in historians. Authors originally trained as scientists, physicians and engineers were increasingly replaced by those trained in the humanistic disciplines of history. Although many historians in that generation had substantial technical backgrounds, they were more willing than their forebears to see change forming and evolving spontaneously rather than taking steps somehow destined to generate modern science.

In the USA, for instance, scholars of medical history increasingly were educated and employed outside the medical schools. That directly involved them in the shift of orthodoxies from 'internalism' (the study of ideas without reference to their social formation) to 'externalism' (the study of the interaction between scientists and society without regard to the content of science). Externalism reflected, among other things, a growing awareness that science could be used toward heedless and destructive ends. This new orthodoxy was also attractive because scholars without technological training or experience could make important contributions.

Internalism and externalism faded away as innovation focussed on interrelations. Through the 1980s, the most influential historians of science, and those historians of medicine close to them, acknowledged that a dichotomy of ideas and social relationships made it impossible to see any historical situation as a whole. In this effort they were greatly aided by tools and insights borrowed from anthropology and sociology. To take the most obvious example, the notion of culture provides a unitary view of concepts, values and social interactions. Out of this synthesis has emerged a history of medicine that is precise about the large role of bluff in the Greeks' presentation of their theories as rational and demonstrative, and about the competition for livelihood that made bluffing productive; well informed about the astrological and religious as well as the anatomical side of +17th-century English medicine; and attentive to the therapies and social statuses of the 'quacks' who provided most of the medical care in France *ca.* +1800.[44]

Sinology

Over the last half of the 20th century, occidental scholars have transformed their methods of approach to the academic study of China. The Sinological branch of Orientalism

---

[42] Shryock (1936) is an important exception to this general shying away from social history. Although physician-centred, the book's overview of British, French, German and American medicine remains unequalled.

[43] Prominent examples are Porter (1985), Porter & Porter (1988, 1989) and, for Japan, Johnston (1995).

[44] Lloyd (1987), pp. 15, 28; MacDonald (1981); Ramsey (1987).

began at the confluence of two traditions. It was formed not only by the enterprising philology of the European Renaissance but by the Chinese tradition of evidential research (*khao-chêng* 考證), which, culminating 2,000 years of critical scholarship, had developed equally powerful methods for testing and interpreting textual evidence.[45]

Naturally enough in view of these patrimonies, Sinologists tended to concentrate on explaining, with the greatest possible care, what ancient books said, recapitulating their gist in topical or chronological order. They paid little attention to the motivations, interests and social circumstances that made such sources a good deal less than objective testimony. The European specialists' taste for early classics and for conventional topics reflected those of the orthodox Chinese scholars who taught, advised and assisted them.

Here as in the history of medicine, the multiplication and expansion of universities and the introduction of new subjects following the end of World War II opened secure careers for large numbers of scholars who could show their proficiency according to established standards. The Cold War of the 1950s multiplied governmental support for studies of China, mostly but not only in the USA. The scale of research thus rapidly expanded. At the same time the added investment encouraged two shifts that turned out to be fundamental.

As Sinologists established a noticeable presence in university departments, colleagues in History and other departments, who had come to think of philology as a tool, challenged reliance on it as an end in itself. Advances in language teaching in this period stressed intensive oral drill with native speakers to provide an active command of the modern language. That enabled the best programmes to give students a sound grasp not only of modern but also of classical Chinese in three years or less. Language ceased to be a career-long fixation. Young scholars could be expected to learn a discipline before they began dissertation research. No occidental specialist today, however experienced, finds reading ancient Chinese texts effortless or free from uncertainty, but the same can be said of their East Asian colleagues. With remarkably ample classical dictionaries and concordances, recently published, one text after another has become fully accessible to those systematically trained.[46] This is now merely one among the many perennial difficulties of scholarship.

Another important change, natural enough when so much of the volition (and funding) for Chinese studies came from enmity toward Communist China, is that their centre of gravity shifted toward the present. By now, outside China there are many more academic experts on the period after +1800, which hardly existed as an academic subject in 1950, than on the period before. The literature on early China, once largely concentrated in the late Chou dynasty, is coming to be evenly spread across at least the major dynasties.

RESEARCH IN ASIA

This sketch of the background against which Lu and Needham worked would be seriously flawed if it did not point out that the most valuable contributions to the factual

---

[45] On evidential research, see Elman (1984).

[46] The most notable are the splendid classical dictionary by Lo Chu-fêng 羅竹風 (*1987–94*) and the new series of concordances edited by Liu Tien-chüeh 劉殿爵 & Chhen Fang-chêng 陳方正 (*1992– *).

documentation of Chinese science and medicine have originated in China and Japan. The circumstances in which this work has been done differ considerably from those I have outlined above.

In 1937, when Chhen Pang-hsien 陳邦賢 published the first modern-style history of Chinese medicine, the Rockefeller Foundation's Peking Union Medical College, one of the leading medical schools of the world, was being abandoned to the Japanese occupation. The influence of this research-oriented medical school, and the effect of biomedicine generally on China's health care system, remained negligible until the second half of the 20th century.[47] Nor did modern medicine greatly influence Chhen's book. His work was characteristic of the best historians of the time in its great erudition, its tendency inherited from the evidential research tradition to juxtapose summaries of primary sources rather than to explain change, and its silence about the social context of medical practice. Also characteristic, the sole exception to this silence was detailed sections on the organisation of the palace medical bureaucracy in each major dynasty. There is no evidence that palace officials were responsible for much of the significant change in health care over the centuries, but in the 1930s a career in the civil service remained, as it had been for two millennia, the one occupation that scholarly families urged on their offspring.

The Sino-Japanese War, the end of the Kuomintang government on the mainland, and the vicissitudes of the People's Republic led to remarkably little net change in the writing of medical history, at least once authors were no longer required to apply (or even quote) the thought of Marx, Engels, Lenin and Mao Tse-Tung. But the scale of publication has expanded greatly, due to the growth of support in China for specialised research. In addition to the growth of full-time researchers in central institutes of the history of medicine and of science, teachers and physicians have responded to the prestige that strong government support gave the field. A survey of 1987 counted a total of roughly 500 historians of medicine publishing the results of their enquiries into original sources and artefacts.[48]

This swelling band has poured forth monographic studies on individual authors and sources, reprints, critical editions, translations into modern Chinese, and reference books.[49] Although the chaotic book distribution system makes this literature difficult to find inside China as well as elsewhere, with access to a good library it is possible for the first time to find accurate information about a great deal of the ancient medical landscape.

There has been no corresponding revolution in methods or conclusions. Almost all the historians of medicine in China were trained in medicine rather than in history. The

---

[47] By far the best study is Bullock (1980).   [48] Sivin (1988), pp. 49–50.

[49] On reference books, including those that provide access to other recent publications, see note 46, p. 18, above, and Sivin (1989). Among the most useful publications since those listed in the latter are Chungkuo Chung-i Yen-chiu-yüan 中國中醫研究院 & Kuang-chou Chung-i Hsüeh-yüan 廣州中醫學院 (1995), a comprehensive dictionary of medical terminology and related matters; Chung-kuo Chung-i Yen-chiu-yüan Thu-shu-kuan 中國中醫研究院圖書館 (1991), a national union catalogue of primary sources; Li Ching-wei 李經緯 (1989), a biographical dictionary; Ho Shih-hsi 何時希 (1991), a compendium of biographical notices; Yen Shih-yün 嚴世芸 (1990–4), a vast collection of prefaces; and Chung-kuo Chung-i Yen-chiu-yüan. Chungkuo I-shih Wên-hsien Yen-chiu-so 中國中醫研究院中國醫史文獻研究所 (1989), an index to publications on the history of medicine.

training of Ph.D.s in the field began in the late 1980s, but the curriculum does not yet provide a substantial introduction to the modern humanities and social sciences. Students are still being rigorously trained in evidential studies.

In China before the late 1980s, history remained a risky occupation. In a consciously totalitarian society, reinterpretation of the past is an essential means of control. Any issue may become politically charged, any position ideologically incorrect. Although such threats have largely abated, intellectuals who grew up earlier have not forgotten that they were periodically terrorised in public, and that periods of calm have regularly been succeeded by new terrors. This is not the sort of atmosphere that encourages methodological innovation.

Authors aware of the risks and sincerely eager to be correct, even when there were no government bans or party persecutions, have to some extent censored themselves, shunning risks in subject matter and interpretation. At the same time, high officials who aim to build national pride have tended to support history only to the extent that specialists can identify discoveries or inventions that appeared first in China. Such exercises fight with its own weapons the Western parochialism that believed the Chinese mentally incapable of technical priorities. Thanks to the work of Needham and others, that parochialism has become less assertive, but few Chinese policy-makers are aware of the change. It is unlikely that 'leaders' would stop exerting pressure to claim technological priorities even if they knew that who did what first has ceased to be what the history of technology is about.[50]

A scholarly tradition that explicates texts without much conscious interpretation, historians mostly trained as scientists or physicians and thus inclined toward positivism,[51] recognition and praise awarded mainly for claiming scientific priorities, and self-censorship to avoid being linked to the 'bourgeois liberalism' of colleagues abroad have all conspired to inhibit qualitative change proportionate to quantitative growth. This picture has begun to change over the past decade as economics has taken the helm. Pride has come increasingly to depend on wealth, and campaigns against 'spiritual pollution' have begun fading into the featureless past. But it will be some time before iconoclasm becomes nothing more than an intellectual issue. At the same time, many colleagues in China consider the system in crisis, because it has become practically impossible to recruit first-rate graduate students. Exceptional university graduates want a high income quickly.

Work in Japan and Taiwan is also methodologically very conservative, for different reasons. In Japan, where the public esteems the indigenous tradition of Chinese-style medicine, there are well-organised traditions of historical research. Some of the world's most comprehensive and painstaking studies have been carried out by Japanese scholars, to the point that a reading knowledge of modern Japanese is essential for historians of Chinese medicine.[52] The political constraints that operate in China are of course absent

---

[50] See, as an example of official presentation, China Science and Technology Museum & China Reconstructs (1983).

[51] On this point, see Sivin (1991).

[52] This volume cites the works of Okanishi Tameto, Miyashita Saburō, Akahori Akira and many others.

in Japan. But Japanese academic careers depend on maintaining the traditions of one's mentor, on keeping cultural barriers to iconoclasm and foreign influence high, and on maintaining the evidential research tradition to shape method.[53]

In Taiwan, lack of government support in a culture focussed on central approval and initiative has made specialisation in technical history rare. A handful of enthusiastic young scholars, many of them educated abroad, reflect new insights and methods in their work, and have begun to train successors.[54] But the impact on the educational system remains negligible. Few institutions are interested in hiring anyone whose qualifications must be evaluated by any other criterion than skill at evidential research.

This survey of local proclivities perhaps makes it easier to understand what has happened, and what has not happened, as the study of Chinese medicine has evolved. It is now appropriate to consider the future.

CURRENT AND FUTURE RESEARCH ISSUES

The intersection of a changing Sinology and a changing history of medicine must itself be a quickly moving point. From new successes in the study of early Europe, initiatives in the last generation and scattered innovations of the last decade or so, we can identify some research questions that can serve as first steps toward a comprehensive history of Chinese medicine:

(1) What were the traditions of health care? How were they related? To what extent did medicine compete with religious and other popular forms of therapy? How was each tradition situated in society? What was the economic base of each? How were their intellectual content and social forms related?

(2) What were the social settings of medical practice and scholarship? Were these separate activities? If so, how were they related? If not, how were they articulated to form a single complex?

(3) What were the sources of new ideas, techniques and problems? To what extent did these stimuli come from outside medicine, and from where? What role did foreign influence, and the influence of non-Han people within the Chinese realm, play in their formation?

(4) What were the directions of change in the long term? What caused medicine to move in those directions? What were the constraints on change?

(5) To what extent have changes in technology affected medicine? Have innovations in medical technique found non-therapeutic uses?

(6) How were medical ideas related to other kinds of thought and activity? The interplay with politics, philosophy and religion obviously played out differently. How can concrete instances illustrate the diversity of these links?[55]

---

[53] An important exception is the work of Yamada Keiji, an original and polymathic scholar who from time to time contributes to medical history.

[54] See especially Leung (1987) and Hsiung Ping-chen 熊秉真 (1995).

[55] For previous discussions see Sivin (1977), pp. xv–xx, and *ibid.* (1988), 42ff.

It is obvious that in order to investigate these issues they must be resolved into a multitude of concrete questions, of which I now give some examples. Some of this work has been done long since for occidental medicine, but there is very little for that of China.

European and American medicine

More than half a century ago Henry Sigerist, who greatly influenced the formation of medical history, observed that 'we know much about the history of great medical discoveries but very little on whether they were applied or to whom they were applied'. He tried to redefine his field as 'the history of healers and sick people seen within the actual context of their interaction (social and intellectual)'. He was not calling for social history, but for something broader.[56]

Historians of Western medicine have begun to study all healers, not only the eminent. They have learned to gather the very scattered records from which they can reconstruct the experience of patients. 'Medical pluralism' (see p. 28), they realise, is found to some extent in every society, from ancient Sumeria to modern France. It implies, among other things, that the study of physicians cannot of itself lead to an adequate picture of health care.

Historians now understand that most patients who choose modern medicine do not passively accept a physician's diagnosis and therapy; medical care still involves negotiation. The doctors may feel that their special knowledge gives them the last word in therapeutic encounters, but the multiplicity of healers greatly limits that power.[57]

Historians no longer choose between social, cultural and intellectual contexts, but see them as part of one picture. They now examine aspects of this picture that their predecessors overlooked. An important aspect is how differences in gender, ethnic group and social class affect health, sickness and therapy. Historians no longer analyse therapy only as curative technology that is either effective or ineffective. They see it instead as a kind of social interchange in which the drugs and other measures may be less important than ritual or other symbolic acts,[58] and the doctor less important than other people present at the medical interview, in affecting patients' concrete conditions. They are attempting to see the many dimensions of disease in a unitary way. Here is a recent definition of 'disease':

at once a biological event, a generation-specific repertoire of verbal constructs reflecting medicine's intellectual and institutional history, an occasion of and potential legitimation for public policy, an aspect of social role and individual – intrapsychic – identity, a sanction for cultural values, and a structuring element in doctor and patient interactions.[59]

---

[56] Sigerist, cited in Marti-Ibañez (1960), pp. 25–6; the second quotation is a paraphrase from Leavitt (1990), p. 1473.

[57] Chang Che-chia (1997), ch. 3, analyses the consequences of this competition in the collective behaviour of physicians. See also Li Yun (1995).

[58] Rosenberg (1997), p. 12, speaks of traditional medicine as 'a ritual legitimated . . . by a rationalistic model of pathology and therapeutic action'. This description also holds for many aspects of modern clinical therapy.

[59] Rosenberg in Rosenberg & Golden (1992), p. xiii.

Sigerist's prescription is now widely taken. It has given the history of medicine, as practised at the research frontier, great vitality and scope. It has widened the focus beyond medical ideas to incorporate the whole spectrum of health care, establishment practice and alternative practice, public practice and private practice. Scholars of European history now know something about the costs of therapy in +1650; the enormous differences between practice in cities and in small towns around +1750; the rivalry between the many kinds of healers, respectable and marginal, around +1800; and the unimportance of medical advances in increasing lifespans before about 1900.[60] By throwing light on questions of political policy and social change, historians are helping people to make sensible decisions about today's urgent health care problems.

Chinese medicine

If we turn our attention to the study of Chinese medicine, we see, in contrast, an approach to forming problems that has hardly changed in fifty years. When its specialists speak of the history of medicine they seldom mean, as historians of the West do, explaining change. Histories of Chinese medicine continue to be summaries of primary sources in chronological order. This sometimes gives a good idea of what changes took place in the interval between two texts, but does not explain why. When we find attempts at explanation, they tend to be guesses based on crude generalisations. The history of East Asian medicine has never evolved a methodology to account for change, and has ignored that of other fields. This somnolence has begun to dissipate as a number of newly qualified scholars begin to publish, but is still the norm.

We know nothing about the economics of medical practice in traditional times, nothing about occupational organisation, nothing about the careers and practices of ordinary practitioners. The histories of medicine do not investigate physicians who were not élite authors.[61] They ignore the health care that was performed by priests, healers not educated in the classics, and laymen. Although non-classical healers greatly outnumbered physicians, the historians dismiss their therapy as worthless superstition or quackery instead of studying it. They say nothing concrete about the experience of patients. When they mention such topics, they usually give offhand opinions rather than the conclusions of research.

A well-known example is the question of when medicine came to be widely considered a respectable livelihood for the offspring of élite families. Nearly everyone who asked it has given a different answer. We are told that this was a very gradual development, or that it never happened, or that it happened in one dynasty or another from the Sung on. The only systematic study, based on the rich documents of one Chiang-nan prefecture, proves that it happened there in the Yüan period. As Hymes put it, this was one result of 'a permanent shift among elite families from exclusive concentration on the

---

[60] Among many other sources, MacDonald (1981), Ramsey (1987), O'Boyle (1994), Loudon (1986) and McKeown (1979) deal with these issues.

[61] For important new discoveries of popular writings, especially about Western science and medicine, see Chung Shao-hua 鍾少華 (1996). Charlotte Furth (1998) studies healers marginal to the élite in the Ming and Chhing. Chia-feng Chang (1996) discusses what she calls 'alternative' practitioners.

achievement of high office to a broader strategy focusing on local position'. This shift took place around the fall of the Northern Sung. The Yüan transition brought other changes, including fewer opportunities to teach, that accelerated the move into medicine.[62]

Conventional work on traditional topics will continue to be necessary and valuable. We have learned a great deal about Chinese scientific and medical concepts in the past couple of decades. We need no longer rely on vague, undefined notions such as Confucianism and Taoism to explain intellectual change.

Studies of recently excavated manuscripts earlier than the Inner Canon have fundamentally changed our view of the beginnings of medicine.[63] New studies have led to a redating of most classics that used to be considered pre-Han, and to a new view of how the earliest books came together.[64] This has led us to reconsider how science separated from philosophy to form distinct bodies of knowledge, and how the high tradition of medicine emerged from a combination of popular religion, philosophy, techniques of self-cultivation and lore about effective cures.[65] Historians are clarifying technical issues that used to be dismissed without examination, for instance the importance of bloodletting in early medicine.[66] We have learned a good deal for the first time about health care among the minority peoples of China and about military medicine.[67]

There remains plenty of work of familiar kinds for specialists in textual study (*khao-chêng*) to do. The problem is that studies of medicine in the rest of the world, even Africa, no longer rest on such a narrow methodological base.[68] Their scope has changed rapidly as a result of new analytical tools adapted from history, sociology, anthropology, folklore studies and other disciplines. Ignoring this larger perspective has isolated East Asian history and made its influence on the history of medicine smaller than it should be.

A few enterprising young scholars of East Asian medicine have already begun the necessary broadening of skills and research questions. They have begin to draw freely on new sources of insight, among them the sociology of knowledge, symbolic anthropology, cultural history and literary deconstruction. I will not pause over the strengths and weaknesses of subaltern studies, ethnomethodology, discourse analysis and other methods of approach that they are learning. I will simply call attention to Chinese issues, already mentioned, on which such methods cast light. The topics discussed below range from the relations of theory and practice, through therapy as the patient rather than the doctor sees it, to the economics and social organisation of medicine.

---

[62] Hymes (1987), pp. 57, 65; see also Yamamoto (1985). This may not be the end of the story. Leung (1987) has argued that, in the Ming, families of the lower Yangtze region began to shift their children away from therapeutic careers. She explains the change by more attainable official careers and growing competition between élite and lower-class practitioners. This argument would benefit from a prosopographic analysis, since, as the Ming wore on, appointments expanded but central government posts that yielded a high income became less accessible. Again the notion of a 'changing opportunity structure' needs evidence of actual competition.

[63] I have surveyed recent research studies and reference works in Sivin (1989) and (1995b). See the important publications by Harper, especially (1998a).

[64] Loewe (1993) summarises recent studies; see also Brooks (1994). On the character of early books the most important study is Keegan (1988).

[65] Harper (1990), Sivin (1995c).   [66] Epler (1980), Kuriyama (1995).

[67] E.g. I Kuang-jui 伊光瑞 (*1993*), Li Kêng-tung 李耕冬 (*1988*), Tshai Ching-fêng 蔡景峰 (*1982*), Yü Yung-min 于永敏 (*1987*) on the former topic and Chu Kho-wen 朱克文 *et al.* (*1996*) on the latter.

[68] On Africa, see Feierman (1985), Pryns (1989).

Integration of intellectual and social history

Benjamin Elman's *From Philosophy to Philology* demonstrated in 1984 that it is possible to overcome the split between intellectual and social history, and that the rewards of doing so are rich. It showed that in the Lower Yangtze region from the Sung on, official position and commercial wealth tended to run in the same families for many generations, that such families in the Chhing sponsored a system of education not aimed at the imperial examinations, and that these schools were at the centre of major intellectual transitions. The book paid due attention to astronomy, showing that such a broad approach could throw fresh light on technical history.

As for medicine, Ma Po-ying and Paul Unschuld have raised questions that are neither purely conceptual nor purely social.[69] Ma looks closely and impartially at the role of popular religious practices in the beginnings of medicine, examines the connections of medicine and population growth, and takes up other previously neglected topics. There is still no history of Chinese medicine in any language that draws on current medical anthropology, sociology and general history, but both Unschuld and Ma have raised important questions that carefully educated young scholars can hope to answer.

Theory and practice

The changing relation between medical theory and practice has been an important theme in the history of Western medicine. It regularly appears in studies of China, but has not led to rigorous research and firm conclusions. Needham assumes that medicine is a body of abstract knowledge derived directly and unproblematically from empirical findings. Unschuld avers that yin–yang, Five Phases and other basic concepts were a collective delusion that distracted physicians from empirically sound results.[70] Both positions are redolent of an earlier generation. Both are philosophically questionable and uninformed about the use of theory in practice.

The anthropologist Judith Farquhar has broken through this impasse with a study based on prolonged and systematic field observation in a school of traditional Chinese medicine and its clinic in Guangzhou. At the same time it is informed by familiarity with history. It makes a strong case that one cannot speak of theory and practice as separate activities which may or may not be related. Joanna Grant has studied the interaction of practice and doctrine in the early +16th century.[71]

Much of the confusion has arisen because in Europe, after the time of Galen, the near collapse of classical culture separated the men of religion, who preserved the books but did not use them for practice, from the healers, who had no access to the books. What had previously been a unity, in which concepts arose from therapy and served it, was decisively sundered. It took many centuries to overcome this break.

---

[69] Ma Po-ying 馬伯英 (*1994*), Unschuld (1985); see also Li Liang-sung 李良松 & Kuo Hung-thao 郭洪濤 (*1990*).

[70] E.g. Needham (1980), p. 15, and Unschuld (1985), p. 197. These points of view pervade the writing of both.

[71] Farquhar (1994); see also *ibid.* (1995), Grant (1997).

In China no such break ever took place. Foundational works from the Inner Canon on state clearly their concern with practice. It is remarkable how many authors of very abstract medical books, even philological commentaries, were practitioners, and explicitly address the needs of praxis.

I have found it heuristically desirable to avoid the word 'theory' in connection with medicine. One is less likely to confuse the very different European and Chinese experiences if, with respect to China, we speak instead of the 'doctrines' that underlie practice, and resist the temptation to trace their history without reference to therapy.[72]

The complexity of this issue has become clear in a new study of the Heat Factor Disorder (wên-ping 瘟病) movement, which historians generally describe as the last major new school of theory. They trace its origins as a school to the Wên i lun 溫疫論 (Treatise on Warm-factor Epidemic Disorders, ca. +1642) of Wu Yu-hsing 吳有性, and explain that it was active by the middle of the +18th century.

Marta Hanson's careful reading of the early writings associated with this 'school' indicates that at the time no one associated them intellectually or institutionally. She has discovered that physicians of the Lower Yangtze region invented this school in the 19th century.[73] The novel issue around which they formed it was not, as historians usually claim, the intricate conceptual relation between the various diseases in the Cold Damage (shang han 傷寒) group. It was rather a method of approach to therapy for epidemic fevers suitable for practice in the south, where classical approaches, doctors believed, were not appropriate.

The late fabrication of a Heat Factors school, putatively based on a new doctrine, is not just a medical issue. It was part of a much broader social movement that created local identities for Soochow and other cities in the Lower Yangtze region, and for the region as a whole – a process that was bound to challenge earlier assumptions about the universal scope of classical medicine.

Local studies

The examples just given are typical of the growing importance of local traditions in research. Work of this kind raises an important point peculiar to China.

Anyone who writes on European medicine is aware of the great diversity of that subcontinent. The category often refers to a culture that élites of many countries who were educated in Latin shared. But if we focus on the full diversity of practice and activity up and down the social scale rather than on the learning of a minority, on health care rather than medical doctrine, 'Europe' is too large a unit to be of any use.

---

[72] Historians of Europe are also finding it useful to resist this temptation. See, for instance, Garcia-Ballester et al. (1994).

[73] Hanson (1997). This is a complex story; key works were Yeh Kuei's 葉桂 Wên je lun 溫熱論 (Treatise on Heat Factor Disorders, +1746?), tractates collected in Thang Ta-lieh's 唐大烈 Wu i hui chiang 吳醫會講 (Collected discourses of Soochow physicians, +1792), Wu Thang's 吳塘 Wên ping thiao pien 瘟病條辨 (Systematic manifestation type determination in Heat Factor Disorders, +1798) and Wang Shih-hsiung's 王士雄 Wên je ching-wei 溫熱經緯 (System of Heat Factor Disorders, 1852). See the conventional accounts in medical-school textbooks on 'schools of medicine', e.g. Kuo Tzu-kuang 郭子光 (1988).

China, a yet larger unit, is even less suitable. G. William Skinner long ago made this point in great detail. He proposed nine distinct regional systems, which he called physiographic macroregions. They are divided by the drainage basins of rivers. Each macroregion is environmentally distinct. Each has its own social, cultural, economic and political resources, which gave it a history more or less independent of the others. Skinner showed that much of what we call Chinese history holds only for the region in which the capital was located. Officials gathered and aggregated some of the data in the standard histories so unreliably that accurate studies must use local records instead.[74]

This is hardly a trivial issue. Anyone who reads a few hundred medical books from the Sung on is aware that Skinner's Lower Yangtze region (parts of Kiangsu, Anhwei and Chekiang), and parts of his Middle Yangtze (Kiangsi) and Southeast Coast (Fukien) regions have consistently provided leadership in medicine for the past thousand years. It is no doubt pertinent that they have also provided leadership in trade, agriculture, science, technology and, certainly not least, politics. On the other hand, considering the importance of Szechwan as a source of drugs, it is remarkable that no important figure in medical practice from there, aside from legendary ones, is known from before the Sung, and that Szechwan medical authors do not appear regularly before the Ming.[75]

Another pertinent example is the work of Wu Yiyi (1993), who has challenged the looseness with which historians use the word 'school'. He has reconstructed the lineages of transmission of medical writings that began with the famous Liu Wan-su 劉完素 (+1120? to +1200). Wu was able to find biographical information on fifty-three physicians who claimed to be in this line of transmission. For each he determined what the nature of the linkage was. Although the textbooks describe this 'school' in terms of theoretical tendencies, that is not what held it together. Nor was it united by a single style of practice. First Chang Tshung-chêng 張從正 (or Tzu-ho 子和), a generation after Liu, and then Chu Chen-heng 朱震亨 (or Tan-chhi 丹溪, +1281 to +1358) championed practices quite different from those of Liu. That did not, however, undermine the unity of the lineage.

In the early phase, what defined it was precisely its locality, Ho-chien 河間, the home town of Liu Wan-su in Hopei province, in or near which all the main figures lived. When after a century the centre moved decisively to the Lower Yangtze region, its relationships changed just as decisively. It was largely the widespread distribution of printed medical books that again changed those relationships to make local, personal ties less important than links through textual study.

Local studies have also brought precision for the first time to the history of epidemiology in China. Carol Benedict has used gazetteers and other local materials to reconstruct in remarkable detail the transmission of bubonic plague from one locality to the next in the 19th century. Paul Katz has used a similar technique to document the

---

[74] The most important sources are the introductory essays in Skinner (1977; map of macroregions on p. 214) and the compendious statement in *ibid.* (1985); on local data, see *ibid.* (1987).
[75] Jen Ying-chhiu 任應秋 (*1946*); see the data collected in Chhen Hsien-fu 陳先賦 & Lin Sen-jung 林森榮 (*1981*).

spreading cult of an epidemic god in 19th-century Chekiang. Studies of this kind offer a pattern for many new kinds of investigation.[76]

Explaining change

The few projects that so far have sought to trace change rigorously have been productive. They point the way to much more comprehensive historical explanations.

As an example, it has been clear for some time that the Northern Sung was a major time of transition. Many aspects of medicine changed, including its role in the civil service, the role of government in medical education, the classics people studied, the kinds of medical innovation, special attention to the disorders of women and children, and the economics of the drug trade. In a series of classic studies, Miyasita Saburō has shown that there was a widespread shift in the drugs most often used for particular ailments.[77]

There have been many guesses, ranging from the influence of neo-Confucianism to effective government control, to explain why medicine was so decisively transformed. There has even been some research, bearing on such factors as the innovative and flourishing economy centred along the southeastern coast, the growth of medical publishing, and the changing social circumstances of women. Rather than continuing to expand the list of unrelated, unweighted factors, a comprehensive analysis is needed.[78]

Medical pluralism

Therapy does not begin with doctors. It was normal in China, as in every other society, for people to rely first on themselves for treatment, and then on their families and other people around them. When that did not work, people chose in turn from a great variety of curers, medical, religious and so on. Anthropologists call this 'medical pluralism'. They have been exploring the topic in China for much of this century. We find great overlaps with magical and religious curing from the very beginnings of medicine to the present day.[79] My own work in progress on the diversity of health care in the last thousand years reveals a consistently rich picture, and indicates that élite physicians provided only a small part of the health care of the whole population. It shows not mutually exclusive realms of 'scientific' and 'superstitious' medicine but an unceasing flow of methods and understandings up and down the social scale, constantly translated back and forth between the vitalistic and god-centred view of popular medicine and the cosmological and secular assumptions of the various high traditions. Early sources for studying practitioners who did not belong to the élite, although seldom noted, are not rare. This work can progress rapidly as more young historians learn the skills of current anthropology.

---

[76] Benedict (1996), Katz (1995).   [77] Miyasita (1976, 1977, 1979, 1980).
[78] Asaf Goldschmidt is completing such an analysis in a dissertation.
[79] See notably Kleinman's (1980) anthropological study of contemporary Taiwan. On early texts, Harper (1990, 1998b); for occult traditions coexisting with 'official' medicine in the late 1980s, see Hsü (1992).

Gender, ethnic group and social class

Medical history has usually been written as though the same therapy were available to everyone in China. This has not been true of any society. Inequality of treatment based on social and ethnic distinction has not yet been studied, but it needs attention if we are to understand fully the spectrum of health care in China.[80]

In the history of medicine generally, questions of gender have ceased to be simply a feminist theme. They have to do with the most fundamental characteristics of health care. Although the body as defined by modern biology is much the same everywhere, people think about their own bodies and those of others, and their ailments, differently in every society and even every subculture. The diseases peculiar to women are not only physiological conceptions, they are also tools of social control, species of deviance and justifications for low status. Insights about gender cast light on every aspect of medicine, for men as well as women.[81]

Gender issues illuminate Chinese medicine. Charlotte Furth has presented evidence that in the +17th through 19th centuries physicians drew the boundaries between genders to emphasise the physical deficiency of women and define their potential almost exclusively as a matter of childbearing and rearing of children. She argues that distinct views of female bodies and their disorders emerged in the Sung period and evolved in the direction that she has already described. As Grant (1997) has shown, a picture drawn from medical case histories of men as well as women is considerably more nuanced than one derived from writings on childbearing and women's disorders.

Furth also takes up the neglected topic of women as practitioners, not only midwives but doctors. A number of female physicians were not limited to gynaecology, but practised across the whole range of medicine. At the same time, their own seclusion restricted them to treating women and children.[82]

Experience of patients

The voice of the patient used to be silent in the history of medicine. In the last decade, however, it has become a central theme in the study of Europe. The writings of patients offer a wealth of sources. We can now reconstruct the experience of surgery before anaesthetics, and understand why patients until late in the 19th century endured the purging, bleeding and other drastic remedies upon which doctors depended. We have also learned from patients how inadequately physicians recorded the roles of other healers, especially those of midwives, in health care.[83]

---

[80] Dikötter's book, *The Discourse of Race in Modern China* (1992), although it cites more perceptions of ethnicity than of race, demonstrates that prejudice existed and was freely expressed in the early 20th century.

[81] On cultural definitions of the body, the classic anthology is Feher (1989); see also Gallagher & Laqueur (1987) and Laqueur (1990). The essays in Rousseau (1990) approach the relation of mind and body from a historical point of view. Feher contains a few essays on China and Japan; see also Zito & Barlow (1994). On the dimensions of disease, I paraphrase Rosenberg & Golden (1992), pp. xv–xvi.

[82] Furth (1986, 1987, 1988, 1998). On the last point see Furth (1996). An earlier survey is Liu Hai-po 劉海波 *et al.* (*1982*). A detailed study of Ming paediatrics is Hsiung Ping-chen 熊秉真 (*1995*).

[83] Porter (1985), Porter & Porter (1988), Rosenberg (1977), Ulrich (1990).

In China we do not find the intimate autobiographies and diaries that have been mined so successfully for the West.[84] Still, there are more than enough sources of other kinds to open up the subject. Doctors have occasionally recorded their own medical experiences among their case records and other writings. They and articulate laymen such as Yuan Mei 袁枚 (+1716 to +1798) have had a good deal to say on the subject in prefaces and informal writings, but historians have not yet surveyed them for this purpose.[85] On the other hand, for half a century scholars have been studying the rich depictions of therapy in literature and drama. These are usually written from the patient's point of view, and often make the physician a figure of fun.

Finally, we have massive information about a very exceptional class of patient, namely emperors. Chêng Wên 鄭文 (1992) has reconstructed a grotesque incident in the +1060s involving the death of Jen-tsung, followed by what seems to have been the mental breakdown of the newly enthroned Ying-tsung. Officials discredited and savagely punished doctors inside and outside the palace medical service as an outlet for frustration over the choice of Jen-tsung's successor.[86]

Negotiation in medical encounters

Records in the palace archives make it clear that imperial patients, at least, had a great deal to say about the treatments they received, and did not hesitate to say it, whether politely or abusively. Their absolute power made their encounters with their doctors exceptional. Still, there is a parallel to their negotiations in the tug of war between the ordinary patient's economic status and the physician's occupational authority.

Medical textbooks and handbooks routinely give the impression that doctors could count on less-exalted patients to do what they were told. But informal writings of doctors, and what we can learn directly from patients and historical anecdotes make it clear that therapeutic relationships were negotiated. As early as we have detailed information on cases, from the Thang onward, we find doctors complaining about patients who ignore their instructions and people in the sickroom who argue with the physician. When patients and their families found that their wishes were not taken seriously, or when they lost confidence in their doctor, they did not hesitate to consult someone else. Because patients chose freely between different types of practitioner, in no set order, this issue is inseparable from that of medical pluralism.[87]

---

[84] Wu (1990) has shown, however, that autobiography is a richer genre than previous authors have imagined. For an exploratory essay on doctor–patient relationships, see Li Yun (1995).

[85] Recent reference works such as the anthologies of Fêng Han-yung 馮漢鏞 (1994) and Thao Yü-fêng 陶御風 et al. (1988) greatly ease such work.

[86] See Sun Ssu-mo's records of his own illnesses in Sivin (1968), Appendix A. There are much richer materials on physicians' own medical experiences in later sources, such as the Chê kung man lu 折肱漫錄 (+1635). On medical encounters in literature, see Li Thao 李濤 (1948), Idema (1977) and Cullen (1993). Some anecdotes in Sivin (1995g) are pertinent.

[87] For emperors' attitudes toward their physicians see Chhen Kho-chi 陳可冀 et al. (1982a), Chhen Kho-chi (1990) and Chang Che-chia (1997). Sun Ssu-mo 孫思邈 mentions patients who defied him, for example between +627 and +640, an old man who died rather than give up sexual intercourse; Chhien chin fang 千金方, ch. 27, pp. 28b–29a. Among the many essays in which doctors complain about the recalcitrance of patients and bystanders are I tsung pi tu 醫宗必讀 (+1637), ch. 1, pp. 8–9, Ching-yueh chhüan shu 景岳全書

Economics of medical practice

Early primary sources are seldom explicit about the costs of goods and services, with obvious exceptions such as the price of grain. Sinologues have made some progress over the past generation in working out a history of commerce. We still know very little about the costs of therapy and the incomes of doctors.[88]

One reason is that élite patients and their physicians preferred to treat therapy, like art, as a matter of give-and-take between gentlemen rather than commerce. The more esteemed the painter or physician, the fewer fees and the more gifts he received. Both careers, in later dynasties, offered exceptional social mobility. Doctors concerned about respectability wanted to be thought of in the same class as officials, not as mercenary artisans or tradesmen.

The sources do not openly discuss the cost of treatment by the cream of physicians, and no one bothers to provide information about the majority of practitioners who did not belong to the élite. One might, of course, begin relieving our ignorance by paying close attention to the gifts that recovered patients gave to their doctors. A second obvious move is to identify patients named in case records and thus determine who the clients of a given physician were. I have even found it possible to link the practices of doctors by determining that two of them record treatments for the same patient.

Patronage was a special social form with complicated but generally understood rules. Historians of Europe have recently shown that it was the major source of support for innovators in science and medicine in the +16th and +17th centuries. At times, in the upper social reaches of Chinese medicine, it was equally important. It is well known that, among others, Thao Hung-ching (mentioned earlier) and Sun Ssu-mo 孫思邈 (alive +673) were supported for long periods by the largesse of emperors, and Ko Hung 葛洪 (+283 to +343) by that of high officials.[89]

From the Sung onward, this sort of patronage becomes less visible. Instead, the support of officials and later of local gentry for printing medical books grows in importance. In some Ming and Chhing books, we find lists of subscribers with the sums that they contributed.[90]

There were many other forms of patronage. For instance, in the period +1600 to 1850, the central and local medical academies supported temples to medical deities as a means of 'maintaining symbolic control over the provision of medical care'. This support, like many other kinds of government subvention for popular religion after the Sung, was justified precisely because only symbolic control was possible.[91]

(+1624), ch. 3, pp. 76a–77a, and *I hsüeh yüan liu lun* 醫學源流論 (+1757), p. 131. Li Yun (1995), pp. 68–72, translates the first essay, and Unschuld (1989), pp. 385–7, the second (not very reliably).

[88] A large proportion of the best work is Japanese, but see Wilkinson (1972) for an early example of Western research. For an admittedly 'crude guess' about the number and average income of 'gentry doctors', based on a small collection of biographies from the first half of the 19th century, see Chang (1962), pp. 117–22.

[89] See, for instance, Moran (1991), a collection of studies on patronage, Biagioli (1993), an innovative study of Galileo, and Sarasohn (1993). Patronage has been applied to the history of Chinese thought in Sivin (1995f). On Ko see *Pao Phu Tzu wai phien* 抱樸子外篇, ch. 50, p. 8a, tr. Ware (1966), pp. 15–16.

[90] Hanson (1997) has reconstructed the networks of patrons and supporters of *Wên i lun* from all the Qing prefaces to over 100 editions of this canonic book, as well as from related texts.

[91] Chao (1995), pp. 131–40.

Publishers not only printed medical books but wrote new ones, compiled popular editions of classics, and invented treatises by combining old ones. Their complex activity will obviously play an important part in clarifying the economics of medicine.[92]

Another area of medical economics ripe for exploration has to do with medicines. There have been some Chinese and Japanese studies of the drug trade, although they have not paid much attention to prices. There has been more useful scholarship on the commerce in perfumes and other aromatics, which overlapped considerably with that in drugs. There are a few accounts of long-lived apothecary shops. The famous Hall of Shared Benevolence (Thung-jen-thang 同仁堂) of Peking catered to the Palace Medical Service for a large part of the Chhing dynasty, but the palace controlled the prices it paid for drugs.[93]

Finally, an important issue in the economic history of medicine is the growth of what we might call a national drug market, and the shifts in where particular drugs came from. Here, too, there have been a few useful studies.[94] Easily available primary sources can cast a broad light on this subject.

Occupational organisation

Sociologists dissatisfied with the conventionally vague usage of the term 'profession' distinguish the traditional professions from other occupations by the autonomy that society grants them. This autonomy lets them control who enters and leaves their ranks and set their own compensation.[95] By that definition, Chinese medicine was never a profession. At no point in the past can we speak of a single autonomous occupational group.

I do not know of any organised attempt by élite doctors in pre-modern China, no matter how much they complained about quackery, to prevent practice by people unlike themselves. The government attempted briefly in the Sung dynasty to set unified standards and examine physicians for local service (see Section (c)). There is no reason to think that this initiative had any lasting effect, and it was not revived. Such a move, had it lasted, might have led to a sense of occupational identity, but hardly, given its origin, to autonomy.

The first associations, in the early part of the 20th century, could hope for no more than to protect traditional medicine against abolition. They learned to organise lobbies

---

[92] Among the most interesting publishers are Tou Chieh 竇杰 (often called Tou Han-chhing 竇漢卿) of Fei-hsiang 肥鄉, Hopei (+1196 to +1280), Hsiung Chün 熊均 (or Hsiung Tsung-li 熊宗立) of Chien-yang 建陽, Fukien (fl. between +1437 and +1465), and Wang Chhi 汪淇 of Hangzhou (fl. +1665). Hsiung led the move toward popular editions of classics, as well as the popularisation of Phase Energetics (*wu yün liu chhi* 五運六氣), a widely used method of diagnosis. On publishers' concoctions, much information is scattered in Ma Chi-hsing 馬繼興 (*1990*). See Widmer (1996) on Wang, and Chia (1996) on Yuan and Ming publishing.

[93] Chang Phei-yü 張培玉 *et al.* (*1993*). See Ni Yün-chou 倪雲洲 (*1984*) and Mei Khai-fêng 梅開豐 & Yü Po-lang 余波浪 (*1985*) for local studies of the drug business, and Lin Thien-wei 林天蔚 (*1960*) and Kuan Lü-chhüan 關履權 (*1982b*) for Sung aromatics and medicinals. There is also much of interest on this and other medical topics in the reports of the Chhing Imperial Maritime Customs.

[94] E.g. Li Ting 李鼎 (*1952*) and Wang Yün-mo 王筠默 (*1958*).

[95] Freidson (1970) provides the best critical discussion for historical purposes. Chao (1995) makes professionalisation a focus of her study of Soochow physicians in the Chhing. She uses 'profession' to mean 'livelihood'.

only when the Nationalist government threatened to abolish the legal practice of their art. Since 1949 doctors have been civil servants, until private practice began recently to re-emerge on a limited scale with official approval.

If we want to understand the inability of physicians to organise before modern times, the first place to look is guilds (usually called *hang* 行). We know that Chinese guilds were local associations to restrain competition. They were not broad organisations, claiming from government or society authority to regulate themselves. Despite a few good studies,[96] we do not know much more than that. We have no idea how medical guilds differed from those of tradesmen.

The obvious next step is to look at local medical practitioners and their networks. A good deal of the recent scholarship on that topic has to do with 'schools' (as usual, a term used so vaguely as to be meaningless) defined by books, not by interactions between people that create permanent institutions. Still, there are now a few good biographical compilations by province or, in the case of Huichow, by prefecture.[97] These do not analyse their materials for signs of medical organisation, but they do facilitate this work. For one thing, it is now possible to trace the movement of physicians in and out of official positions, as they traded in their court status for local eminence. This pattern becomes important in the Ming and Chhing periods.

Finally, doctors sometimes mention in their case records other physicians who were consulted at the same time as they, or who had failed to cure the patient earlier. A careful biographical analysis of these data will let us establish local links of cooperation and competition between doctors.

At the same time, the issue of competition will need to be handled subtly. In China, lacking effective organisation or government regulation, élite physicians could not hope to monopolise even the care of the gentry. Most collections of case records leave a clear impression that literate doctors were generally not eager to treat their inferiors. The overwhelming majority of the population, in the countryside, were barely involved in the money economy before the 1980s. They could hardly shop around for services.

The doctors' frequent complaints about quacks, itinerants, priests of the popular religion, spirit mediums and Taoist masters were clearly meant to impress prospective patients.[98] These complaints were neither part of an organised campaign nor tools that a professional group adopted to compete against rival guilds. There was no such cohesive group.

I would suggest that individual physicians were trying by this rhetoric not to protect a profession from incompetents, or to drive their competitors out of business, but to assert their personal standing in the company of the Yellow Emperor's successors. No matter what their birth, they wanted to be identified with conventional upper-class

---

[96] E.g. Niida (1950), Katō (*1953*).

[97] Chhen Tao-chin 陳道瑾 & Hsüeh Wei-thao 薛渭濤 (*1985*), Li Chi-jen 李濟仁 (*1990*), Liu Shih-chüeh 劉時覺 (*1987*) and Yü Yung-min 于永敏 (*1990*).

[98] The gentry and their physicians called doctors of whom they did not approve *yung-i* (庸醫), and lumped together the last three categories as *wu* 巫. Neither of these terms stands for a social category. Both were epithets.

medicine, not with the popular therapy that orthodox male patients despised, or the many varieties of practice in the undefined area between the two. The issue, in other words, was social mobility.

Once we look at the Chinese data in the light of recent European studies (e.g. Ramsey, 1987), we see linked to these social issues a political one. In the late Ming and Chhing, profits from business easily bought official status. At the same time, fewer descendants of the gentry than before could hope for good careers. The borders between simple respectability and no prospects at all seemed vague and chaotic. Pronouncements about the need for strict stratification of health care *at the same time* expressed nostalgia for a stable society that seemed to be quickly disappearing.

TERRA INCOGNITA: THE SCIENTIFIC VALUE OF THERAPIES

The clinical evaluation of traditional techniques is too important to ignore, but no one has come to grips with it, and so far it appears to be intractable. It is probably impossible to form a view of early medical practice without considering its usefulness in curing medical disorders. Given the limits of curing before the last couple of generations, it might be more apposite to speak of its utility in supporting rather than interfering with the patient's recovery.

It is hard enough to measure the curative power of a broad range of therapies, understood from the purely technical point of view. One must also ask whether their power was in fact purely technical. If it is equally necessary to consider the social and ritual circumstances in which each therapeutic encounter took place, how does one weigh their role? One must then admit that, although medical case records from the Sung on are often detailed and concrete, they are practically never detailed and concrete enough to indicate exactly what the patient's condition was according to modern diagnostic criteria. Any practitioner is aware of how seldom a mature patient presents a simple and classical form of a single disorder. One also encounters claims of cures so clearly at odds with modern knowledge that one has to consider them sceptically, or at least to admit that what brought about the result more likely lay in the realms of emotion, ritual and symbol than in those of chemistry and physics.[99] But the case records rarely mention emotion, ritual and symbol. These obstacles make the prospect of reasoned evaluation seem unattainable.

Medical historians seldom reflect on this problem. Many in East Asia accept any assertion about efficacy that they find in the sources. Conversely, many outside East Asia refuse to take any such assertions seriously, or even to contemplate them.

Neither of these positions is tenable. The second position falls into one of the most elementary traps of historiography, namely assuming that the people one studies were less alert and capable of critical reflection than oneself. The first assumes that ancient physicians were able to determine cures and cure rates more precisely than modern doctors can do today. This too is not an informed position. One looks in vain for a clear

---

[99] Cooper & Sivin (1973).

exposition of how early practitioners defined cure, and by what means they determined when it had taken place.

These questions can be answered. Unlike most ancient biographers and historians writing about medicine, some physicians admitted that their therapy did not always succeed, and that not all virtues recorded for drugs could be validated. Aside from explicit statements, one can assess how often a given therapy was passed down from one written source to another – always asking whether the reasons for its popularity were technical, ritual or a habit of copying on trust.

If we examine therapies one by one, we discover that it is no simple matter to assess their efficacy. The concentration of active substances in animal and herbal drugs is far from constant, depending on where they come from, in what season they were gathered, how they were processed, and so on. It is extremely difficult to untangle the pharmaco-dynamics of constituents in compound formulae. Modern clinical trials of prepared formulae (conducted for commercial, not regulatory or historical purposes) have not ordinarily met current biomedical standards.

Ancient claims for therapy using acupuncture exclusively are rare and practically impossible to evaluate scientifically. No one has yet found a way to apply double-blind testing to its evaluation. We have no basis for measuring the frequency with which it was used for a given disease. It was in any case only one therapy among many, seldom used alone. In some periods its practice was decidedly out of fashion and, in at least one, forbidden to be taught in the Imperial Academy of Medicine.[100]

Blithe claims about curative efficacy also commonly overlook the fact that ancient disease entities do not correspond to those of biomedicine. There was no disease closely equivalent to typhoid, cancer, tuberculosis or, despite the many recent assertions of entrepreneurs, the acquired immune deficiency syndrome (AIDS). To take the first of these as a single example, *shang han* 傷寒, the modern term for typhoid, has an entirely different meaning in traditional medicine even today. It designates both a particular disorder, Cold Damage (its literal meaning), which may correspond to a number of acute infectious febrile diseases including the common cold, and a large group subsuming it.[101] It is also likely that any classical discussion of acupuncture for *shang han* ignores other therapies that the physician brought to bear – and the most important element, the recuperative powers of the patient. In traditional medicine, early or contemporaneous, as in European medicine over most of its history, the conventional wisdom credited the doctor with the cure of every patient who recovered.

But there were unquestionably failures. Cullen (1993) has made the point that, although imperial officials often failed in their duties, that frequently mattered less than whether they said the right thing. The issue was whether orthodox political and ethical convictions backed their actions. In the same way, a physician accomplished his

---

[100] Andrews (1995), p. 13. Historians occasionally misread this internal order of 1822, which simply terminated the department of acupuncture in the palace's medical school, as an edict generally promulgated to outlaw the practice of acupuncture throughout the empire. There is no evidence that its use diminished even slightly as a result. There seems, however, to have been a marked diminution about half a century earlier; see *I hsüeh yüan liu lun* (+1757), p. 98, tr. Unschuld (1989), p. 244.

[101] Sivin (1987), p. 84.

expected task when he set his patient's disorder within the framework of medical cosmology, even if he could not bring about a cure. The same might be said of European physicians in the late +18th century, although they quoted Galen rather than the Yellow Emperor.

These many uncertainties are genuine. We can only hope that future students will overcome them. That is not optional once we acknowledge the centrality of practice in Chinese medicine.

EDITING CONVENTIONS

In revising this volume, my main concern has been to maintain its character as a diverse group of important contributions that, mostly in the 1960s, laid the foundation for a broader and more rigorous approach to the history of Chinese medicine over the last generation. I have taken care to maintain the authors' general interpretations, discussed above, intact. On the other hand, it is obviously important to reflect the results of recent research in the essays when this can be done without interfering with the main arguments.

With this in mind, I have revised many translations, explanations, dates and references. In doing so, it has occasionally been advisable to add a brief clarification, or to drop a statement that conflicts with what is clearly known today, and is not indispensable to the larger theses. I have not labelled such changes when they are minor. At the same time, I have tried to meet the needs of readers concerned with the state of the art by pointing out in footnotes where newer publications have challenged judgements about meaning and importance. Whenever there was danger that a minor revision would make the original discussion incoherent or confusing, I have merely summarised current understanding or important alternatives in a footnote, marked to indicate that it comes from the editor and not the authors.

I have also found it advisable to remove repetition and in a few instances to move a discussion to an earlier essay than the one in which it originally appeared. In a couple of cases I have introduced material from an unpublished essay of Lu and Needham in place of a useless reference to it.

I was able to discuss this *modus operandi*, as well as many concrete changes, with Joseph Needham in the last two years of his life. I have tried throughout this volume to incorporate the understanding that we reached.

Official titles are now translated according to Hucker's authoritative *A Dictionary of Official Titles in Imperial China* (1985). It was difficult to choose between maintaining consistency with previous volumes and introducing consistency with almost all recent scholarship. Hucker carries the day because his article on each title provides detailed and reliable comments on rank, duties, and historical changes in duties and meanings. The articles in the *Dictionary* are arranged by title in Chinese, but a good index makes it easy to find them by looking up the English translation of the title. Many of the authors' designations for artisans and functionaries are not recorded as official titles, and thus are not found in Hucker; they remain unchanged. I have retained a very few titles that typify Needhamite panache or British usage: 'Astronomer-Royal' instead of 'Grand Astrologer',

'Regius Professor' for 'Erudite', and the English 'Marquess' instead of Hucker's French 'Marquis'.

ACKNOWLEDGEMENTS

I am grateful to Bridie J. Andrews, Christopher Cullen and Marta Hanson for suggestions that improved this Introduction, to Francesca Bray and Judith Farquhar for ideas that were most helpful in assembling this book, and to Christine F. Salazar, Sally Church and Asaf Goldschmidt for research assistance. The Needham Research Institute kindly supported a stay in Cambridge, and as always provided a delightful place to work. I was pleased to present an excerpt from this Introduction at the Eighth International Conference on the History of Science in East Asia, Seoul, Korea, 26–31 August 1996, and to publish a section of it in revised form in *Positions*.

44 MEDICINE

(a) MEDICINE IN CHINESE CULTURE

First of all it is necessary to consider the relations between the great medical systems of humanity and the cultures or civilisations in which they arose. It is surely a hopeful circumstance that Europeans are now giving up their rather self-satisfied parochialism and are eager to look at other systems of medicine, not only in the past before our modern civilisation came into being, but also in other parts of the Old World which have highly continuous and complex civilisations paralleling our own.

The attachment of Chinese medicine to its own culture is so strong that it has not yet entirely come out of it. All the sciences of ancient times and the Middle Ages had their distinct characteristics, whether European, Arabic, Indian or Chinese. Only modern science has subsumed these ethnic entities into a universal mathematised culture. But while all the physical and some of the simpler biological sciences in China and Europe have long ago fused into one, this has not yet happened with the medical systems of the two civilisations.[1] As we shall later see, much in Chinese medicine cannot yet be explained in modern terms, but that means neither that it is valueless, nor that it lacks profound interest. We hope that this volume may lead to greater mutual understanding in the inter-cultural and intercivilisational confrontations of our times.

We shall consider in this introductory essay a number of topics indispensable to an overview of classical medicine: its doctrines and early history, the influence of China's characteristic forms of government and religion, acupuncture as a quintessential therapy, the differences between traditional and modern medicine, and the prospects for their integration.

(1) THE GENERAL POSITION OF MEDICINE AND MEDICAL DOCTORS IN TRADITIONAL-CHINESE SOCIETY

In order to understand the position of medical men within Chinese society through the ages, it is first of all indispensable to recognise that the thinkers and the experimenters, the inventors and the physicians, age after age, came from every stratum of society.

It is appropriate to divide the sciences into the 'orthodox' and the 'unorthodox'. Chinese regarded the latter as slightly sinister, and certainly *outré*. Thus in a Confucian society there were limits beyond which a gentleman could hardly go without turning into a Taoist naturalist, not beyond the pale of course but one who had definitely turned his back upon the establishment career of worldly wealth and station.

---

[1] We speak of modern-Western systems of medicine and of traditional-Chinese systems. Concerning the fusion of Chinese and European science see p. 65.

Fig. 1. Consultation with a physician. A section from a copy of the handscroll painting by Chang Tsê-tuan, *Chhing-ming shang ho thu* 清明上河圖 (Going up the river on the Spring Festival, +1125). The doctor, whose shop sign dignifies him with the official title Chief Administrative Assistant Chao, is seen examining a child at the centre. To the right, people are drawing water from a well, and in front a scholar rides by with a servant behind him carrying his lute. The city depicted is Khaifêng just before its fall to the Chin Tartars. From Needham (1970), Fig. 95, facing p. 441.

Mathematics belonged to orthodoxy because the results of computations were essential for the public planning of Confucian officials, though a special skill of this kind would never take one beyond the back rooms of provincial governors. Astronomy was rather more gentlemanly because there were a limited number of jobs in the Bureau of Astronomy attached to the imperial court. Engineering activities, as in hydraulics, bridge building and poliorcetics, were also not unsuited to a magistrate, who might well be wise enough to listen from time to time to the experienced advice of an illiterate foreman grown old in the service. But on the other hand, alchemy and proto-chemistry were distinctly unorthodox, very close to divination, genethliacal astrology, palmistry, physiognomy and various more or less black arts, such as glyphomancy and chronomancy.

Gentility extended only so far. There was nothing wrong in a learned man consulting the *I ching* 易經, and many highly regarded scholars were practical mutationists. The aura of respectability wore rather thin with the readers of geomantic compasses; the geomantic 'Masters of Ganchow' (Kan-chou hsien-shêng 贛州先生) were not often as highly regarded as their title pretended.[2]

Agriculture again was orthodox. Seeing that the main wealth of the whole empire grew up green from out of the earth, writing about farm management was perfectly in order for gentlemen, just as it had been in ancient Rome. This often included rural engineering. Botany and zoology never became distinct fields of study, but were included in the pharmaceutical natural histories and the agricultural literature. The latter comprised a great many books and monographs on horticulture, even of specific genera of flowering plants.

Medicine and the studies which supported it, such as pharmacy and the anatomy of the acupuncturists, constituted a fully borderline case. They were not situated at a particular point inside or outside the conventional social scheme. Therapists could be found at every level from the imperial court to the most isolated mountain temple, and the varieties of healing were correspondingly manifold. Our concern is learned medicine and its written traditions, but it is advisable to keep in mind that this was only the tip of the iceberg of ancient health care.

Pride of place in any sociological investigation of physicians must go to the problem of their social position. Greek appreciation of doctors is well known, as witness the quotation from the *Iliad* (XI.514) which my father was always citing to me:

> A good physician skilled our woes to heal
> Is worth an army to the public weal.

The whole history of the social position of doctors in China might be summarised as the passage from the *wu* 巫, a sort of technological servitor, to the *shih* 士, a particular kind of scholar, clad in the full dignity of the Confucian intellectual, and not readily converted into anyone's instrument. As it is said in the Confucian Analects, 'the gentleman does not act as an instrument (*chhi* 器)'.[3] During the −2nd and −1st centuries, in the Former Han period, there were many men of an intermediate sort called *fang shih* 方士; these were magicians and technologists of all kinds, some of them pharmaceutical and medical. Some Sinologists have translated this expression as 'gentlemen possessing magical recipes' and this, if somewhat stilted, is certainly not wrong.[4]

Their origin in the *wu* connects physicians (*i* 醫), despite their status as *shih*, with one of the deepest roots of Taoism. Far back at the dawn of Chinese history in the −2nd

[2] Vol. 4, pt 1, pp. 242, 282.   [3] *Lun yü*, 2, 12, tr. Legge (1861), p. 14.

[4] [But it is not revealing, since *fang* implies technical methods that are in no sense necessarily magical. *Fang shih*, like *wu* (discussed in the Introduction), is an epithet, not the name of a social grouping; see Sivin (1995d), pp. 29–34. − Editor]

millennium, probably before the beginning of the Shang kingdom, Chinese society had its 'medicine-men', something like the shamans of the North Asian tribal peoples. During the course of the ages these differentiated into all kinds of specialised professions, not only physicians but also Taoist alchemists, invocators and liturgiologists for the ouranic religion of the imperial court, pharmacists, veterinary leeches, priests, religious leaders, mystics and many other sorts of people. By Confucius' time, the end of the −6th century, the differentiation of physicians had not yet occurred. He himself designated healers by a term that did not distinguish *wu* from *i* when he said that 'a man without persistence will never make a good healer (*wu i* 巫醫)'.[5]

Physicians of these ancient times are mentioned in the *Tso chuan* 左傳 (Master Tso's tradition of the spring and autumn annals, between −400 and −250), the greatest of the three commentaries on the *Chhun chhiu* 春秋 (the 'spring and autumn' annals). More than forty-five consultations or descriptions of diseases occur in these celebrated annals of the state of Lu, which cover a span from −721 to −479 and were compiled at the end of that period. Among the older ones is the incident when Huan the Physician (I Huan 醫緩) diagnosed correctly in −580 the illness of a king of Chhin. But the most important, to be discussed below, is the consultation dated −540 which another king of Chhin had with an eminent practitioner, Ho the Physician (I Ho 醫和).[6]

Before the Thang, élite writers tended to think of physicians more or less as a kind of artisan, but most doctors from the Chou to the Six Dynasties about whom we have any biographical record came from aristocratic backgrounds.[7] Still they were hardly typical. There was a general move throughout the Middle Ages to raise the intellectual standing of physicians in general. From the Thang on, as medicine entered the civil service, the literature becomes increasingly diverse. As early as +758, one can find the beginnings of an important development, the examination of medical students in general literature and the philosophical classics. We shall say more about medical qualifying examinations, but here we are concerned with their component of general education.[8] From about the time when printing began to spread in earnest, scholars distinguished for *belles lettres* or statecraft began to write most learnedly on medical matters. In Hangchow from about +1140 onwards the candidates were expected to pass tests in the literary and philosophical classics as well as in medical subjects. An imperial decree of +1188 ordered that unqualified medical practitioners must pass provincial examinations. These included the general classical writings as well as sphygmology and other medical techniques. Anyone who did really well could gain an opportunity of joining the Artisans' Institute (Han Lin Yüan 翰林院). This palace institution was divided into sections for astronomy, painting and calligraphy besides the medical one. Each member was regarded as supreme in his own specialty, but was usually a specialist rather than a regular official.

Although a medical career seldom led to great income and never to high rank in the civil service, from the +13th century scholars of high degree did not hesitate to join the

---

[5] *Lun yü*, 13, 22, tr. Legge (1861), p. 136.   [6] See p. 43.
[7] Based on an unpublished survey by the Editor.   [8] See Section (*c*).

ranks of the physicians. This shift began when Mongol rule greatly reduced the opportunity for the sons of the learned gentry to become officials. It continued as opportunities for advancement in officialdom again diminished in the Ming, and more markedly in the Chhing. Nevertheless, hereditary curers and learned scholars continued to practise side by side until modern times.[9]

These gradual changes in educational and career patterns are important, for they show that considerable numbers of physicians were well educated in general literature, and with greater culture than their predecessors had possessed. Such men called themselves *ju i* 儒醫 (literally, 'Confucian physicians') as opposed to those they belittled as *yung i* 庸醫, mediocre practitioners or quacks.[10] Prominent among those so denigrated were the wandering medical pedlars frequently seen in late imperial China, jingling their special kind of bell on a staff and handing out herbal remedies for the smallest of fees. Indeed the grandfather of the greatest pharmaceutical naturalist in all Chinese history, Li Shih-chen 李時珍 (+1518 to +1593), was a medical pedlar. We ourselves have often met with them, and recall with particular pleasure a brilliant impersonation of the type in a revolutionary opera which we had the pleasure of witnessing at Taiyüan in Shansi in 1964.

We can exclude at the outset any idea that the profession as a whole was despised in Chinese civilisation.

(2) THE PRINCIPAL DOCTRINES OF CHINESE MEDICINE

Now we will say something about the doctrine, the fundamental philosophy of Chinese medicine. We like the saying of Keele (1963) that 'it would seem probable that the first civilised people to free themselves from the purely magico-religious concepts of disease were the ancient Chinese', but we cannot follow him in his belief that this liberation was achieved only briefly until the acceptance of Buddhist thought from India. Nor can we agree with him that the ancient Chinese substituted 'metaphysical' modes of thought for primitive magico-religious conceptions and practices. Everything depends, of course, on what one means by metaphysical, but if we use the term in its generally accepted sense in modern-Western philosophy as meaning ontology, the problem of Being, and the dispute between realists and idealists, it is surely not applicable here. We have to deal with an ancient philosophy of Nature, a set of hypotheses about the Universe and the world of man.

The natural philosophy current among the ancient Chinese was based upon the idea of two fundamental forces, the yang 陽 and the yin 陰, the former representing the bright, dry, masculine aspect of the Universe, the latter the dark, moist, feminine aspect. This conception is probably not much older than the −6th century, but it was certainly

---

[9] Yamamoto (*1985*), Hymes (1987). For details on Ming medical practitioners with the highest examination degree, see Chang Tzhu-kung 章次公 (*1948*).

[10] [The epithet *yung i* seems to have been used from the Sung period on, and became common in the Ming. For general discussions of terms for 'physician' see Yamamoto (*1983*) and Chang Tsung-tung 張宗棟 (*1990*). – Editor]

dominant in the minds of the early royal physicians whom we mentioned just now. In the short lecture given by Ho the Physician in −540 to his patient the king of Chhin we can see Chinese medical thought *in statu nascendi*. Especially important is his division of all disease into six classes derived from excess of one or another of six fundamental, almost meteorological, pneumata (*chhi* 氣). Excess of yin *chhi*, he says, causes *han chi* 寒疾 (cold illness), excess of yang, *jê chi* 熱疾 (hot illness), excess of wind, *mo chi* 末疾 (afflictions of the extremities), excess of rain, *fu chi* 腑疾 (afflictions of the belly), excess of twilight, *huo chi* 惑疾 (illnesses involving confusion) and excess of the brightness of day, *hsin chi* 心疾 (illnesses of the heart and mind). The first four of these are subsumed in the later classifications under *jê ping* 熱病, diseases involving fever; the fifth implies psychological disease, and the sixth cardiac disease.

This classification into six types is of extreme importance because it shows how ancient Chinese medical science grew up to some extent independently of the theories of the Naturalists, which classified all natural phenomena into five groups associated with the Five Elements or Five Phases (*wu hsing* 五行). These ideas were first systematised by Tsou Yen 鄒衍 in the −4th century. The doctrine of the Five Elements became later universally accepted in all branches of traditional science and technology.[11] As is well known, these elements differ from those of the Greeks and other peoples in that they comprised not only fire, water and earth but also wood and metal. Chinese medicine, however, never lost entirely its sixfold classification. The yin and yang viscera (*tsang-fu* 臟腑) were mustered as six of each, although physicians and laymen spoke of them collectively as *wu tsang* 五臟. In view of the duodecimally based mathematics and world outlook of the Babylonians, one cannot but suspect an influence from ancient Mesopotamia on early China in this respect.[12]

It is not the only example of such an effect. The twelve double hours of the Chinese day and night, which go back to the beginning of Chinese culture, have long been thought to be Babylonian in origin, and some evidence has been brought forward also for close parallels in State astrology.[13]

As far as medicine is concerned, we can look for connections between cultures in another direction, namely in the very prominent part played by the conception of *chhi*, closely analogous to the Stoic *pneuma* (πνεῦμα). Both words are almost untranslatable but their significations included 'life-breath', 'subtle influence', 'gaseous emanation' and the like. Somewhat later, Chinese medical theories also dealt much with another word of very similar meaning, *fêng* 風 (wind).

Now Filliozat, in a classical monograph (1949), has shown that the *pneuma* of Greek medicine can be matched word for word, and statement by statement, with the *prāṇa* of the great Indian medical writers. Thus we see, as in perhaps hardly any other science except astronomy, a widespread community in high antiquity between the peripheral

[11] Vol. 2, pp. 232–53. [On the early character of the *wu hsing* concept and its systematisation, see Sivin (1995e). − Editor]

[12] [The most common enumeration totalled eleven, not twelve, and both *tsang* and *fu* were enumerated in various sources as fivefold. See, for example, the modern Chinese source translated in Sivin (1987), p. 213, and the historical analysis in *ibid.*, pp. 124–33. − Editor]

[13] Vol. 2, pp. 351ff.

areas of the Old World. From Greece, through India, round to China, there is 'pneumatic medicine'.

We are well aware that until now, so far as studies of the cuneiform texts have unravelled it, Babylonian medicine has been largely magico-religious in character. Still one cannot help feeling that there must have been some schools of proto-scientific medicine in Mesopotamia which bequeathed their ideas about the subtle breaths, both of normal function and pathological condition, with which the physician must contend. One cannot help feeling that some civilisation older than either Greece, India or China must have originated such conceptions and sent them out in all directions. The Iranian culture area can hardly qualify on account of its relative youth, so that Mesopotamia must have been their home.

Another doctrine prominent in ancient Chinese thought was that of the macrocosm and the microcosm. It envisaged a great interdependence of the State on its people, and of the health of the people on the cosmic changes of the Four Seasons. The Five Elements were associated in 'symbolic correlations' with many other natural phenomena in groups of five. These conceptions were applied in a remarkably systematic way to the structure and function of the living body of man.[14]

As might be expected, a society which was developing the characteristic form of bureaucratic feudalism attached great importance to preventing trouble, in both the political realm and the life of the people, rather than waiting until it arose to control it. And thus in the field of medical thought, prevention was considered better than cure. In spite of all the outside influences which may have acted on Chinese medicine from the beginning onwards, it retained an extremely individual and characteristic quality, still clearly present.

We must, of course, willingly grant to Keele that the practice of using charms, incantations and invocatory prayers to deities persisted through most of Chinese history, particularly among the poorer strata of society and in the exorcistic activities of Taoist adepts and Buddhist monks. But we can find them in the palace as well. In +585, for example, under the Sui dynasty, the Directorate of Medical Administration included, besides two Erudites (or professors) for General Medicine and two Erudites for Massage, two Erudites for Exorcism; thus there was official sanction for magico-religious techniques.

But Keele (1963) is absolutely right in giving the impression that all these phenomena were 'fringe activities' of Chinese traditional medicine. They were quite peripheral to the practice of medicine as such, kept far indeed from the centre of the stage, and it can confidently be asserted that, from the beginning, Chinese medicine was rational through and through.[15] 'The transmutation from magico-religious to metaphysical pathology was an achievement', writes Keele, 'but it was not enough to provide a basis for progress in medicine, for it was not scientific, either in its method of observation or in its reasoning, in that it entirely failed to make use of the method of induction.' Re-writing this in our own language we should say that the advance from magic and

---

[14] On symbolic correlations see Vol. 2, pp. 273ff. Sivin (1995f) analyses the origins of macrocosm–microcosm doctrine.

[15] See especially *HTNC/SW*.

religion to primitive scientific theory was an immense achievement, but that for a wealth of reasons, which we cannot go into here, Europe was the only civilisation in which ancient and mediaeval science could give way to modern science. We would not say that the old Chinese scientific theories gave no basis for progress in medicine, nor that they were unscientific in observation or reasoning. Undoubtedly they did make use of the method of induction, but they remained pre-Renaissance science and never became modern science.[16]

(3) THE FATHERS AND THEIR HISTORY

So much for social position and perennial philosophy; now a word about the fathers of medicine and their history. A comparison between the early classical periods of Chinese and Greek medicine is of much interest. In China there is a figure paralleling Hippocrates (−460 to −379), but not quite so much is known of his personality and he is not directly connected with the Chinese counterpart of the Hippocratic corpus. This was Pien Chhüeh 扁鵲, for whose life we have an authoritative source in the *Shih chi* 史記 (Memoirs of the Astronomer-Royal, *ca.* −100) of Ssu-ma Chhien 司馬遷, the first of the wonderful series of Chinese dynastic histories.

Pien must have been of the generations preceding Hippocrates, for we have a firm date for a famous colloquy of his, −501. When passing through the state of Kuo just after the heir of its ruler had died, he listened to an amateurish analysis of the mortal ailment by a palace cadet. Pien claimed that he could save the king. His interlocutor, taken aback, retorted that, unless, like the legendary physician Yü Fu 俞跗, Pien could cut open the body and manipulate, repair and cleanse its components (an unthinkable idea in a society that did not resort to surgery), this was an infantile boast.

Pien Chhüeh, looking up and sighing, replied 'your ideas of medicine are no better than viewing the sky through a narrow tube or reading a piece of writing through a narrow crack. In my practice of medicine I need not even feel the pulse, look at the complexion of the patient, listen to him, or visually examine his physical condition, in order to say where the disease is located.'

He went on to make his point by diagnosing without even seeing the patient, and restoring him to life.[17]

The passage shows that at this early time the four important diagnostic observations (*ssu chen* 四診) typical of Chinese medicine were already in use. These comprised: first, the inspection of the general physical appearance of the patient, including colour and glossoscopy (*wang* 望); secondly, primitive forms of auscultation and osphristics (*wên* 聞); thirdly, interrogation, including eliciting the patient's medical history (*wên* 問); and finally palpation and sphygmology (*chhieh* 切). Pien's biography also shows that, at the time of Confucius himself, the physicians were using acupuncture needles, cauterisation with the pulp of various leaves (moxa), counter-irritants, aqueous and alcoholic

---

[16] [On this issue, see the Introduction. – Editor]   [17] *Shih chi*, ch. 105; see also Yamada (*1988*).

Fig. 2. A physician applying moxibustion. From a painting by Li Thang 李唐, *ca.* +1150. From Needham (1970), Fig. 82, facing p. 441.

decoctions of drugs, medicated plasters, massage and gymnastics. It is striking to find so many therapeutic methods already elaborated before the time of Hippocrates.[18]

Now what corresponded in China to the Hippocratic corpus? We know that the books in that great collection were written during a period of time covering much more than

---

[18] [Few scholars today consider the Pien Chhüeh anecdotes in this and other sources historically reliable or markedly pre-Han in origin. It is far from certain that they are about the same physician or the same period. See the summary in Takigawa (*1932–4*), ch. 105, p. 7, and the quite different versions of this anecdote, earlier and later than *Shih chi*, in *Han shih wai chuan* 韓詩外傳 (*ca.* −150), ch. 10, pp. 6b–7b, and *Shuo yüan* 説苑 (−17), ch. 18, pp. 13a–14a. Chang Tsung-tung (*1990*), p. 144, discusses 'Pien Chhüeh' as not the name of a person but a title of esteem for physicians.

the life of Hippocrates himself, i.e. from the beginning of the −5th century down to the
end of the −2nd. Only a few of them are now considered 'genuine', in the sense of having come from the pen or the dictation of Hippocrates himself.

The Chinese counterpart is the *Huang ti nei ching* 黃帝內經 (The Yellow Emperor's
manual of corporeal [medicine]), often referred to as the *Nei ching*. The extant forms
appear to be large books divided into separate chapters, but like the Greek corpus
each is a compilation of tractates. The Manual deals indeed, just as the Hippocratic
corpus does, with all aspects of the normal and abnormal functioning of the human body,
with diagnosis, prognosis, therapy and regimen. It was, we think, approximately in its
present shape by the −1st century, in the Former Han dynasty. No one disputes that it
systematised the clinical experience and the physiopathological theory of the physicians
of the preceding five or six centuries.[19] The attribution of the work to the mythical Yellow
Emperor (a favourite Taoist figure) is of little significance.[20] The book contains some
practical knowledge of the ancients and elaborates the *philosophia perennis* of Chinese
medicine. All later writings in this field derive and develop from the *Nei ching*. It is quite
natural that such a compendious treatise should have been made in the Chhin and Han
periods, for the institution of the first unified empire in the Chhin brought about not
only a centralisation of government but also a standardisation of weights and measures,
even down to the gauge of carriage wheels: in sum, a general systematisation of Chinese
practices.

A minor difference from the Hippocratic tractates is that in the *Nei ching* a great deal
of the text is cast in the form of dialogues between the legendary Yellow Emperor and
his biological–medical preceptors and advisors (equally legendary) such as Chhi Po 歧伯.

The *Huang ti nei ching* studied by physicians over the past thousand years consists of
two books, the *Su wên* 素問 (The plain questions [and answers]) and the *Ling shu* 靈樞
(The vital axis), both probably compiled in the −1st century. Wang Ping 王冰 edited
the surviving recension in +762, considerably altering (not for the first time) the form
that the corpus had in the Han period. About a hundred years earlier than Wang Ping,
Yang Shang-shan 楊上善 compiled another recension of the *Huang ti nei ching*, known
as *Thai su* 太素 (The great innocence), which has come to light only in recent times.
Parts of it are nearer the original text of the Han. It is incomplete, but probably contained, organised in a different order, roughly the same range of material as the two later
compilations.[21]

Christopher Cullen has pointed out (private communication) that the 'Inner Canon' and 'Outer Canon' of
Pien Chhüeh listed in the bibliography of the Han history (but, unlike the *Huang ti nei ching*, later lost) make
him an excellent candidate for comparison with Hippocrates. Both are historically shadowy figures to which
large compilations of medical writing were eventually attributed. – Editor]

[19] [On this contentious question see the review in Sivin (1993), pp. 199–201. It is likely that the constituent
parts of the *Nei ching* reflected developments over less than a century. Two particularly useful collections
of research reports on the *Nei ching* are Maruyama (1977) and Jen Ying-chhiu 任應秋 & Liu Chhang-lin
劉長林 (1982). – Editor]

[20] It has taken on great significance in the past decade as scholarly attention has focussed on the Huang-
Lao movement of the Han dynasty. See Yü Ming-kuang 余明光 (1989) and Csikszentmihalyi (1994).

[21] [The *Thai su* reached its final form after +656, and the *Su wên* and *Ling shu ca.* +762; see Sivin (1993).
Several chapters of the *Su wên* were interpolated from late sources at that time or later. Keegan (1988) has
given evidence that none of the recensions is consistently closer to the original or more reliable than another.

The *Nei ching* contains the fundamental principles of traditional Chinese medicine. The *Su wên* recognises and describes many specific disease entities, noting the regular associations of symptoms which permit their diagnosis. It traces their aetiology in terms of the classical physiological theories which it enunciates, having due regard to external influences. As for therapeutics, it concerns itself chiefly with acupuncture.

Unfortunately, although the basic principles of the *Nei ching* are not difficult, its language is archaic and hard to understand. Nor were the ancient commentaries on it easy to understand. Hence, during the centuries, only scholars of high quality could master the corpus and become truly learned physicians. The difficulty of the *Nei ching* is that the technical terms are often ordinary words that have been given special meanings. Sometimes they occur along with the same word used in its ordinary sense in the same passage. Much confusion about Chinese medicine has arisen from misunderstanding of the *Nei ching*.[22]

The *Nei ching* scheme of diagnosis (systematised *ca.* +200 in the *Shang han tsa ping lun* 傷寒雜病論) classified disease symptoms into six groups in accordance with their relation to the six circulation tracts (*ching* 經) through which the pneuma (*chhi*) coursed through and around the body. Three of these tracts were allotted to the yang (*thai-yang* 太陽, *yang-ming* 陽明, *shao-yang* 少陽) and three to the yin (*thai-yin* 太陰, *shao-yin* 少陰, *chüeh-yin* 厥陰). In feverish illness each of them was considered to preside over a 'day', one of six 'days', actually stages, following the appearance of the disease. In this way physicians established differential diagnoses and decided upon appropriate treatments. These tracts were essentially similar to the tracts of the acupuncture specialists, though the acupuncture tracts were composed of two sixfold systems, the cardinal (*ching* 經) and decumane (*lo* 絡). They crossed each other like the streets of a city laid out in rectangular grid arrangement.[23]

By the time of the *Nei ching* the physicians had fully recognised that diseases could arise from purely internal as well as from purely external causes. The ancient 'meteorological' system explained by Ho the Physician had therefore been developed into a more sophisticated sixfold series, namely *fêng* 風, *shu* 暑, *shih* 濕, *han* 寒, *tsao* 燥 and *huo* 火.[24] As external factors they could be translated as wind, humid heat, damp, cold, aridity and dry heat; but as internal causes we could name them blast (cf. Van Helmont's blas), fotive *chhi*, humid *chhi*, algid *chhi*, exsiccant *chhi* and exustive *chhi*. It is interesting to

'Great innocence' seems to extrapolate from 'pristine' as a meaning of *su* 素. This interpretation is not pertinent to the book's contents. A gloss by Chhuan Yüan-chhi 全元起, author of its first known commentary, supports a more obvious understanding: 'Grand basis'. – Editor]

[22] [For additional perspectives see the essays by Chinese and Japanese authors in Unschuld (1988). The papers by Western writers in this conference volume should be read with caution, since the authors' grasp of the classical medical literature and its language varies greatly. – Editor]

[23] [European acupuncturists in the 19th century fell into the habit of calling the *ching* 'meridians' by mistaken analogy to another sense of *ching*, namely longitude. Some occidental writers on acupuncture now call them 'conduits', unaware that *ching* was only one of a number of ways, each with distinct implications, to designate circulation pathways. Early physicians used *sui* 隧 when they wished to imply a conduit, since *ching* has no such associations. Lu and Needham appropriately call *ching* tracts or acu-tracts. See Sivin (1987), pp. 135–7. – Editor]

[24] See p. 43.

notice the partial parallelism with the Aristotelian–Galenic qualities, which were part of a quite different fourfold system.

It will have been noticed that we translate the title *Huang ti nei ching* as 'The Yellow Emperor's manual of corporeal [medicine]'. This raises an extremely interesting question. The first English version translated the title as 'The Yellow Emperor's manual of internal medicine', but this is indisputably wrong, and other scholars have not accepted it. Not only did it introduce a modern conception where it has no place to be, but it also entirely mistook the significance of the word *nei*. The two last words of the title mean literally 'inner manual'.[25] The bibliography of the Western Han history also contains a *Huang ti wai ching* 黃帝外經, literally the 'outer manual', which we prefer to entitle 'The Yellow Emperor's manual of incorporeal (or extra-corporeal) medicine', as well as other pairs of 'inner' and 'outer' medical manuals from other traditions. All disappeared during the early centuries of our era, and no explicit statement of what the titles meant survives.

Many other ancient Chinese books are grouped in 'inner' and 'outer' chapters (*phien* 篇). For example, the 'Inner chapters of the Preservation-of-Solidarity Master' (*Pao Phu Tzu nei phien* 抱樸子內篇), written by Ko Hung 葛洪 about +320 and accompanied by a separate book of 'outer chapters' (*wai phien* 外篇) on other topics, is the greatest of Chinese alchemical books.[26] One might be tempted to translate 'inner' and 'outer' by esoteric and exoteric, respectively, the former being a secret doctrine not to be revealed to people in general, the latter being the overt publicly preached system. But this would involve just as serious a mistake as that which we are trying to correct.

The key to the real meaning for which we are seeking is to be found in the classical statement of the Taoists that they 'walked *outside* society'. Again, the *Chuang tzu* 莊子 book says: '*outside* time and space (*liu ho* 六合, literally, "the six directions") is the realm of the sages, and I am not speaking of it here'. In other words, *nei* or 'inside' means everything this-worldly, rational, practical, concrete, repeatable, verifiable, in a word, scientific. Similarly, *wai* or 'outside' means everything other-worldly, everything to do with gods and spirits, sages and immortals, everything exceptional, miraculous, strange, uncanny, unearthly, extra-mundane and extra-corporeal or incorporeal. Let it be noted in passing that we are not here using the term supernatural. In classical Chinese thought nothing, however strange it may happen to be, is outside Nature. This is why we propose the translation 'The Yellow Emperor's manual of corporeal [medicine]'. The fact that the *Wai ching* was lost so early emphasises once again precisely the quite secondary character of the magico-religious aspect of medicine in China; for cures

---

[25] It is normal to translate *ching* in early book titles as 'canon' or 'classic'. The word implies an authoritative text, often considered to have been revealed by a sage, and transmitted by a lineage of scholars. Joseph Needham preferred 'manual' to emphasise its Thang and later derivative usage in the titles of various technical writings. It would be hard to argue that Han physicians thought of the *Nei ching* primarily as a handbook. See Henderson (1991) and Sivin (1995c).

[26] Vol. 5, pt 3, pp. 75–113. [In Ko's autobiography he lists the outer and inner chapters as separate books, and they have been so treated by bibliographers. See the 'outer chapters', ch. 50, pp. 9a–9b. Ware (1966) includes the autobiography in his translation of the 'inner chapters'; see *ibid.*, p. 17. – Editor]

effected by charms, cantraps and invocations must certainly have been included in the 'outside' corpus.[27]

Before leaving this subject we should like to refer to certain other uses of the terms *nei* and *wai* which might be made to explain the title of the Chinese Hippocratic corpus but in fact cannot serve that purpose. Modern medicine in China incorporates the Western distinction between *nei kho* 內科 and *wai kho* 外科 (or *yang kho* 瘍科), the former meaning internal and general medicine, and the latter external medicine. In traditional medicine *wai kho* included such surgery as the Chinese carried out, but the term was much wider than surgery in the modern sense, for it included the treatment of fractures and dislocations, boils and eruptions, and dermatological and other conditions of the outer surface of the body. This internal–external distinction in medical practice, however, does not go much further back than the Sung period (+10th to +13th centuries), where the Imperial Academy of Medicine began with three specialties (*kho*) and added others until in the Ming a classification into thirteen became usual. But neither internal nor external medicine was restricted to one of these thirteen departments. All this can have nothing to do with the *Nei ching*, the text of which recognises no such distinctions.

Another *nei–wai* differentiation occurs in historical writing associated with the words *shih* 史 (historical account) or *chuan* 傳 (tradition, often biography). Oddly enough, 'inside history' and 'outside history' both denote unofficial versions, with somewhat different shades of meaning. This does not apply in any way to the *Huang ti nei ching*.

In alchemy there is an important distinction between *nei tan* 內丹 and *wai tan* 外丹, the 'internal' and 'external' elixirs. The former term refers to 'psychophysiological alchemy', in which the elixir used to be made from the juices and organs of the body itself. *Wai tan* on the other hand refers to the elixirs of longevity or immortality prepared from chemicals by manual operations in the laboratory. Here, as in the case of esoteric and exoteric, the meanings are almost diametrically opposite, for the 'inner' was the physiological and the 'outer' was the practical and proto-scientific.[28] Lastly, the words could be used in a perfectly straightforward and unsophisticated way, as in the title of

[27] *Chuang tzu*, ch. 2, l. 56, tr. Fêng Yu-Lan (1933); Legge (1891), p. 189. [This argument is extremely difficult to follow, since the two books of Pao Phu Tzu are counter-examples of the distinction that the authors propose. The *Pao Phu Tzu nei phien*, the 'inner chapters', is explicitly concerned with Taoist esotericism, including 'cures effected by charms, cantraps and invocations'. Ko Hung describes his own 'outer chapters' as 'concerned with success and failure in the world of men, and what is propitious and unpropitious in secular affairs; they belong to the Confucian school'; *Wai phien*, ch. 50, p. 9b.

The authors may have been tempted to take this approach by the unsatisfactory translations of the *Chuang tzu* available when they began to investigate ancient medicine. The standard-setting version of Graham (1981), p. 57, renders the cited sentence and the one that follows as 'What is outside the cosmos the sage locates as there but does not sort out. What is within the cosmos the sage sorts out but does not assess.' That does not have the connotations that the authors propose, and is irrelevant to book titles. Given the lack of evidence for exactly what the Yellow Emperor titles meant, and the inconsistency in the usage of *nei* and *wai* elsewhere, historians of medicine now translate the titles literally (as 'Inner Canon' and 'Outer Canon') rather than offer an interpretation. – Editor]

[28] [This is an evaluation from the standpoint of modern science, not a distinction meaningful at the time. Internal alchemy, as adepts thought of it, was not physiological manipulation but meditative visualisation. External alchemy was not proto-science, if this involves cognitive goals. It aimed, like internal alchemy, at spiritual perfection, personal immortality in the flesh, and appointment in the bureaucracy of the gods – not, in Chinese thought, contradictory goals. See Vol. 5, pt 4, pp. 210–98. – Editor]

a lost book of acupuncture diagrams written by Chu Kung 朱肱 in +1118 – the *Nei wai êrh ching thu* 內外二景圖. The 'Illustrations of internal and external views' included diagrams of the circulation near the surface of the body and the organs at its centre. We think that this sort of philological excursus into the proper nuances of words is abundantly worthwhile for the prevention of serious misunderstanding.

It is indeed fortunate for the historian of Chinese medicine that we have an extremely important physician's biography dating, we believe, from roughly two generations before the *Huang ti nei ching* was put together. The biography of Shun-yü I 淳于意 by Ssu-ma Chhien follows that of Pien Chhüeh in the same chapter of the *Shih chi*. The second part is by far the more important, because it contains twenty-five clinical histories related by Shun-yü I as well as his replies to eight specific questions, all on the occasion of an imperial decree of not long after −167 seeking information about the prognostic abilities of doctors.[29]

Born *ca*. −214 in the old State of Chhi (now Shantung), Shun-yü I, although he held the post of Director of Granaries, practised medicine among kings as well as officials and common people. In about −176 he was accused of an unspecified crime and imprisoned, but released after the supplication of his daughter. This was the famous occasion when mutilative punishments were revoked, alas only temporarily. Shun-yü practised until about −150.

It is possible to explain the nature of nearly all the cases attended by Shun-yü I in modern terms. Though a few of these interpretations may be subject to revision, the majority are perfectly clear. We have thus a unique record of medical practice and knowledge in the −2nd century.

On the occasion of the imperial enquiry, Shun-yü I referred to more than twenty books that had been handed down to him by his teachers or that he had taught his disciples. As Keegan among others has shown, many of these titles are mentioned, and some of the texts appear, in the present *Huang ti nei ching*. This implies, although it does not prove, that Shun-yü had at his disposal some of the materials that were later incorporated into the *Nei ching*.[30]

Bridgman in 1955 ended his medical assessment of Shun-yü's clinical histories by making a weighty comparison with Greek medicine. Far from being an assembly of magical practices and inexplicable fantasies, he says, it appears that in China the examination of the sick person, the investigation of the clinical history, the comparison of data from different examinations, and the therapeutic deductions all formed part of a discipline which constituted a valid and valuable precursor of contemporary clinical science. In this light ancient Chinese medicine can fully sustain any confrontation with Greek or Roman medicine of the same period. With this we wholeheartedly concur.

The Later Han, Three Kingdoms and Chin periods brought a number of outstanding physicians and medical writers roughly corresponding to Aretaeus, Rufus, Soranus

---

[29] [Bridgman (1955) has consecrated to the life and times of Shun-yü I a monograph that provides valuable medical assessments, although it frequently misunderstands the text. Keegan (1988), Sivin (1995c) and Loewe (1997) have studied aspects of Shun-yü's career. – Editor]

[30] See the translation in Sivin (1995c), pp. 179–82, and the discussion in Keegan (1988), pp. 226–31.

and Galen. The life and work of Chang Chi 張機 (Chang Chung-Ching 張仲景), who probably lived from +152 to +219, closely paralleled that of Galen (+131 to +201). One could hardly say that in China the influence of this younger contemporary of Galen over the ages was less than Galen's in the Western world. For his *Shang han tsa ping lun* (Treatise on febrile diseases), produced about +200, was one of the most important medical classics after the *Huang ti nei ching* itself, and more important from the viewpoint of drug therapy.

Next came Hua Tho 華佗 (+190 to +265), about whom many stories subsequently clustered. Little of what he wrote has come down to us, but Chinese physicians trace back to him the great developments of medical gymnastics, massage and physiotherapy.[31] The third century brought two more men of the highest importance. First, Huangfu Mi's 皇甫謐 (+215 to +282) *Huang ti chia i ching* 黃帝甲乙經 (The A–B manual [of acupuncture], named after its method of designating chapters) was a most influential work. No less important, however, was the *Mo ching* 脈經 (Pulse manual), compiled by Wang Hsi 王熙 (usually called Wang Shu-ho 王叔和) about +300 from the *Nei ching*, *Shang han tsa ping lun* and other early classics. It became the foundation of all later works on the pulse.[32] As Wang was born about +265 and died in +317, we have come down to the time of Oribasius, and the classical period of Chinese medicine draws to a close.[33] Its vast developments in later ages we cannot follow further here.

(4) INFLUENCES OF BUREAUCRATISM ON CHINESE MEDICINE

Let us now turn to how developing within a society based on bureaucratic feudalism affected the medical profession. Westerners seldom understand that for some 2,000 years Chinese society was constructed in an entirely different way from anything known in the West. The principles of aristocratic-military feudalism are familiar to all educated Europeans, though historians are aware that its practice was far more complex and diverse than laypeople usually imagine. Broadly speaking, traditional China (at least after the −3rd century) lacked the occidental apparatus of fiefs and feudal ranks, of primogeniture and inherited lordships. Instead of all this the culture was governed by what became in the Thang and Sung dynasties a non-hereditary bureaucracy. The members of this immensely elaborate civil service were drawn from the ranks of the educated gentry. Instead of earls and barons there were governors and magistrates. Access to this 'mandarinate' was the result of passing the official examinations, to an extent that varied greatly over history. In this sense Chinese invented the 'career open to talent' a

---

[31] [Equally important in the second century is the *Huang ti pa-shih-i nan ching* 黃帝八十一難經 (Manual of 81 problems [in the Inner Canon] of the Yellow Emperor, usually now referred to as the *Nan ching*). Unschuld (1986b) has emphasised the book's importance as a first attempt to construct a synthesis out of the frequently inconsistent discussions in the *Nei ching*. – Editor]

[32] [More than half of the text can be traced to extant sources. Some scholars, e.g. Kosoto *et al.* (1981), vol. 8, pp. 333–402, argue that, since diagnosis is not primarily by the pulse, the title should be translated 'Canon of the circulation vessels'. – Editor]

[33] [One might argue that the classical period ends with the *Nei ching*, and that *Nan ching*, *Mo ching* and *Chia i ching*, all of which are attempts at synthesis based on it and other early canons, represent the beginning of classicism. – Editor]

millennium before it appeared in France.[34] Knowledge of Chinese customs, we know, greatly influenced +18th-century France.

The Chinese society of the −1st millennium, the time of the Spring and Autumn and Warring States periods, is justly characterised as feudal or proto-feudal. But it is certain that with the passage of time all feudal elements persistently declined and were replaced by the non-hereditary bureaucratic society.

As might be expected, the influence of this very different form of society upon medicine was profound. Instruction began early in those sciences important to the State, for example, medicine, astronomy and hydraulic engineering. So we find that the government established medical professorships, colleges and examination ranks between the end of the +5th and the middle of the +8th century. These activities aimed primarily at meeting the need of the palace for medical care, but government sponsorship extended gradually to military and provincial medical administration. We tell this story in Section (c), with special attention to qualifying examinations.

At first the dates may seem remarkably precocious. They are less so once the distinctively bureaucratic-feudal character of Chinese society is understood, together with the age-old respect of the Chinese for learning and for a learned, non-hereditary civil service. It would hardly be possible to imagine a deeper effect of the environing culture on medicine than this 'bureaucratisation' of medical knowledge, which had the extremely happy effect of protecting people at large from the activities of ignorant physicians.[35]

In a bureaucratic society it was quite natural that, as the conception of hospitals formed, religious and governmental initiatives should, from time to time, contend together. The idea of the hospital in China first arose in the Han before the introduction of Buddhism. During the Six Dynasties period, religious motives led to the foundation of many institutions, always by Buddhists. Then, when Confucianism regained strength towards the end of the Thang and especially during the Sung dynasty, the national medical service more and more took over the hospitals. Under the Yüan dynasty, at the time of the Mongol conquest of Persia and Iraq, medical organisations of Arabic type and tradition were added, just as a Muslim Bureau of Astronomy was set up as an auxiliary to the age-old department of the Astronomer-Royal. Finally, however, under the Ming and Chhing dynasties social organisms of many kinds, hospitals among them, decayed. When Westerners first began to visit China in considerable numbers early in the 19th century they gained an altogether wrong idea of the history of medical administration in China. Nevertheless, many interesting hospitals and public charities did continue in these late times.[36]

---

[34] See Sections 48 and 49 in Vol. 7. [The examinations were a remarkable vehicle for recruiting widely, but their role in social mobility was limited. A large proportion of initial civil service positions, varying through history, were filled by directly appointing the sons or other close relations of high officials. As for the rest, Hartwell's demographic studies (1982, pp. 419–20) indicate that much of the winnowing took place before the candidates entered the examination rooms. A system of recommendations and control of examination quotas made it possible for a small number of local gentry families of the Yangtze region in the Sung to 'perpetuate their political position', largely interdicting even the initial examinations to any but members of their own lineages. This made birth or marriage into such lineages an indispensable step toward a great career. – Editor]

[35] [See the Introduction. – Editor]

[36] Leung (1987) argues that, as government medical charities decreased in this period, private support grew.

As in so many other fields, the beginnings of the hospice go back to the troublous but venturesome times of Wang Mang 王莽 (r. +9 to +23). On the occasion of a severe drought and locust plague in +2 'commoners stricken by epidemics were accommodated in empty guest-houses and mansions, and medicines were provided for them'.[37] But this, it seems, was only a provisional measure, not the foundation of an institution.

The first permanent hospice with a dispensary is that founded by Hsiao Tzu-liang 蕭子良, a Buddhist king of the Southern Chhi dynasty, in +491. Characteristically, the first government hospital followed very soon afterwards when in +510 Tho-pa Yü 拓跋余, a king of the Northern Wei dynasty, ordered the Court of Imperial Sacrifices (*Thai chhang ssu* 太常寺) to select suitable buildings and attach a staff of physicians for all kinds of sick people who might be brought there. This hospital, called merely *pieh fang* 別坊 or 'separate buildings', had a distinctly charitable purpose, being intended primarily for poor or destitute people suffering from disabling diseases. Severe epidemics were again the background of the initiative. Later in the same century we have a good example of the pattern of semi-private benefactions by officials which afterwards became widespread. Hsin Kung-i 辛公義, one of the generals who had conquered the house of Chhen and helped to unite the empire under that of Sui, encountered a violent epidemic in the province where he had retired to be governor. He turned his own residence and offices into a hospital and provided drugs and medical attendants to thousands of people (*ca.* +591). The classical example of such a benefaction no doubt was the action of the great poet Su Shih 蘇軾 (or Su Tung-pho 蘇東坡). When Prefect of Hangchow in +1089, he founded and richly endowed a government hospital that formed a model for other provincial cities.

It is in the Thang that we can best study the conflict between religious and governmental control of hospitals. In +653 Buddhist and Taoist monks and nuns were forbidden to practise medicine. In +717 the minister Sung Ching 宋璟 memorialised the throne saying that ever since Chhang-an had been the capital (i.e. since the beginning of the Western Wei in +534), hospitals there had been supposedly controlled by government officials, but because of neglect the Buddhist religious leaders had taken over these functions more and more. By +734 action was taken, at least in the capital, to establish government-supported orphanages and infirmaries for the destitute. By +845, as part of the great dissolution of the monasteries under Wu Tsung 武宗, the hospices long called Compassion Pastures 悲田 (*Pei thien*) were transferred to lay control under the name of Patients' Buildings 病坊 (*Ping fang*). At the same time, much temple property in land and buildings was expropriated by the emperor and allocated to these hospitals. Meanwhile, since the beginning of the dynasty in +620, there had been a special hospital and clinic, the Affliction Buildings or *Huan fang* 患坊, within the imperial palace, with its own medical stores, under the control of a special superintendent. The Medical Supervisors, Principal Practitioners and Master Physicians of the Imperial Medical Office served in turn at this institution.

The regularisation of hospital services carried out in the Thang bore great fruit in the Sung, when we find (*ca.* +1050 to +1250) a wide variety of State institutions at

---

[37] *Chhien Han shu* 前漢書, ch. 12, p. 353.

work both in the capital and the provinces. There were infirmaries for the care of the aged and the sick poor (*Chü yang yüan* 居養院 and *An chi fang* 安濟坊, from +1102, and the *Fu thien yüan* 福田院), as well as a hospital mainly for foreigners (*Yang chi yüan* 養濟院, from +1132), another for sick officials (*Pao shou tshui ho kuan* 保壽粹和館, from +1114), and even one for Chin Tartar prisoners of war (also called *An chi fang*, about +1165). Besides all these there were orphanages (*Tzhu yu yüan* 慈幼院, from +1247, and *Yü ying thang* 育嬰堂) and subsidised government apothecaries (*Mai yao so* 賣藥所, *Hui min yao chü* 惠民藥局, and other titles, from +1076).

Comparative data suggest that in hospital organisation Chinese practice was not so far ahead of the rest of the world as it was in the matter of qualifying examinations and government medical services. Hospitals of some kind are attested by the +1st century both for India (as in the *Carakasamhita*, or at Mihintale in Ceylon) and for the Roman Empire (the *valetudinaria* of legionaries, gladiators, etc.). More exact studies are needed to elucidate their nature. The Chinese Buddhist pilgrim Fa-hsien 法顯 described the Indian facilities in the +5th century. At this period too there arose the great hospital of Jundi Shahpur in Persia, heir of the former University of Edessa and precursor of the splendid foundations of Iraq, especially Baghdad, from the +8th to the +12th centuries, which correspond to the institutions we have mentioned in Thang and Sung China.

For a bureaucratic society there is also interest in examining the beginnings of quarantine regulations. As early as +356, the Chin Emperor, on the occasion of a disastrous epidemic, applied what were called the 'old rules', which prohibited officials whose families had three or more cases from attending court for a hundred days.

Another question arising is the isolation of lepers. Though we are as yet uncertain when this started, it is sure that the Indian monk Narendrayasas, who died in China in +589, established leprosaria for men and women at the Sui capital. During the Thang these institutions continued. A Chinese monk, Chih-yen 智嚴, acquired much fame by his preaching and nursing in a leper colony, where eventually he died (+654).

Whatever may be said against bureaucratic systems of society, they do at least go in for rational systematisation. This is certainly relevant to that wonderful series of pharmacopoeias, or rather pandects of natural history, which followed each other throughout the centuries between the Later Han dynasty and the Chhing. The first of these, the *Shên Nung pên tshao ching* 神農本草經 (Pharmacopoeia of the Divine Husbandman), was not produced under imperial auspices, but a number of later ones were.[38]

Such treatises, some of them vast in size, go under the generic name of *pên tshao* 本草, and most have these characters in their titles. Perhaps the best translation of the phrase would be 'the fundamental simples'. The term first appears in the History of the Former Han (*Chhien Han shu* 前漢書) for +5, when Wang Mang, shortly to become emperor

---

[38] We have described the *pên tshao* literature in detail in Vol. 6, pt 1, pp. 220–328. [See also Ma Chi-hsing 馬繼興 (*1990*), a survey of the full range of medical sources, Chang Ju-chhing 張如青 *et al.* (*1996*), Okanishi (*1958*), the excellent overview in Okanishi (*1974*), and Unschuld (1986a), largely summarising Japanese reference works.

Because, among other reasons, the Shên Nung compilation used early Eastern Han place names, was not included in the bibliography of the Western Han history (completed after +50), and was not quoted until the mid +3rd century, it is now generally dated to the late +1st or +2nd century. On its contents and history see the massive scholarly analysis in Ma Chi-hsing (*1995*). – Editor]

of the short-lived Hsin dynasty, called what might be described as the first national scientific and medical congress. They also appear in the biography of Wang's friend Lou Hu 樓護, an eminent physician. In later centuries the *Hsin hsiu pên tshao* 新修本草 (Newly revised pharmacopoeia, +659) was a striking example of an imperially commissioned pharmaceutical natural history.[39] In the Sung there followed Su Sung's 蘇頌 *Pên tshao thu ching* 本草圖經 (Illustrated pharmaceutical natural history) of +1062, and the many successive recensions of the great *Ching shih chêng lei pei chi pên tshao* 經史證類備急本草 (Pharmaceutical natural history for emergency use, classified and verified from the classics and histories) from +1097 on.

Books of standard drug formulae, like those of materia medica, were frequently motivated by official connections. For example, in +723 the Emperor Hsüan Tsung 宣宗 and his assistants composed the *Kuang chi fang* 廣濟方 (Formulae for widespread benefaction) for what the title announces were charitable ends. The emperor then published it and sent it out to each of the provincial medical schools. Officials actually wrote up some of its prescriptions on notice-boards at cross-roads so that the ordinary people could take advantage of them. The Arabic traveller Sulaimān al-Tājir, who was in China in +851, observed and described this practice. In +796 the Emperor Tê Tsung 德宗 disseminated throughout the country his *Chen-yüan kuang li fang* 貞元廣利方 (Medical formulae for widespread benefit from the Chen-yüan emperor). These compilations set a precedent for massive official compilations in later centuries.[40]

The systematisation of drugs and prescriptions was extended to diseases at the beginning of the +7th century. About +610 Chhao Yüan-fang 巢元方 produced, under government sponsorship, the *Chu ping yüan hou lun* 諸病源候論 (Aetiology and symptoms of medical disorders). The great interest of this large work is that it systematically classified medical disorders according to the ideas of the time, only occasionally mentioning therapeutic methods. It was thus essentially a natural history of disease, a thousand years earlier than the time of Felix Platter (+1536 to +1614), Sydenham (+1624 to +1689) and Morgagni (+1682 to +1771).

One cannot but feel that the bureaucratic mentality of 'pigeon-holing', and routing things 'through the right channels', had something to do with this early appearance of systematisation in medical science. Indeed, the classificatory sciences as a whole were strong in traditional China. The very word for science itself in modern Chinese, *kho-hsüeh* 科學, adopted (probably from Japan) at the end of the 19th century to translate the foreign word, means literally 'classification knowledge'. Of course the bureaucratic world outlook affected many other things besides medicine. As we have shown elsewhere, it is in China that one must look for the filling up of prearranged forms, the

---

[39] On the tangled history of this reconstituted work, see Ma Chi-hsing (*1990*), pp. 269–75, and Shang Chih-chün 尚志鈞 (*1981*). The first 'officinal' pharmacopoeia in the Western world, the *Pharmacopoeia Londiniensis* of +1659, was produced just 1,000 years later. There has been some dispute about what constitutes an officinal pharmacopoeia, commissioning by an emperor or king or enforceability at law. We prefer the former criterion, applicable to the Chinese case; Unschuld (*1986a*), p. 47, admits only the latter, no matter how great the extra-legal authority of a given book.

[40] On the formulary literature see Yen Shih-yün 嚴世芸 (*1990–4*) and, for works before the Yüan, Okanishi (*1958*).

beginnings of filing and card-indexing systems, and the differentiation of texts by different coloured inks.[41]

(5) INFLUENCES OF THE CHINESE RELIGIOUS SYSTEMS ON MEDICINE

As is generally known, the three great religious systems or doctrines, the *san chiao* 三教, were Confucianism, Taoism and Buddhism. Only the first two were autochthonous, for the latter came in from India beginning in the Later Han. The thought of these religious philosophies affected all aspects of medicine, and they must have influenced entry into the profession.

A great many medical men throughout the Middle Ages in China were trained at the government's expense, and often became civil servants, even Imperial Physicians. In addition, there must always have been a host of auxiliary practitioners who learned medicine as apprentices and treated the poor. Physicians tended to come from the families that had produced medical men for several generations. Indeed, the Record of rites (*Li chi* 禮記), contemporary with Confucius himself (early in the −5th century), has been interpreted to say that one should not take the medicine of a physician whose family had not been physicians for three generations.[42]

From what we have already said it is clear that the class structure in mediaeval China was quite different from that of Europe, because of the non-hereditary bureaucracy of the scholar-gentry. Social mobility was great. Families rose into office-holding, and sank out of it, within a few generations. The medical profession, as we have emphasised, was not wholly looked down upon after its early beginnings, for as the centuries went by more and more Confucian scholars tended to enter it.

One interesting reason why men of scholarly families took up medicine was because Confucian filial piety enjoined them to attend upon their parents. This made the great Thang medical writer Wang Thao 王濤, among many others, embark upon the studies that issued in the *Wai thai pi yao* 外臺秘要 (Arcane essentials from the imperial library, +752). Cases are also known of men who became physicians on account of the challenge of an illness from which they themselves suffered.

We must not forget here the role played by Buddhist compassion. The forbidding aspect of Buddhism which may be epitomised in the word *Śūnya* or emptiness, i.e. utter disillusionment with this world and the need to escape from the wheel of rebirths, was modified in all varieties of Buddhism by a limitless compassion for all created beings, which may be epitomised in the word *karuṇā*. Thus it came about that no Buddhist abbey was likely to be without its medical specialists. For many centuries, as we have seen, the Buddhists were active in the foundation and maintenance of hospitals, orphanages, etc.[43] The Taoists also participated in this movement, because as an organised religion Taoism tended more and more to imitate Buddhist practices. But they were not so important in the field of medical organisation.

[41] Needham (1964), p. 13.
[42] Ch. 2, p. 18. [Jeffrey Riegel, in Loewe (1993), pp. 293−5, argues that this chapter originated early in the −1st century. − Editor]
[43] On the influence of Buddhism see Ma Po-ying *et al.* (1993), pp. 350−89.

The profound influence of Taoism on Chinese medicine was exerted in quite a different direction. At an earlier stage (p. 41) we had occasion to speak about the primitive shamans of Chinese society, the *wu* 巫. There can be no doubt that Taoist philosophy and religion took its origin from a kind of alliance between these ancient magicians and those Chinese philosophers who, in ancient times, believed that the study of Nature was more important for man than the administration of human society, upon which the Confucians so much prided themselves.[44] At the heart of ancient Taoism there was an artisanal element, for both the wizards and the philosophers were convinced that important and useful things could be achieved by using one's hands. They did not participate in the mentality of the Confucian scholar-administrator, who sat on high in his tribunal issuing orders and never employing his hands except in reading and writing.

This is why it came about that wherever in ancient China one finds the sprouts of any of the natural sciences the Taoists are sure to be involved. The *fang shih* 方士 or 'gentlemen possessing magical recipes' were certainly Taoist, and they worked in all kinds of directions as star-clerks and weather-forecasters, men of farm-lore and wort-cunning, irrigators and bridge-builders, architects and decorators, but above all alchemists. Indeed the beginning of all alchemy rests with them if we define it, as surely we should, as the combination of macrobiotics and aurifaction.[45]

These words are a little unusual but they are carefully chosen. The ancient Alexandrian proto-chemists in the West were aurifictors, i.e. they believed that they could imitate gold, not that they could make it from other substances. Though they had a spiritual side to their endeavours, it was not a predominating one.[46] Macrobiotics, on the other hand, is a convenient word for the belief that, with the aid of botany, zoology, mineralogy and alchemy, it is possible to prepare drugs or elixirs which will prolong life, giving longevity (*shou* 壽) or immortality (*pu ssu* 不死). Similarly, aurifaction is the belief that it is possible to make gold from other quite different substances, notably the ignoble metals.

These two ideas came together first in the minds of the Chinese alchemists from the time of Tsou Yen in the −4th century onwards.[47] Europe had no alchemy in the strict sense until this combination had made its way from China through the Arab culture-area to the West. The macrobiotic preoccupation made Chinese alchemy, as it were, iatro-chemistry, almost from the first, and many of the most important physicians and medical writers in Chinese history were wholly or partly Taoist. One need only mention Ko Hung about +300 and the great physician Sun Ssu-mo 孫思邈 (fl. +673). There was never any prejudice against the use of mineral drugs in China such as existed long in Europe. Indeed the Chinese went to the other extreme. They prepared elixirs containing metallic ingredients which must have caused a great deal of harm.[48]

---

[44] Vol. 2, pp. 3ff.   [45] See Vol. 5, pt 2, pp. 8–126, on which the following paragraphs are largely based.

[46] [Although the positivist view of alchemy as primarily proto-chemistry was widespread when the original version of this chapter was written, few historians of early Western alchemy adopt it today. On the gnostic roots of the Alexandrian art see the writings of Sheppard, esp. (1962) and (1981), and on the context of alchemical operations in mystery cults, Wilson (1984). Sivin (1990) reviews recent research on Chinese alchemy. – Editor]

[47] On the association with Tsou see Vol. 5, pt 3, pp. 7, 13. [Sivin (1995e) offers a different view of the role of Tsou Yen; on Taoists as physicians see the Introduction. – Editor]

[48] On elixir poisoning, see Vol. 5, pt 2, pp. 282–94.

The object of the devout Taoist was to transform himself by all kinds of techniques, not only alchemical and pharmaceutical but also dietetic, respiratory, meditational and sexual, into a *hsien* 仙, in other words an immortal, purified, ethereal and free, who could spend the rest of eternity wandering as a wraith through the mountains and forests to enjoy the beauty of Nature without end. These are the beings that one can discern, tiny against the immensity of the landscape, flitting across remote ravines in many beautiful Chinese paintings.

The elixir concept was characteristic of China, and only of China. In the West, the alchemists were more interested in metallurgical exercises, in imitating gold, not so much in actually believing that they had made real gold from other substances, but in China there was a close connection between alchemy and medicine from the very beginning. Whether it was *wai tan*, elixirs produced from external sources, chemical substances such as minerals and metals; or whether it was *nei tan*, elixirs produced within the body for the advancement of longevity and perhaps material immortality; whichever it was, the idea that knowledge of chemistry would increase man's life by many decades was undoubtedly a Chinese conception.

That conception reached the Arabs by about +700; it got through to the Byzantines by about +1000. Then about +1250, Roger Bacon, the English Franciscan, was the first European to talk like a Taoist. In his book *De retardatione accidentium senectutis*, he said that if only we knew more about chemistry we could lengthen life enormously. Later came Paracelsus, at the end of the +15th century, with his statement that 'the business of alchemy is not to make gold but to make medicines'. With that the beginnings of all modern medical chemistry were achieved.

As time went by, the hope of developing into an immortal receded somewhat, and from the Sung onwards external alchemy shaded imperceptibly into iatro-chemistry. What Chinese iatro-chemistry was capable of can be seen by the extraordinary fact that the mediaeval Chinese chemists succeeded in preparing mixtures of androgens and oestrogens in a relatively purified crystalline form and employing them in therapy for many hypo-gonadic conditions.[49]

In connection with the possible influence of religious systems upon medical science we ought perhaps to take up a very different matter, namely the question of the mental health of the mass of the people in the culture. This opens many wide perspectives. In the absence of adequate statistical analyses we can only give our impression that in traditional and indeed in contemporary Chinese society, while the incidence of psychoses is about the same as in the West, that of the neurotic conditions is considerably less.[50] The incidence of suicide may have been about the same in the past, but for different reasons. There is much here that needs further thought and investigation, but it is generally agreed that neither of the three Chinese religions gave rise to a sense of sin

[49] See Vol. 5, pt 5, pp. 301–37. [No attempt to duplicate any of the ancient procedures has provided a substantial yield of purified hormones. See, for instance, Chang Ping-lun 張秉倫 & Sun I-lin 孫毅霖 (*1988*) and Huang *et al.* (1988). During the discussion of the latter at the Fifth International Conference on the History of Science in China, San Diego, California, 5–10 August 1988, the authors reported a lower concentration of active hormones in the product than in the urine used as raw material. – Editor]

[50] [This is not a widely shared impression. – Editor]

and guilt as Christianity did in the West. Perhaps China was a 'shame society' rather than a 'sin society'. Other facts are interesting in connection with the low incidence of neurosis, e.g. the general acceptance of Nature and natural phenomena inculcated by Taoism, and the extreme permissiveness of Chinese parents in the house-training and home life of young children.

If Chinese mentality was on the whole better balanced than that of the West, this was in spite of great uncertainty of life. Since capitalism did not spontaneously develop in China, and there was no bourgeois revolution, policed bourgeois society did not develop either. Even as late as the end of the 19th century public life could be quite dangerously at the mercy of bandits, bullies, loafers, corrupt magistrates and family tyrants. We dare not follow any further the sociological avenues opened up here except to say that the universal squeeze, graft and corruption, of which the 'Old China Hands' complained in the last century, was simply the way in which the bureaucratic medi-aeval society had always worked. It seems strange only because Western society, having passed through the stage of 'serving God in the counting-house', had left that level some time before.

Of course in making sociological comparisons between Chinese and Western society one must take all periods as well as all aspects into account. To the credit of the Chinese side must be placed an almost total absence of persecution for the sake of religious opin-ion. No such phenomenon as the Holy Inquisition can be found in all Chinese history, nor was there anything corresponding to the witchcraft mania which makes so great a blot on European history between the +15th and +17th centuries.[51] Chinese psychology and psychotherapy remain as yet a closed book to the Western world, but there are many texts available which could be drawn upon to outline it, not least some extremely inter-esting books of the Middle Ages and later on the interpretation of dreams. A great work remains to be done in this direction.

(6) ACUPUNCTURE

The armamentarium of the Chinese physician by the Middle Ages was on the whole similar to that of his Western counterpart. It is possible to show that all the active prin-ciples known and used by Western physicians in the mediaeval centuries were also known in China. In some cases the Chinese had a clear advantage; for example ephedrine, which was described in the oldest of the pharmaceutical natural histories. In others, the Chinese introduced drugs later than elsewhere. In some instances the active principle was the same in China and the West, though it came from different sources. In still other cases the active principle produced a similar effect though it was chemically different.

What sets Chinese therapy most fundamentally apart from that of Europe is of course acupuncture.[52] It has been in constant use throughout the Chinese culture-area for some

[51] [In China it was secular authorities who persecuted religious institutions and fomented witchcraft scares. On the former see, for example, Weinstein (1987), pt 2; on the latter, Kuhn (1990). – Editor]

[52] For details see Lu & Needham (1980), which this Subsection largely summarises.

2,500 years. The labours of thousands of learned and devoted men through the centuries have turned it into a highly systematised department of medical doctrine and practice. Briefly, this system, as is well known, consists of a large number of *hsüeh* 穴 on the surface of the body. Europeans frequently refer to them as 'points', but since they are not at all miniscule points, we call them 'loci'. The physician, in order to affect the branches of the circulation system, penetrates the loci in different specified manners with needles of varying lengths and thicknesses.

The oldest catalogue of these loci occurs in the part of the *Huang ti nei ching* corpus called the Divine Pivot, or *Ling shu*. When the precursor of this Thang book was written, probably in the −1st century, these loci were 360 in number, possibly because of equivalence with the fancied number of bones in the body, in turn connected with the round number of days in the year. Each locus has a distinctive technical name which has developed through the ages, but there is a good deal of synonymy. The total number of loci which have been identified by distinct names is about 650. In the late 20th century, about 450 are recognised, but those most commonly employed are much fewer in number, not exceeding perhaps about 100. Before the Sung period (+11th century) we know the titles of some eighty books on the system of acupuncture loci, but the majority of these were lost. We have already taken note (p. 52) of one of the earliest systematic treatises on this art, the extant *Huang ti chia i ching*, about +280.

If this were the whole story, the system would be indeed purely empirical, but it is far from that. Practitioners connected the loci to form a complicated reticulate system quite resembling a map of the London underground railways, the *ching-lo* 經絡 (cardinal and decumane tracts). The analogy can be carried somewhat further because the tracts are indeed invisible, like modern anatomy's principal blood vessels and nerves, running along as though under the surface of the city. It is as if one had in *ching* and *lo* two transport companies with exchange points for the public, well defined at their junctions. We call these junctions anastomotic loci (*hui hsüeh* 會穴).

The classic names of the loci did not long antedate the system of tracts. We find loci designated only by location in the Ma Wang Tui documents buried in −186 and probably written a little before −200.[53] Only a few loci are named in the case histories of Shun-yü I (*ca.* +150), perhaps two generations or so earlier than the *Ling shu*. The tractates in this book outline, far from consistently, the system of tracts connecting the loci, and add correlations with the yin–yang forces and the six *chhi*. All of this was systematised in the *Chia i ching*.

There is no doubt that in the *ching-lo* system we have to deal with a very ancient conception of a traffic nexus with a network of trunk and secondary channels and their smaller branches. It seems as if from the beginning these were thought of in terms of not only civil but hydraulic engineering, for there are greater and lesser reservoirs of *chhi*. We

[53] [The picture has been additionally complicated by the excavation in Szechwan in 1993 of a lacquered wooden figure probably made between −179 and −141, painted with what appear to be circulation tracts, but no loci. Whether this reflects an early stage in the history of medicine or a distinct local tradition is not yet clear. See Ma Chi-hsing (*1996*) and He & Lo (1996). – Editor]

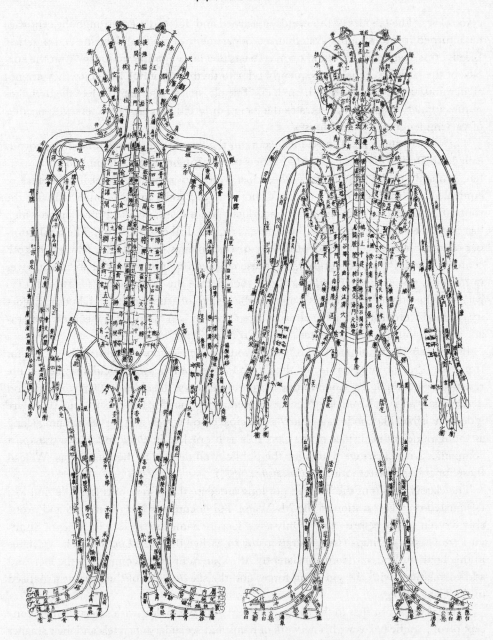

Fig. 3. General views of the circulation tracts and the most important loci for acupuncture and moxibustion, superimposed on a simple skeleton. An exceptionally fine version of an illustration often reproduced in books on acu-moxa therapy, from *Chen fang liu chi* 針方六集 (Six collections of acupuncture procedures, +1618).

are thus in the presence of an important doctrine arising from the idea of the Chinese double microcosm, the body of man corresponding to the State, since both reflect the order of the physical Universe. The basic idea of circulation, which originates unmistakably in the Former Han period, may also be derived in part from a recognition of the meteorological water-cycle – the exhalations of the earth rising into the clouds and falling again as rain.

The question of the origin of acupuncture is surely one of extraordinary interest. Therapists must have closely observed symptoms, especially pain, and their relief by various methods. We suspect that the profound conviction of the organic unity of the body as a whole that was reflected in the acupuncture system may have arisen challengingly out of the phenomenon of referred pain. Perhaps some passages in the ancient Chinese texts not yet noted will justify recourse to this as part at least of the explanation. The relation of transitory pain in the extremities or trunk with passing malfunctions of the viscera is so common an observation of everyday physiopathology that it may well have struck the ancient Chinese physicians with particular force. The secular accumulation of clinical experience too must have convinced Chinese physicians that acupuncture is effective.

Millions of people for something like twenty centuries have received and accepted acupuncture. No one will really know the effectiveness of it, or of other characteristic Chinese treatments, until accurate clinical statistics have been kept for several decades. Western physicians commonly express the view that acupuncture acts purely by suggestion, like many other things in what they often call 'fringe' or 'alternative' medicine. Some medical authors are convinced that it has a basis in physiology and pathology. In matters so uncertain, which belief is the most difficult? This is, we believe, a question of what one might call relative credibility (or perhaps a calculus of credulity). For our part we find the purely psychological explanation of acupuncture much harder to credit than an explanation couched in terms of physiology and pathology, without forgetting the subsidiary role of the mind in all somatic healing.

The practices of phlebotomy and urinoscopy in the West may seem analogous. They had exceedingly little physiological and pathological basis on which to sustain their long-enduring popularity, but neither had the subtlety of the acupuncture system. Perhaps blood-letting had some slight value in hypertension, and extremely abnormal urines could aid in diagnosis, but neither contributed much to modern practice.

Chinese, Japanese and Western laboratories have for decades actively pursued physiological and biochemical acupuncture experiments on animals, where the psychological factor is ruled out. Chinese have conducted clinical tests for a large part of this century, and Westerners since *ca.* 1960. The results so far support this opinion.

The suggestion that the action of the needles may stimulate the production of antibodies by the reticulo-endothelial system is now being tested. It has been a matter of surprise for Western biologists that the acupuncturists claim their treatment to be effective, at least in some degree, not only in diseases such as sciatica or rheumatism where no treatment in any part of the world can be considered very successful, but also in cases of infectious disease where an external causative agent is fully recognised. For example, it is difficult to believe that in such an entity as typhoid fever acupuncture could be effective; nevertheless that was the claim of the traditional physicians.

However, if the reticulo-endothelial system could be stimulated to produce antibodies in great quantity, possibly by indirect stimulation through the autonomic nervous system, that could explain the results. Alternatively, there may be a neurosecretory effect mediated through the autonomic and sympathetic systems upon the suprarenal cortex, inducing a rise in cortisol production. Again, there may be a neurosecretory influence upon the pituitary gland. A wealth of experimental approaches lies open. It is quite likely that acupuncture results in production of several groups of biochemical agents in the tissues near and far, prostaglandins, histamines and antihistamines, interferons and other antibiotic-like substances, kinins, etc. Whether it acts on gating centres in the central nervous system or in the spinal cord, so that 'all lines are busy' and the pain impulses cannot get through; or whether it is a matter of mobilising the endogenous opiates, those morphine-like derivatives such as the enkephalins and the endorphins, in the brain, or both, we still do not entirely know. This is merely to point out that a typically Western notion of one-to-one cause-and-effect may be too one-dimensional to properly appreciate the complexity of the 'esoteric' ancient Chinese medicine.

Acupuncture as practised in the late 20th century has two main branches: modern analgesia and traditional therapy. It is no good looking for anatomical correlates of the acu-tracts, because they seem not to exist. It is the stimulation at the loci that produces analgesia sufficient for major surgery.[54] The authors have themselves witnessed in China many examples of acupuncture adapted to modern surgery. Although it is no longer used as widely as in the late 1950s, in the heat of its discovery, nevertheless it is extremely important. It was the very first Chinese therapeutic technique that made Western physicians and neurophysiologists sit up and take notice. They had previously not been willing to manifest an interest in anything Chinese, or indeed Asian, but the fact that major surgery had been successfully performed under acupuncture analgesia could not be denied.

The term *ching* 經 has been known in the West only as the name for the linear arrays of acupuncture loci on the surface of the human body that we call circulation tracts. But the word has a much deeper meaning than this, denoting a basic physiological conception in ancient Chinese medicine founded on the theory of the Two Forces (yin and yang) and the Five Elements (*wu hsing*), which recognised six patterns of physiological function and pathological dysfunction. This sense emerges in the ancient and mediaeval Chinese system of diagnosis now called *liu ching pien chêng* 六經辨證 (differentiating the syndrome in accordance with the six *ching* patterns).[55] During the course of a disease these patterns became abnormal in diverse succession according to the character of the pathology.

---

[54] [Porkert (1974), pp. 197–8, has argued that, although the concept of loci is based on empirical data, the tracts, as the pathways that connect them, 'are only the result of systematic speculation'. Medical school acupuncture textbooks since 1960 have tended to avoid discussing the tracts as physical structures. This change has not generally affected books written for foreigners. On this point see Sivin (1987), pp. 142–5. – Editor]

[55] [See the Chinese account translated in Sivin (1987), ch. 7–9, and the analysis based on intensive clinical observation in Farquhar (1994), pp. 154–61. The translation above follows her usage. – Editor]

(7) The contrast between traditional-Chinese and modern-Western medicine

The time has now come to draw up a balance-sheet of the merits and demerits of traditional-Chinese as opposed to modern-Western medicine. In the first place, modern-Western medicine is based upon the modern sciences of anatomy, physiology, pathology, pharmacology, immunology and so on. None of these sciences was available anywhere in the Middle Ages, when traditional-Chinese medicine was forming its developed system; but as the modern sciences became reductionist they have found history irrelevant. This decisive break with the past has made it feasible for modern-Western medicine to recognise specific disease entities, whereas traditional-Chinese medicine still seldom sharply distinguishes symptom, syndrome and disease.

Modern-Western medicine is generally recognised to be particularly good for acute diseases, as in the case of antibiotics. One regrettable effect of modern-Western medicine is that the active principles in certain drugs, as identified by modern pharmacology, are administered as simple agents, producing side-effects on the patient. These are sometimes very serious.

When we turn to look at traditional-Chinese medicine, we have to recognise at once that the concepts with which it works – the yin and the yang, and the Five Elements – are all more suited to the times of Hippocrates, Aristotle and Galen than to modern times. They are unquantifiable, and indeed Chinese did not attempt to quantify them. Traditional-Chinese medicine knows nothing of the atomic and molecular conceptions so characteristic of modern biochemistry. Its background is very inadequate and indefinite.

A feature in which traditional-Chinese medicine is extremely good is its organic approach to illness. Two patients with identical symptoms may be given quite different treatments, depending on their backgrounds, which the physician has enquired about, and the general pictures of their body processes as ascertained in the examination. Another excellent feature of traditional-Chinese medicine is its notion of disease as a process that passes through various stages. This can lead to some very sophisticated cures. Generally speaking, a strength of traditional-Chinese medicine lies in curing chronic diseases. Thus we can see how valuable it would be if the insights of traditional-Chinese medicine and modern-Western medicine could be united.

(8) The possible integration of traditional-Chinese and modern-Western medicine

A topic of general concern is the possible unification or integration of the medical systems of the East Asian peoples with modern medicine. It has to be admitted that traditional-Chinese medical theory is mediaeval in character, because the yin and yang, the Five Elements, and innumerable other concepts are not congruent with modern scientific medicine. They are really parallel with the four elements of Aristotle, which cannot be put into modern scientific terms at all, or the four humours of Galen. They are neither verifiable nor falsifiable. Nevertheless Chinese physicians, whose clinical insights were truly profound, used these concepts as a trellis-work on which to hang their understanding of diseases.

The fact that the terminology and the concepts are mediaeval, whereas the concepts of Western medicine are essentially modern, does not mean that we cannot look forward to an ecumenical medicine of the future, which should embody all the clinical insights as well as the techniques characteristic of East Asian medicine, while remaining firmly based on modern biological science.

The Chinese government has for decades strongly supported research to assess the therapeutic achievements of traditional medicine by the standards of modern science, and to explore the possibilities of synthesis. It is equally important to accommodate modern scientific medicine to the practices and ideas of traditional-Chinese, traditional-Japanese and traditional-Indian medicine. Modern-Western medicine must be prepared to learn as well as to teach.

For example, medicine could become much more organicist or holistic than it is, and it could avoid active principles that are too powerful when used alone. It could use active principles in natural form, just as is done today with Chinese-style (*kanpōyaku* 漢方藥) prescriptions in Japan, and products of modern pharmacognosy elsewhere. There has been considerable progress in using modern biochemical and immunological methods to test their effects.[56]

This then is as far as we can go in our account of Chinese medical ideology and what ought to be done about it. One might feel that if any type-case were needed to demonstrate the moulding of medicine by the culture in which it grew up, Chinese medicine would be such a case. But, on second thoughts, is there any reason for regarding it as more 'culture bound' than Western medicine?

To think of the latter as self-evidently universal in application may be an illusion of those who happen to have been educated in that occidental Semitic–Hellenistic culture. True, it was destined by a series of historical accidents to give rise in the later stages of the Renaissance to specifically modern science. Western medicine became modern only in the 19th century as it was rebuilt upon the assured results of modern scientific physiology and pathology. The traditional medicines of the Asian civilisations are facing this transition only in our own time. Western medicine cannot become truly and ecumenically modern until it has subsumed all the clinical experience, special techniques and theoretical insights achieved in the non-European medical systems. Then will have occurred that fusion of Eastern and Western medicine to which we have referred above.

In the last resort, all medical systems have been 'culture bound'. Modern medicine is separating from its local historical roots only insofar as it partakes of the universality of modern mathematised natural science. Everything that the Asian civilisations can contribute must and will, in due course, be translated into these absolutely international terms. Only thus will medical science be able to free itself from connections with particular cultures and to minister universally to a united humankind.

---

[56] [For a typical modern guide to occidental pharmacognosy see Tyler (1988) or Trease & Evans (1989). Liu Shou-shan 劉壽山 (*1963–92*) is a massive collection of abstracts on scientific research in China. – Editor]

(b) HYGIENE AND PREVENTIVE MEDICINE

(1) INTRODUCTION

Ancient Chinese medicine was closely associated with the beliefs of the philosophers who may broadly be termed Taoist. In contrast to the Confucians who were interested primarily in human society alone, the Taoists devoted themselves to the study of Nature, believing that life should be lived in conformity with her, and they developed a system of religious mysticism, which Fêng Yu-lan has termed the only one ever known in the world that was not essentially anti-scientific.[1] The Taoists believed in the possibility of attaining a specifically material immortality so that they could continue to exist as etherealised beings on the Earth, enjoying the beauties of Nature without end. For this purpose they engaged in the study of alchemy, sought drugs which would confer longevity or immortality, and practised all kinds of techniques (some ascetic, some not) which they thought might contribute to this end. Their relation with preventive medicine was therefore particularly intimate. They spoke of the art of nourishing or preserving life (*yang shêng* 養生, *shê shêng* 攝生 or *wei shêng* 衛生), concentrating special attention on the inner causes of illness, which they pictured as the result of an improper balance between the yin and the yang. It was necessary to harmonise the two in order to remove the causes of disease.

Hygiene was their aim, reached by a large repertory of methods and techniques. To attain longevity, and perhaps immortality, it was necessary to cultivate the mind and the emotions, and to train the body to accord with the cyclical changes of the four seasons. One who thus pursues a normal and regular way of life should live at least a hundred years. Besides the search for the elixir of immortality, and for drugs (often mineral) which would promote longevity, the other Taoist techniques (*shu* 術) thought likely to contribute to that desirable end included various forms of gymnastics, special sexual practices, fasting and abstinence, and meditative visualisations involving the forces of Nature and the gods. Many of the techniques both of the Taoists and the ancient physicians were necessarily mixed with beliefs in what we should today call magic. Nevertheless the practical benefits to health of what they recommended are undeniable.

(2) EARLY CONCEPTS OF PREVENTION

We shall now present a series of texts illustrating concepts of safeguarding life from the end of the Chou period or just after, down to the the beginning of the Thang. It will be seen that the philosophers and the medical writers are at one on the subject. First of all, in a −3rd or −2nd century commentary on the *I ching* 易經 (Book of changes), we have the fundamental statement that 'the *chün-tzu* 君子 always meditates on trouble

---

[1] Personal communication; see Vol. 2, p. 33.

in advance and takes steps to prevent it'.[2] In the *Tao tê ching* 道德經, the great poem attributed to Lao Tzu 老子 but probably of the late −3rd century, we have the following verse:

> To know when one does not know is best;
> To [think] one knows when one does not know is a disease.
> Only when one recognises this disease as a disease
> Can one be free from the disease.
> The sage is free from the disease;
> He recognises this disease to be a disease
> And therefore is free from it.
> The best physicians always treat disease when it is not [yet] a disease
> And so [their patients] are not ill.[3]

Here disease is a metaphor for wilful ignorance, but perhaps we can see in this verse the recognition of abnormal functioning at its small beginnings and of causes bodily as well as mental.

Next is a celebrated saying in the *Huai nan tzu* 淮南子 book, compiled just before −139 by a group of scholars and proto-scientific magicians under the patronage of the king of Huai nan, Liu An 劉安: 'A skilful doctor always treats illness when there is no sign of disease, and thus the disease never comes; a sage (ruler) always restores crises to order when there is no sign of crisis, and thus the crisis never comes.'[4]

Not long after, the greatest of all the Chinese medical classics, the *Huang ti nei ching, Su wên* 黃帝內經素問 (The Yellow Emperor's treatise on corporeal [medicine]: plain questions [and answers], probably compiled in the −1st century), similarly juxtaposes the work of the physician and the ruler: 'the sage treats illness before, not after, it has arisen; he imposes order on disorder before, not after, it has arisen'. The chapter goes on to say 'to administer drugs after the illness has developed, or to impose order after disorder has developed, is like digging a well when one feels thirsty, or casting daggers in the midst of battle. Is that not too late?'[5]

The first indication of how this idea was applied in clinical reasoning appears about +200 in *Shang han tsa ping lun* 傷寒雜病論 (Treatise on febrile and miscellaneous diseases). When a disciple asks 'What does "the highest type of practitioner treats illness before it has arisen" mean?' his teacher explains:

[2] Wilhelm (1924/1950), vol. 1, p. 261, on hexagram 63. This statement occurs in the 'Commentary on the images' (*Hsiang chuan* 象傳, mid −3rd to early −2nd century).

Like '*tao* 道' and '*chhi* 氣', we prefer to leave '*chün-tzu*' untranslated. Though the word originally meant the ruler or the lord in Bronze-Age proto-feudalism, it came eventually to have all the aura of Aristotle's 'great-souled man', or what is implied by the term 'gentleman', even 'knight': of aristocratic birth but not necessarily, a scholar but not necessarily, an officer of state but not necessarily. For Westerners, Sir Thomas More might be cited as the type of all *chün-tzu*.

[3] Ch. 71, tr. Chan (1963), p. 225, mod. The laconic and epigrammatic quality of this famous text may be appreciated from the text of lines 5–7: *Shêng jen pu ping, i chhi ping ping, shih i pu ping* 聖人不病。以其病病。是以不病. [See the variants in the Ma-wang-tui MS as translated in Lau (1982), p. 251. Lau's understanding of the passage is quite different. On the strong case made by Ku Chieh-kang and Lau for dating the contents of the *Tao tê ching* in the −3rd century, see W. Boltz in Loewe (1993), pp. 269–92. – Editor]

[4] Ch. 16, p. 4b.   [5] *HTNC/SW*, ch. 1 (*phien* 2), 3.

treating illness before it has arisen means that when he observes an illness of the liver, he knows that it will be transmitted to the spleen, so that the [vitality of the] spleen should first be restored. But when, in the [cycle of the] seasons, the spleen [in its turn] has assumed kingship [over the body's functions], it cannot accept pathological agents, and so should not be replenished.[6] A mediocre practitioner will not know about transmission. When he observes an illness of the liver, he does not know about restoring the spleen, and will only treat the liver.[7]

There is an interesting passage in the *Ho kuan tzu* 鶡冠子 (Book of the Pheasant-Cap Master), a very mixed compilation of a philosophical character. Many parts date from as early as the −3rd century; when it reached its present form is still under debate. One of the discussions there recorded, not necessarily historical, is between Chao 趙, the king of Cho-hsiang 卓襄 and son of Duke Hsiao 孝 of the State of Chao, and his general, Phang Hsüan 龐煖.

Phang Hsüan said to the king of Cho-hsiang, 'Have you not heard that Duke Wên 文公 of Wei 魏 asked the great physician, Pien Chhüeh 扁鵲, "of your three brothers, which is the best physician?" Pien Chhüeh answered "The eldest is the best, then the second, and I am the least worthy of the three." Duke Wên said, "Might I hear about this?" Pien Chhüeh replied, "My eldest brother, in dealing with diseases, is attentive to the spirit (*shên* 神). Before [any symptoms] have formed, he has already got rid of it. Thus his fame has never reached beyond our own clan. My next brother treats disease when its signs are most subtle, so his name is unknown beyond our own village. As for myself, I use stone needles on the blood vessels, prescribe strong drugs, and fortify the skin and the flesh. Thus my name has become known among all the feudal lords." '[8]

Aphorisms of this kind may be found *ca.* +320 in the *Pao phu tzu nei phien* 抱樸子內篇 (Inner chapters of the Preservation-of-Solidarity Master) of the great alchemist and physician, Ko Hung 葛洪. He says: 'The adept disperses troubles before they begin, and cures diseases before the illness appears. If he treats in advance of the problem, he will not be trailing behind as the patient dies.'[9]

The Prescriptions worth a thousand (*Pei chi chhien chin yao fang* 備急千金要方, between +650 and +659) of the great physician Sun Ssu-mo 孫思邈, outlines a preventive regimen. For instance, he says:

After enjoying excellent health for ten days, one should employ moxibustion at a few points in order to drain out [noxious] wind and *chhi*. It is well, day by day, to harmonise the *chhi*, adjust the circulation, massage oneself, and practise gymnastics. Do not take good health for granted. [Just as one] should not forget danger in time of peace, try to prevent the coming of disease beforehand.

Sun suggested keeping what we should now call first-aid kits. He recommended that readers, whether at home or travelling, have artemisia pulp ready for moxibustion, keep

---

[6] 'Kingship' refers to the paramount status of one of the visceral systems during the part of the year that corresponds to it. During this phase it is not subject to pathogenic *chhi*, according to the old principle that the ruler is not directly subject to evil. The splenetic functions, because they correspond to earth, rule, according to the schema, either in midsummer or in the transitional periods between seasons. For an example of 'kingship' see *HTNC/LS* (*phien* 41), 5.

[7] *Chin kuei yao lüeh*, ch. 1, p. 1a.

[8] Ch. 16, pp. 10bff.; cf. Defoort (1997), p. 215. Pien portrays himself as relying on the full range of therapy instead of preventing disease or stopping it as it begins.

[9] Ch. 18, p. 4b.

rhubarb, liquorice root, cassia bark, ginger, mercury and a few other simple medicines handy for therapy, and store certain emergency formulae in compounded form. Sun urged every family to possess one or two treatises on emergency medicaments.[10]

(3) ANCIENT LITERATURE

Before proceeding further, it will be desirable to have a look at preventive medicine in certain interesting specimens of ancient literature in the Chinese tradition. We shall examine first the *Chou li* 周禮 (Rites of the Chou dynasty), and secondly the *Shan hai ching* 山海經 (Classic of the mountains and rivers). The *Chou li* is a large and interesting work compiled by scholars of the early Han dynasty, about the −2nd century. It purports to be a register of the officials and their duties under the High King of the Chou dynasty, in the middle of the −1st millennium. It cannot be a record of that date, but constitutes rather a system showing what the Han Confucian scholars considered the ideal form for a unified imperial bureaucracy. Some parts of the book, especially the *Khao kung chi* 考工記 (Artificers' record), are probably genuine documents dating back to the −4th century or a little before, in this case from the State of Chhi 齊, but the greater part of the work is essentially Han.[11]

Pertinent to our topic is the detailed account of the medical and health officials attached to the imperial court. We find a Master Physician (*I shih* 醫師) who superintends the imperial medical staff. Next in order is a Dietician (*Shih i* 食醫). His duties and those of his assistants were to take charge of the diet of the emperor and the court. They bore in mind the nature of the various foods to be combined into menus that were balanced and at the same time adapted to the cycle of the seasons. Besides, there were the Physicians-in-Ordinary for Diseases (*Chi i* 疾醫), who dealt with the infections and epidemics characteristic of the four seasons, certifying and recording causes of death as well as applying remedies to the sick, and the Physicians-in-Ordinary for External Disorders (*Yang i* 瘍醫). These latter officials cannot quite be called surgeons because they rarely operated. Their domain included war wounds and fractures as well as ulcers, oedemas and skin disorders. Lastly came the Veterinarians (*Shou I* 獸醫).

Elsewhere the *Chou li* enumerates among the complements of official departments 2 Senior and 2 Junior Master Physicians, with 2 each Administrative Officials of the 5th (*Fu* 府) and 6th class (*Shih* 史) and 20 of the 8th class (*Thu* 徒).[12]

In addition to all these groups there are a number of sanitary officials whom we might call 'fumigators' or 'vermin exterminators'. For example, the Worm Specialist (*Chu shih* 庶氏) is charged with combating poisonous creatures of all kinds (*tu ku* 毒蟲). Similarly, the Exterminator (*Chien shih* 翦氏) is charged with combating insect pests (e.g. silver

---

[10] Ch. 27, p. 481.

[11] [Recent scholarship makes it unlikely that any part of the *Chou li* long pre-dates the compilation of the work. Its date remains contentious within a range from the early +3rd to the mid +2nd century. Some parts of it do quote extant earlier writings, but the Artificers' record is not prominent among them. Its place of origin is also uncertain. See R. Fracasso in Loewe (1993), pp. 25–9. – Editor]

[12] Ch. 1, p. 5a; ch. 5, pp. 1a–5b.

fish, moths, boring beetles, ants, termites and so on).[13] The source does not specify the methods used by either, but in later times their counterparts used not only physical and chemical agents but also exorcistic rites.

Among the drugs used for fumigation was the plant *mang* 莽 (*Illicium religiosum*, the bastard anise).[14] This poison for insects and fish was well known in antiquity. We find, for example, in the *Huai Nan wan pi shu* 淮南萬畢術 (The inexhaustible arts of the king of Huai nan, *ca.* −120), a collection of formulae attributed to the group of experts surrounding Liu An, the king of Huai nan, the statement that *mang* is a powerful fish poison (*mang tshao fu yü* 莽草浮魚).[15] Returning to the *Chou li*, the Extirpator (*Chhih pa shih* 赤犮氏) is charged with combating vermin in walls and houses. For this purpose he uses lime prepared from mollusc shells and 'scattered ashes', words which may conceal the use of caustic alkali prepared by burning wood and 'sharpening' its lye. Then the Master of Crickets (*Kuo shih* 蟈氏) has the duty of combating creatures which make disagreeable noises such as frogs or cicadas. He significantly uses *mu chü* 牧鞠 (*Pyrethrum seticuspe*, the winter aster)[16] both for sprinkling as powder and as fumigatory smoke. It is interesting that such a valuable plant insecticide is mentioned among the paraphernalia of the sanitary officials in the *Chou li*. Lastly, the Water Sprinkler (*Hu cho shih* 壺涿氏) got rid of pests (*chhung* 蟲). He beat an earthenware jug in order to frighten them away, and also performed rites with elm-wood branches and elephant ivory. We are obviously in the presence of a mixture of ancient magic and effective insecticides. It is remarkable that this archaised imaginary reconstruction of the perfect governmental staff includes so many sanitary officials.

Indeed, this was not the end of them. Elsewhere in the same book we meet with the River Patrollers (*Phing shih* 萍氏). Their duty was to warn of dangerous places as well as to protect fisheries out of season. They were concerned with public safety in more senses than one, since not a few species of fish in Chinese waters are poisonous at certain seasons of the year. More directly important for our present subject, perhaps, are the Protectors of Corpses (*Chhü shih* 蜡氏, the word *chhü* here meaning decaying flesh caused by vermin). This official and his staff are charged with the removal of rotting corpses, whether of men or animals. When someone dies on the road, they have to bury the body and report to the magistrate on the day and month of the event, bringing the clothes and possessions of the deceased so that the family can reclaim them. Last of all come the Travel Patrollers (*Yeh lu shih* 野廬氏), whose duty was to inspect communications, direct traffic both on roads and waterways, keep rest-houses in good condition, conduct visitors to and from the frontiers and – again for us important – organise the

[13] Physician, ch. 1, p. 4a, ch. 2, p. 1a; Dietician, ch. 1, p. 4a, ch. 2, p. 1b; Internal Disorders, ch. 1, p. 4b, ch. 2, p. 2a; External Disorders, ch. 1, p. 4b, ch. 2, p. 3a; Veterinarian, ch. 1, p. 4b, ch. 2, p. 4a; Worm Specialist, ch. 9, p. 5b, ch. 10, p. 7a; Exterminator, ch. 9, p. 6b, ch. 10, p. 8b. All the passages referred to in this paragraph will be found in the Biot (1851) tr., vol. 1, pp. 8ff., 92ff. Each official has two references; one chapter musters him and his staff, while another describes their duties.

[14] R/505; Stuart (1911), p. 489.

[15] Item 53. Cit. in *TPYL*, ch. 993, p. 2b. [The text, literally 'the bastard anise makes fish float', is quite ambiguous, but the commentaries imply the interpretation given above. – Editor]

[16] R/26; Stuart (1911), p. 260; Bretschneider (1881–98), vol. 2, nos. 130, 404. Pyrethrum is mentioned again, and in the same context, in the *Huai Nan wan pi shu*, entry no. 107; cit. in *TPYL*, ch. 996, p. 2b.

sweeping of the streets in the capital and other cities. Thus from the point of view of preventive medicine there is a great deal of interest in this ancient survey of the ideal imperial bureaucracy.[17]

It is difficult to be sure how far the organisation imagined in the *Chou li* refers only to the imperial capital and the imperial court itself and how far it was regarded as extending outward in a provincial network throughout the whole of the country. During the early Han period, when the book was compiled, a bureaucratic organisation of prefectures and commanderies was evolving all over the empire, subject to the one universal monarch, head of the Chinese *imperium* (*thien hsia* 天下). Thus we may interpret the scholarly writers of the *Chou li* as outlining a bureaucracy throughout the provinces as well as in the capital itself. Of course, officials such as the imperial medical staff were attached only to the court.

Another literary work of equal interest is the *Shan hai ching*, which has sometimes been described as the oldest geographical book in China. Although it is archaistic in style, recent studies tend to date the composition of its various parts between the late Chou and some time in the Han period.[18] This book is ostensibly a geographical account of all the regions of the Chinese culture-area. It contains indeed a good deal of mythological material about strange beings, gods and local spirits who were worshipped in different places, yet the tone is surprisingly matter of fact. The content includes a large quantity of very rational description, including the trees, animals and minerals found in different places, and the difficulties of communications.

Surprisingly perhaps, the *Shan hai ching* usually recommends particular drugs not for curing diseases, but for preventing their onset. It uses no less than ninety plant, animal and mineral substances to promote health and to prevent illness. The word *fang* 防, which we might translate 'will ward off', is extremely prominent here. The number of measures listed to prevent each disorder or condition is shown in Table 1.

Apart from words for ingestion, the text also speaks of *phei chih* 佩之, meaning to wear on the body an object that repels some evil; and very occasionally, of paying homage to it or worshipping it.

The interest of this analysis is that one can see the kinds of disease which were feared in the Warring States period, especially by travellers.[19] Because more animal substances than plants are mentioned, one can also visualise the great extent of forest country and uncultivated waste land in those days. Table 2 analyses the diversity of materia medica in the book.

Thirty-one items were eaten for protection and only slightly fewer used in other ways, such as being worn or smeared on the body. The idea of promoting general health is in accordance with the Taoist principle of 'nurturing vitality' (*yang shêng*).[20] In certain

---

[17] *Chou li*: Extirpator, ch. 9, p. 6b, ch. 10, p. 9a; Master of Crickets, ch. 9, p. 6b, ch. 10, p. 9a; Water Sprinkler, ch. 9, p. 7a, ch. 10, p. 9b; River Patroller, ch. 9, p. 5a, ch. 10, p. 4a; Protector of Corpses, ch. 9, p. 4b, ch. 10, p. 3a; Travel Patroller, ch. 9, p. 4b, ch. 10, p. 2a. Biot (1851), tr., vol. 2, pp. 380ff.

[18] This book has already been described in the section on Botany, in Vol. 6, pt 1, pp. 255–6.

[19] No doubt Fan Hsing-chun 范行準 (1953), pp. 325–6, is right in believing that '*ku* poison' was various parasitic disorders such as schistosomiasis.

[20] On the notion that this principle is Taoist, see the Introduction, p. 11.

Table 1. *Preventive medicine in the* Shan hai ching

Agent or measure	Instances
Infection by worm or insect parasites (*ku* 蠱)	13
Epidemic or infectious diseases (*i* 疫)	7
Hunger or excessive emotion (fear, jealousy)	33
External diseases of the sense organs	13
External diseases of the skin and limbs	15
Affections of the viscera, etc.	6
Conception	3
Animal diseases	1
Total	91

Table 2. *Materia medica in the* Shan hai ching

Animals		Plants		Minerals	
Mammals	5	Herbs	10	Minerals	1
Fish	13	Shrubs	12		
Birds	9				
Reptiles	2				
Total	29	Total	22	Total	1

descriptions the advantageous effect of the medicament is said to lie in removing fear of storms, thunder, wild beasts, etc., which indicates a concern with promoting mental health.

It is very likely that maintenance and improvement of good nutritional status were an important function of the items in the third group of Table 1. No better means would have existed for keeping in good heart those ancient travellers, official, military or private, through the wild mountains and forests between the isolated and far-scattered urban foci of ancient Chinese culture. Thus in general one can see how the *Shan hai ching*, as a compilation of popular beliefs, represents a more archaic stage in the development of materia medica in China than the *Shên nung pên tshao ching* 神農本草經 (Pharmacopoeia of the Divine Husbandman), the first of the long line of pharmacopoeias (late +1st or +2nd century).

Whether there were any written manuals of public health and the prevention of diseases in the Warring States period, we do not know. None has come down to us. There is, however, the possibility of seeing a reference to such manuals in one of the stories of the *Chuang tzu* 莊子 book, assembled early in the Han from writings of Chuang Chou 莊周 about −320 and others of the −3rd and −2nd century. One of what A. C. Graham has called the 'mixed chapters' purports to record a dialogue between Lao Tzu 老子 and an enquirer, Nan-jung Chhu 南榮趎, who is unable to comprehend how the Way can lead him to long life. More and more frustrated as the interview continues, Nan-jung tries to sum up the problem:

When a bumpkin falls ill and his fellow villagers enquire about his illness, the fellow can describe it. When the patient can speak of it and agree that he has it, he is as if no longer ill. As for my hearing of the Great Way [from my own teacher], it was like taking medicine and the illness worsening. All I really want to hear about are the guidelines for protecting my vitality (*wei shêng chih ching* 衛生之經).

Nan-jung is saying, as scholars have consistently understood the anecdote over the centuries, that since he knows what his problem is, teaching ought to relieve him, but the more he listens to his teacher the further he is from solving it. He is not asking for medical information, but for a set of rules that will lengthen his life. Lao-tzu obliges, with neither medical information nor guidelines, but instead with a reminder that the preservation of life depends on spiritual union with the Way:

You want guidelines for protecting vitality, do you? Can you embrace the One? Can you avoid losing it? Can you recognise what is fortunate and unfortunate without divining? Can you cease? Can you abstain? Can you stop seeking it in others and find it in yourself?

and so on.[21]

In the medical literature of later centuries, *wei shêng chih ching* is more likely to mean technical writings, literally 'manuals for preserving life'. The context makes that meaning in this passage unlikely, but it is not impossible that Nan-jung is asking for oral teaching from actual written manuals, whether on mystical or physical methods.

From the Sung period on, the expression *wei shêng* was employed in the titles of many medical books, generally with a wide implication. The oldest about which we have information is the lost *Wei shêng chia pao* 衛生家寶 (Family treasure for preserving life), by Chang Yung 張永, of the early +12th century.[22] One might also mention Lo Thien-i's 羅天益 celebrated *Wei shêng pao chien* 衛生寶鑒 (Precious mirror for preserving life, +1281). *I hsien* 醫先 (Antecedents to therapy, +1550), by Wang Wên-lu 王文祿 of the Ming, is no mere handbook of preventive medicine. It is concerned with knowledge of cosmic principles that must be applied to forestall disease and make therapy unnecessary.

(4) THE YELLOW EMPEROR'S TREATISE

We now scrutinise the ideas on preventive medicine in the greatest of all the Chinese medical classics, the *Huang ti nei ching* 黃帝內經.[23] The work emphasises throughout the responsibility of the physician not only to attend to the sick but to maintain health. At its very opening the *Su wên* 素問 stresses the need for people to guard their well-being by avoiding agents of disease and remaining tranquil.

---

[21] *Chuang tzu*, ch. 8A (*phien* 23), ll. 23–36. [On the complex relations of mixed chapters like this one to the other strata, see Graham (1979). Whether this fragment is early or late remains in question. – Editor]
[22] Okanishi (*1958*), p. 1067.   [23] See above, p. 47.

What the sages of high antiquity taught their subjects to do, they did. Depleting agents of disease, thievish winds: they avoided them in ample time. They were tranquil and quiescent. Their *chen chhi* 真氣 responded and followed. If vitality is maintained within, from whence can diseases come?[24]

The philosophy of the *Nei ching* is pneumatic. The term *chen chhi* (lit. 'true pneuma') means primarily the innate constitution or endowment of the mind–body organism at birth.[25] Synonyms for this, often found, are *yüan chhi* 元氣 (original pneuma) and *ching chhi* 精氣 (essential pneuma). The expression *chen chhi*, however, has a wider connotation, because it may be regarded also as a a general term for all the bodily vitalities, the *chêng chhi* 正氣 (pneumata proper to the body) as distinguished from the *hsieh chhi* 邪氣 (unbalanced and thus unbalancing pneumata). When the *chen chhi* is adequate, the pneumata of the four seasons act upon the body at the right time and to the right extent. The *hsieh chhi* comprise all agents adverse to health – unseasonable or harmful 'winds' (*hsü fêng* 虛風, *hsü hsieh* 虛邪, *tsei fêng* 賊風), and the poisons of animate or inanimate things. If they invade the body, they may overcome the *chen chhi* and cause disease.

We may find the narrower sense of *chen chhi* well described in the *Huang ti nei ching, Ling shu* 黃帝內經靈樞 (The Yellow Emperor's treatise on corporeal [medicine]; the vital axis): 'the *chen chhi* is [the vitality] endowed by Heaven. It combines with the vitality extracted from food (lit. 'the cereal pneuma', *ku chhi* 穀氣) to fill the body.'[26]

Physicians often referred to the inborn and acquired pneumata by the expressions *hsien thien* 先天 and *hou thien* 後天, literally what is 'prior to Heaven' and 'posterior to Heaven'. These two technical terms are connected with the *I ching* (Book of changes), to which we have already referred, where they have to do with two celebrated arrangements of the hexagrams. The *hsien thien* and *hou thien* arrangements may be pre-Han, but they are certainly associated primarily with two famous mutationist diviners of that dynasty, Chiao Kan 焦贛 and Ching Fang 京房 in the −1st century. These terms occur in many passages in the *Nei ching*.[27] Thus factors 'prior to Heaven' are those of innate constitution plus the effects of adequate nutrition and living conditions, while the factors 'posterior to Heaven' are those of specific external influences which act harmfully or beneficially upon the organism.

The second chapter of the *Su wên* is entitled 'The great treatise on adjusting the *shên* 神 to the *chhi* of the four [seasons]'. *Shên* here is untranslatable; 'spirit' will not do, for *shên* denotes both the body's governing vitalities and the fine *chhi* that makes consciousness possible. The chapter sets out the results of failure to observe this vital adjustment in the summer:

---

[24] Ch. 1 (*phien* 1), 2. [For an analysis see Larre & Rochat de la Vallée (1983). In the first sentence, which is incomprehensible in the received text, I read *wei* 為 for *wei* 謂, following the Northern Sung commentators, who report this reading (now lost) in *HTNC/TS*, as well as in other sources. – Editor]

[25] For a detailed analysis of thirty-two types of *chhi* in *Huang ti nei ching* see Porkert (1974), pp. 168–73.

[26] Ch. 11 (*phien* 75), 4.1.

[27] For example, ch. 69, p. 2a; ch. 70, p. 34a. Further details of the mutationist use of them will be found in connection with the ancient history of the magnetic compass and the compass points in Vol. 4, pt 1, p. 296. [The citation of the *Shuo kua* commentary there should refer instead to the *Wên yen*. Although the terms *hsien thien* and *hou thien* were common from the Han on, recent research indicates that their use for arrangements of the hexagrams is not earlier than the Sung. See, for instance, Smith *et al.* (1990), pp. 110–20. The orders used in connection with the compass were of the trigrams, not of the hexagrams. – Editor]

The three months of summer we call [the time of] burgeoning and bloom. The *chhi* of Heaven and Earth couple, and the myriad things [put forth their] blossoms and fruit. After the night's sleep arise early, and do not be bothered by the [long] days. Then there will be no anger in your thoughts. Your appearance will be flourishing. The [stagnant] *chhi* [in the body] will be able to seep out 'as though its love were outside'. These are [the body's natural] responses to the *chhi* of summer, the way of nourishing and growth. Acting in conflict with it will damage the cardiac function, giving rise in the autumn to *nueh* 瘧 disorders. The basis for [autumnal activities characterised by] contraction will be inadequate, so at the winter solstice there will be serious disease.[28]

The leading idea is that the right regimen in the appropriate season prevents the occurrence of diseases in the following season. Among the broad *nueh* group of disorders, which includes intermittent and other fevers, malaria among them, are types that begin in summer when pathogens invade the body, but become manifest later.

(5) Hygiene, mental and physical

It is remarkable to see how far what we should now call the psychosomatic causation of diseases was empirically recognised in ancient China. The philosophers of all the schools agreed without hesitation that it is necessary to cultivate the mind and control the emotions. A typical statement may be found in *Shên chien* 申鑒 (Extended reflections, +205), by Hsün Yüeh 荀悅: 'In order to nourish the spirit, at all costs one must moderate pleasure, anger, pity, happiness, worry, and anxiety.'[29] He goes on to say that those who are good at controlling their *shên chhi* 神氣 (mental pneuma) can direct it as Yü 禹 the Great (the legendary hydraulic engineer and emperor) regulated watercourses, avoiding excess and deficiency. The same attitude appears in the *Lü shih chhun-chhiu* 呂氏春秋 (Springs and autumns of Lü Pu-wei 呂不韋), that wonderful compendium of natural philosophy compiled *ca.* −239 by the proto-scientists and magicians gathered under the auspices of Lü Pu-wei in the State of Chhin:

Heaven engendered yin and yang, cold and heat, dry and wet, the transformations of the four seasons, the metamorphoses of the myriad things. Every one of these can bring benefit, and every one can bring harm. The sages investigated what is favourable in yin–yang, and what is beneficial amongst the myriad things, for the good of [human] life. [Following their prescriptions] the vitalities remain stable in the body and the lifespan can be prolonged. Prolongation does not mean adding to what is deficient, but completing an allotted span. [One succeeds in banishing harm through avoiding excess of flavours, of emotions and of climate.] For nurturing vitality nothing is as important as understanding these fundamentals. Once one understands them, there is no way in which diseases can enter.

The chapter ends with a splendid passage:

Nowadays people appeal to divination and offer up prayers and worship, yet diseases rampage all the more. It is like an archer at a shooting-match who, failing to hit the bull's-eye, decorates the target. How will this help him to score a bull's-eye? If one adds boiling water to stop something from boiling, the more one adds the longer it boils. When one takes away the fire from underneath, the boiling stops. As for all these healers (*wu i* 巫醫) with their toxic drugs,

---

[28] *HTNC/SW*, ch. 1 (*phien* 2), 1.   [29] Ch. 3, p. 2b.

treating [disease] by driving it out, the ancients esteemed them little, because they dealt with the ramifications [and not with the root].[30]

How can one sum up these fundamentals of a long and serene life? The immediate answer of the philosophers, common to most of the schools and evident in this passage, was an inner balance that comes from avoiding all excess, particularly excessive emotion. The medical writers, such as those who compiled the *Nei ching*, tended to emphasise the correct balance (to use the Greek word, *krasis* ($\kappa\rho\hat{\alpha}\sigma\iota\varsigma$)) between the yin and the yang. Failure to achieve this adjustment arises from proceeding contrary to (*ni* 逆) the correct mode of life. The philosophically important word *ni* throughout ancient Chinese thought refers to actions that oppose the patterns into which the universe spontaneously organises itself. We can see by these examples that the serenity and calmness of mind requisite for health was not merely a stoic ataraxia, an impassibility untouched by the misfortune of others, but part of a cultivated middle way that avoided every extreme.

Turning now to the medical writers, the *Su wên* describes explicitly the effect of mental health upon the resistance of the body to external attack: 'Wind (*fêng* 風, an external pathogen) is the beginning of all diseases. If the mind is clear and calm, then the flesh and the interstitial tissues close the way and resist entry (*chhing ching, tsê jou tshou pi chü* 清靜則肉腠閉拒). Though powerful winds and virulent poisons may be at work, they are quite unable to do any harm.'[31]

The balance of the mind being thus safeguarded, the Taoists and the physicians advocated several kinds of mild gymnastics. They generally called these exercises *tao yin* 導引, that is to say, guiding and leading the vitalities. The terms *nei kung* 內功 in later times, and *kung fu* 功夫 in very recent times, both implying inwardly directed work, were used for such gymnastics.[32] Possibly the exercises derived from the dances of ancient rain-bringing shamans, but in any case they were associated with the idea, as old in Chinese as in Greek medicine, that the circulatory system was liable to become obstructed, thus causing stasis (*yü* 鬱) and disease.

Already in the *Lü shih chhun chhiu* we can find the aphorism that 'Running water does not stagnate, nor does a door-pivot become worm-eaten, because they move. That is also true of the physical form and the *chhi* [of the human body]. If the physical form does not move, the essences do not flow [freely]; if the essences do not flow [freely], the *chhi* becomes static.'[33] This was the basic idea behind the gymnastic exercises, the origin of which we see quite clearly in the biography of Hua Tho 華佗 (fl. +190 to +265), an outstanding physician and surgeon about whose name many legends gathered. In his biography in the *San kuo chih* 三國志 (History of the Three Kingdoms) we read:

Wu Phu 吳普 of Kuang-ling 廣陵 and Fan A 樊阿 of Phêng-chhêng 彭城 both became pupils of Hua. Wu Phu followed exactly the therapies of Hua, so that his patients generally got well. Hua Tho told him 'the human body ought to be exercised until it is tired, but this should not be

---

[30] *Chi* 3 (*phien* 2), pp. 136–7. Cf. Wilhelm (1928), p. 30. On *wu i*, see above, p. 41.
[31] Ch. 1 (*phien* 3), 2.3.   [32] See, *inter alia*, Despeux (1981) and Engelhardt (1987).
[33] *Chi* 3 (*phien* 2), p. 136. [This passage occurs in the same essay as, and falls midway between, the two quoted on the preceding page – Editor]

carried to an extreme. As it is agitated, the digestion improves (lit. "the cereal pneuma is able to disperse", *ku chhi tê hsiao* 穀氣得消) and the circulation through the blood vessels is freed (*hsüeh mo liu thung* 血脈流通), so that disease is unable to arise. This is exactly like a door-pivot never rotting. It was with this in mind that the ancient sages performed guiding and leading exercises, [for example] hanging like a bear and turning the head backward like an owl, flexing the waist and moving the joints, seeking to inhibit old age. I have a method, known as the "play of the five beasts" (*wu chhin chih hsi* 五禽之戲), the tiger, the deer, the bear, the ape and the bird. It serves both to get rid of disease and to loosen the joints of the feet. It can be used as guiding and leading exercises. If the body feels out of sorts, get up and do one of these exercises until the body feels moist and perspiration appears. Then powder yourself. Your body will feel limber and your stomach will feel ready for a meal.' Wu practised [these techniques] himself. When he was over 90 his ears and eyes were sharp and his teeth were intact and strong.[34]

Chinese hygienic and remedial exercises generated a large literature. There is considerable reason to believe that information about practices from Asia influenced the growth of calisthenics in modern Europe.[35] This influence was marked from the +18th century onwards. Earlier traditions in Europe presumably derived from Greek sources. Beyond that, the story of the origins of modern gymnastics remains to be written.

(6) PRINCIPLES OF NUTRITIONAL REGIMEN

(i) *Diet*

The medical classics of the Han fully recognised the importance of a complete and balanced diet. For example, the *Su wên* explains the difference in longevity between archaic times and the degenerate modern age as due to diet and other aspects of hygiene:

Among the people of high antiquity, those who apprehended the Way took yin and yang as their pattern, harmonised themselves through the regularities of [life-sustaining] techniques (*shu shu* 術數), moderated their diets, lived in a regular way, and did not exhaust themselves through undisciplined activity. Thus able to keep their bodies and consciousnesses (*shên* 神) in unison, they lived out the lifespans allotted them by Heaven, enduring a hundred years before they passed on.

The *Su wên* explains bad health by lack of this discipline: 'When eating and drinking are doubled the stomach and intestines are seriously harmed.' And again we read of a certain disease that 'the root of it is the patient's [abnormal] diet and mode of life'.

The same principle governs the use of food in therapy:

Toxic drugs attack agents of disease; the five (i.e. various) cereals provide nutriment [to supplement the drugs]; the five fruits assist them, the five meats aid them, and the five vegetables reinforce them. Thus they are ingested in a way that balances the *chhi* 氣 and the *wei* 味 [i.e. the five-element and yin–yang characteristics of the foods taken with the drugs], so as to replenish the seminal essence (*ching* 精) and the *chhi*.

This passage is discussing therapy which moderates – buffers, so to speak – the aggressive ('toxic') virtues of active medicinal principles by administering, with the drugs, foods that build up the resistance of the body.[36]

---

[34] *Wei shu* 魏書, ch. 29, pp. 6aff.   [35] Vol. 2, pp. 145ff.
[36] Ch. 1 (*phien* 1), 2; ch. 12 (*phien* 43), 3; ch. 7 (*phien* 22), 4. See also the quotation from a lost essay by Chang Chi 張機 in *Chhien chin fang*, ch. 26, p. 1a.

In later times the tradition of dietary medicine developed and expanded fully. The celebrated 'Essay on dietary regimen' in the *Chhien chin fang* (between 650 and 659) quotes several early sources.[37] Diet became a regular topic in writing on the care of elderly people. For example, Chhen Chih 陳直 said in his New book on longevity for one's parents and nurture for the aged (*ca.* +1085):

If it is possible to know the [medicinal] characteristics of foods and regulate them properly, they can be twice as good as drugs. That is because old people are generally averse to taking medicine but enjoy eating, so that treating their complaints through diet is more effective. Furthermore, in geriatric diseases, one must be cautious about purging old people, which makes treatment by diet even more desirable. In general, when an aged person is suffering, the physician should resort to nutritional therapy first, and call for drugs only if this fails. This is the great rule in caring for the elderly.

With this in mind, to be good at curing diseases does not compare with being prudent about [the treatment of] disease. To be good at treatment with medicines does not compare with being good at treatment with food.[38]

The editor of the same work, Tsou Hsüan 鄒鉉 (*ca.* +1300), wrote that 'physicians must first recognise the causes of an illness and understand in what way the body's own vitalities (*chhi*) have become unbalanced. To correct this imbalance, adequate diet is the first necessity. Only if this has failed should drugs be prescribed.'[39]

How large the nutritional literature was in the Chinese Middle Ages may be gauged from a glance at the titles of some of the more important of these lost dietary manuals (*shih ching* 食經), which Shih Shêng-han calls more accurately 'catering guides'. In the San Kuo period (+3rd century) there was a *Shih liu chhi ching* 食六氣經 (Manual for ingesting the six *chhi*), a very Taoist book. In the bibliography that preserves the title, the context suggests that this work had to do partly with quasi-magical heliotherapy and similar techniques of union with Nature. It would also have included principles of dietary regimen, e.g. how to deal with the cereal pneumata (*ku chhi* 穀氣). Of regular practical treatises on diet there was no lack, however. The bibliographical treatise of the *Sui shu* 隋書 (History of the Sui dynasty), compiled by +635, and its later commentaries list no less than thirty-two *shih ching*. A fragment of another survives. One of the most famous of its lost manuals is attributed to, among others, Tshui Hao 催浩, a minister of the Northern Wei dynasty who died in +450.[40]

The Thang bibliographies list several more. From this era we can still peruse a collection of menus (*Shih phu* 食譜) and a handbook of culinary raw materials (*Shan fu ching* 膳夫經). Part of the *Shih i hsin chien* 食醫心鑑 (Essential mirror of nutritional medicine, *ca.* +850) by Tsan Yin 咎殷 survives. By the Sung we hear of a *Shih chin ching* 食禁經 (Manual of dietary prohibitions) by Kao Shên 高伸. Among the pharmacopoeias, several concerned themselves specifically with foods, notably the extant

---

[37] Ch. 26, p. 464.
[38] *Shou chhin yang lao hsin shu* 壽親養老新書, ch. 1, p. 22a, sect. 13.   [39] Ch. 2, p. 21a.
[40] See, for the Three Kingdoms era, *San kuo i wên chih* 三國藝文志, p. 108.3. The *Tshui shih shih ching* 崔氏食經 is listed in *Sui shu*, ch. 34, p. 1043a, and discussed in Okanishi (*1958*), pp. 1390–1. A fragment survives of the *Shih ching* of Hsieh Fêng 謝諷, a Sui author of whom nothing more is known. [There is no evidence that *Shih liu chhi ching* had to do with diet, the preparation of food or Taoism. In the bibliography it is listed under breath disciplines. – Editor]

Shih liao pên tshao 食療本草 (Pharmaceutical natural history: Foodstuffs) by Mêng Shên 孟詵 (+670) and the lost *Shih hsing pên tshao* 食性本草 (Pharmaceutical natural history: Foods) by Chhen Shih-liang 陳士良 in the late +9th century.

We have already made a special study of the empirical discovery of vitamins in medi-aeval China through the investigation of deficiency diseases.[41] The greatest name in this field is no doubt that of the Mongol Hu-ssu-hui 忽思慧, Imperial Dietician between +1315 and +1330. His book, *Yin shan chêng yao* 飲膳正要 (Standard essentials of diet), finished in the latter year, describes in much detail the wet and the dry forms of beri-beri, and advocates foods now known to be rich in vitamins for the treatment of those suffering from deficiency diseases. Hu summarised his empirical recognition of deficiency diseases in the caption 'using foods to cure illness (*shih liao chu ping* 食療諸病)'.[42]

(ii) *Water and tea*

The Chinese appreciated very early the importance of pure drinking water. Already in the *I ching*, probably from the end of the −9th century, we find the commonplace 'Men do not drink water from foul wells.' The *Shih ming* 釋名 dictionary (+200) says, pun-ning, 'A well [*ching* 井] means essentially clear and clean [*chhing* 清], the clear produce of a spring.'[43]

Regular custom in ancient China demanded the periodic cleaning of wells. For exam-ple, the *Kuan tzu* 管子 (The book of Master Kuan), a diverse collection that dates from the −5th to −1st century, says:[44]

In the third month [of the year] they dried out their dwellings and seasoned the stove with heat. [*The commentary says*:] In the third month the yang *chhi* is expanding so that pestilences easily arise. [Therefore it is necessary to] fumigate dwellings with catalpa wood, which takes away bad smells and removes poisonous vapours (*tu chhi* 毒氣). The same wood is burned to dry out new houses, and this forms part of the purification ceremony. [*The text continues*:] They use the wood drill and the burning-mirror to change the fire [in the stove], and clean out [lit. ladle out, *chu* 杅] the well to change the water. This is a way of expelling poisons harboured from the past.

These customs continued right through the Han period. In the *Hou Han shu* 後漢書 (History of the Later Han dynasty) we find that the summer solstice (late June) was then considered the right time for 'changing the water', i.e. digging new wells and cleaning out the old ones. From the solstice until August, making big fires and melting or cast-ing metals were prohibited. This cleaning out of wells in the summer was again analo-gised with the obtaining of new fire by boring of wood or the use of a burning-mirror at other times of the year, especially the winter solstice.[45]

In the +11th century, Shên Kua 沈括, the famous scientific scholar, writing in his *Mêng-chhi wang huai lu* 夢溪忘懷錄 (Record of longing forgotten), records the care taken even at that time to preserve the purity of water sources. He describes 'medicated wells'

---

[41] Lu & Needham (1951).   [42] *Yin shan chêng yao*, ch. 2, p. 214. See also Sabban (1986).
[43] *I ching*, hexagram 48, line 1; Wilhelm (1924/1950), tr., vol. 1, p. 199; *Shih ming*, ch. 17 (palaces and dwellings); Bodman (1954), p. 97, item 693. Originally *jing* and *qing* were homonyms.
[44] Ch. 53, p. 11b; there is a very similar passage in ch. 85.   [45] Ch. 15, p. 5a.

(*yao ching* 藥井), incorporating something like what we call today 'sand filters'. He says that quartz, magnetite and mica gathered in the mountains, as well as stalactites and stalagmites retrieved from caves, were pounded into chips not too fine and filled into the wells for several feet. Some, following lines of thought derived from alchemy, put down cinnabar and sulphur and chipped jade. In the Thang period, there was a celebrated alchemical medicated well in the mansion of Li Wên-shêng 李文勝. Shên adds that such a well 'is provided with a framework that can be locked to prevent insects (worms, reptiles, etc.) and rats from falling into it, and servants and children from polluting it'.[46]

In the −2nd century the *Huai nan wan pi shu* writes of clearing water or fermented beverages by glue or isinglass.[47] Late in the +10th century the Buddhist master Tsan-ning 贊寧, on the opening page of his *Wu lei hsiang kan chih* 物類相感志 (On the mutual stimulation of things according to their categories), mentions purifying well water by filtering it through sand.

The extent of piped water supplies in ancient China has been very much under-estimated by most later writers, both Chinese and Western. It is only now with the great expansion of archaeological investigation and excavation in China that evidences of this are coming to light. Throughout the Chhin and Han periods, palaces and cities were supplied with water piped not only through bamboo tubing, easily installed and replaced, but also in conduits of earthenware. Chinese museums possess abundant examples of these water pipes. There are large earthenware rings about three feet across for lining the walls of wells, and piped lengths usually about two feet long, sometimes with right angles or two-way divisions, and usually with flanges for fitting the lengths of piping together. Fuller particulars of these we have given elsewhere.[48]

An astonishing statement is found in the writings of Chuang Chho 莊綽 (fl. *ca.* +1126): 'Even when the common people are travelling they take care to drink only boiled water.' We have not been able to locate the source of this statement, which Fan Hsing-chun quotes but which seems not to be in any of the books of Chuang available to us. There is nothing in the least improbable about it, since people drank tea infused in boiled water a thousand years before the time of Chuang Chho. In fact, the practice of tea-drinking originated during the Later Han period.[49]

(iii) *Cooking and nutritional hygiene*

One of the most important early developments in hygiene is the recognition that food spoils. Food prohibitions played a prominent part in ancient Chinese life, if not perhaps so extensively developed as in ancient Israel. About −500, we meet with a relevant passage in the Confucian Analects (*Lun yü* 論語). We are told that 'rice which has been injured by heat or damp and has turned sour (*i êrh ai* 饐而餲), one must not eat, nor

[46] In Hu Tao-ching 胡道靜 & Wu Tso-hsin 吳佐忻 (*1981*), pp. 5–6.
[47] Item 84, cit. in *TPYL*, ch. 736, p. 8a.   [48] Vol. 4, pt 2, pp. 127–30, 345–6.
[49] See Bodde (1942) and Goodrich & Wilbur (1942). [In the Han, tea was still not widely grown in the south, and the custom of drinking it did not work its way north until after +494; see Loewe (1968), p. 170, and Shinoda (*1974*), p. 62. On Ming connoisseurship of water for tea see *Yin shih hsü chih* 飲食須知, ch. 1, and comments by F. W. Mote in Chang (1977), pp. 227–30. – Editor]

fish or flesh which has spoiled. That which is discoloured or smells badly one must not eat, nor any food which has been insufficiently cooked (*shih jen* 食飪) or which has been kept too long.'[50]

Safety regulations about food are greatly elaborated in the late +2nd century work of the eminent physician Chang Chi 張機. He tells us that eating and drinking are for the enjoyment of taste and the nourishment of life. Yet, if diet is not correctly ordered, what is eaten can be harmful. There are certain things from which one must abstain. Only if the selection of diet is wise will the body benefit by it; if the selection is unwise, great dangers, illnesses very difficult to cure, may arise.

Chang gives long lists of what should not be eaten. The significance of numerous taboos is not obvious to us today. Some of the understandable ones are hygienic. For example, he forbids all meat and fish that dogs or birds will not eat or that has a disagreeable smell, meat that has on it red or coloured spots, and dirty rice. He also warns against all animals which have died spontaneously. He says that raw meat or milk will turn to white or red parasites (*pai chhung* 白蟲), *hsüeh chhung* (血蟲, lit. 'bloodworms'). Elsewhere he gives a number of antidotes for poisoning by mushrooms, and cautions against certain toxic wild yams. All this is contained in his chapters on miscellaneous disorders that have survived in the *Chin kuei yao lüeh* 金匱要略 (Essentials of prescriptions in the golden casket). The relevant chapters deserve much more study than they have so far received.[51]

No people have paid more attention to the importance of correct cooking than the Chinese. We do not refer here to their gastronomic triumphs, but rather to the principles of Chinese cooking that insisted on using oil at high temperature to fry food in large thin-walled cast-iron vessels (*kuo* 鍋). It seems likely that the disinclination for cold food which seems to have characterised Chinese cooking for twenty centuries was a powerful hygienic factor in preventing the spread of infections. Even under the unsanitary conditions of the old China, those Westerners who knew the country realised that the method of preparing food was extremely hygienic, so that as long as one always ate hot dishes, the risk of infection was slight.

A statement from an apocryphal treatise of the Han period illustrates the importance of cooking. In the *Li wei han wên chia* 禮緯含文嘉 (Apocryphal treatise on the [Record of] rites: excellences of cherished literature), we read that 'It was Sui Jen 燧人 who first drilled wood to obtain fire, and taught the people to cook food from raw materials, in order that they might suffer no diseases of the stomach, and to raise them above the level of the beasts.' This refers to a purely legendary character, a culture-hero, but the recognition that cooking prevents disease is quite clear. From these roots came forth the proverbial saying, 'Anything boiled long enough will not be poisonous (*pai fei wu tu* 百沸無毒)'.

---

[50] *Lun yü*, 10. 8. 1–4, 8; Legge (1885), p. 96; Waley (1938), p. 148. This chapter used to be regarded as describing the behaviour of Confucius himself, but it is now considered to be a ritual text concerned with the rules for well-born men in general, and here particularly with the consumption of sacrificial food. [When it was incorporated into the Analects is uncertain; E. Bruce and Taeko Brooks have recently proposed *ca*. –380; see Warring States Working Group (1993–), note 7. Most specialists today consider the chapter at least partly about Confucius. – Editor]

[51] See especially chs. 24, 25, pp. 89ff., 95, 96.

Closely allied to the recognition of the role of heat went the mistrust of the presence of insects or other small animals. For example, about +75, Wang Chhung 王充 in his *Lun hêng* 論衡 (Discourses weighed in the balance) argues by analogy that good and bad outcomes may depend on chance, not ordained destiny:

Grain is steamed to make cooked rice, and the rice fermented to make wine. As the wine matures it may turn out sweet or bitter; as the rice cooks, it may become hard or soft. This is not because the cooks or the brewers intend it to be that way; the movements of their fingers are subject to chance. The prepared rice is kept in different baskets; the sweet wine is stored in various vessels. If insects [or worms] fall into one of the vessels, the wine in it will be thrown away rather than drunk. If rats run over one of the rice baskets, the rice in it will be thrown away rather than eaten.[52]

The *Chou hou pei chi fang* 肘後備急方 (Handy therapies for emergencies, *ca.* +340) of Ko Hung elaborates on the danger of eating inappropriately. His topic is the '[intestinal] turmoil (*huo-luan* 霍亂)' group of diseases, which includes cholera, the dysenteries and diarrhoea. Ko says:[53]

Intestinal turmoil diseases arise from food and drink, often from raw or cold things, foods that do not go together, greasy food, or uncooked fish marinated in wine; or else because of exposure to draughts and dampness, or by sitting out of doors with insufficient clothing, or sleeping while inadequately covered.

In these sources we see the suspicions aroused by uncooked food and possible traces left on it by animals. In Wang Thao's 王燾 *Wai thai pi yao* 外臺秘要 (Arcane essentials from the imperial library, +752) we find this explicit in his discussion of *lou* 瘻 diseases. The term *lou* originally meant swollen glands, especially chronic, suppurative infections of lymphatic channels or vessels, due to a number of affections which it is hard to be precise about now. The form of the swelling was supposed to indicate by its mimetic shape the animal responsible for the disease. He says:[54]

Lou lesions include the rat, snake, bee, toad, and worm types, which do not greatly differ. All are caused by the essential *chhi* [*ching chhi* 精氣; of the vermin] entering in contaminated food and drink. [Their *chhi*], passing into the flesh, transforms to give rise to these diseases. When the local lesions break down, [the poisons] invade the circulation vessels, and may cause death. The rat and ant types are the most common, for [those pests] come more often into contact with human beings.

More than a century earlier, the *Chu ping yüan hou lun* 諸病源候論 (Aetiology and symptoms of medical disorders, +610), under the *lou* group, enumerates many diseases, mostly named after vermin, and says that one type comes from eating contaminated fruit or melons. In this connection, the book quotes a *Yang shêng fang* 養生方 (Prescriptions

---

[52] Ch. 2 (*phien* 5), p. 1b; Forke (1907), p. 154.

[53] Ch. 2 (*phien* 12), p. 1a. Chhao Yüan-fang 巢元方 quotes Ko on this point, emphasising again the dangers of raw flesh and cold uncooked food, in *Chu ping yüan hou lun* 諸病源候論 (+610), ch. 22, pp. 1aff. In modern medicine *huo-luan* merely means 'cholera'.

[54] Ch. 23, p. 11b (p. 641). One of the most ancient prescriptions for *lou* disease caused by rats must be that in the *Huai nan wan pi shu* of the −2nd century (item 67). Ko Hung's entry on *lou* is solely devoted to therapy (ch. 5 (*phien* 41), pp. 168ff.).

for nurturing vitality) as saying that in the sixth month (i.e. July or August) one should never eat fruit which has dropped on the ground and been left overnight. Furthermore, in winter time one should never eat flesh which has been fouled by dogs or rats.[55]

(7) PERSONAL HYGIENE AND SANITATION

The desirability of frequent and adequate ablutions is a commonplace in the old Chinese medical texts, as no doubt amongst most ancient nations. From the general literature, too, it is possible to collect abundant material for a history of bathing customs in the Chinese culture-area.[56] Such a history would consider baths and washing facilities first in relation to the family, then in monastic establishments and colleges, continuing with the development of public bath-houses and ending by an account of the elaborate and celebrated bathing-pools of the imperial palaces in successive dynasties. A special chapter might be devoted to the hot springs that were numerous in China.

Here we need only say that the terminology of washing and bathing was already stabilised in the late Chou period, indeed in Confucian times (−6th century).[57] The *Li chi* 禮記 (Record of rites) gives a striking impression of the cleanliness which traditional etiquette and ceremonial required of patricians in the late Warring States period.[58] Washing the hands five times a day, with a hot bath every fifth day and a hair washing every third day, was obligatory. This latter operation was especially important, since the hair was worn long. Many a story in the classical books depends upon high esteem for this duty. The *Lun hêng*, as it attacks the superstition that there were lucky and unlucky days for bathing, shows us that the scholars of the Han took frequent baths.[59] We know little about the customs of the plebs in ancient times, but there are many indications in the texts (including poetry) that they took full advantage of natural bathing-places in lakes and rivers.

With the growth of Buddhism after the +2nd century, an Indian emphasis on personal cleanliness powerfully reinforced the indigenous prescriptions. Buddhist abbeys (as in India and Sri Lanka) normally contained a bath-house or a bathing-pool. Its use

---

[55] Ch. 34, p. 3a (p. 179). In the bibliographies of the dynastic histories, no *Yang shêng fang* is to be found, but a *Yang shêng shu* 養生書 (Book for nurturing vitality) was current in the +3rd century. More probably Chhao was referring to the *Yang shêng ching* 養生經 (Manual for nurturing vitality), written somewhere about the +5th century by Shang-kuan I 上官翼.

[56] Schafer has brought together an excellent sample in a learned and interesting paper (1956). Some of the same ground is covered in Fan Hsing-chun (*1953*), pp. 58ff. Still the surface has barely been scratched. There were certainly parallels for those treatments which in Sri Lanka have left such remarkable traces behind them as the stone medicinal baths like crusaders' coffins still found on the sites of mediaeval hospitals at Anurādhapura and Mihintale, the imperial bathing-pools at Anurādhapura, and the remarkable +5th century royal pleasaunce still conserving its original geometrical lay-out at Sigiriya. One of us (J.N.) had the good fortune to visit these places in the spring of 1958.

[57] For a discussion of the terms, see Schafer (1956).

[58] See especially *phien* 12, 13, and 41; Legge (1885) tr., vol. 1, pp. 449ff.; vol. 2, pp. 1ff., 402ff.

[59] *Phien* 70. In the Han and later, every five or ten days officials took a day off for 'relaxation and hair-washing'; *Shih wu chi yüan* 事物紀原, ch. 1, pp. 36a–b, and Yang (1961). Many Chinese books were devoted to bathing and personal hygiene. The bibliography of the *Sui shu*, ch. 34, p. 1037, lists a *Mu yü shu* 沐浴書 (Treatise on the bath), which would thus have dated from before the +7th century. Furthermore, an emperor himself did not disdain to write a Manual of Balneology, the *Mu yü ching* 沐浴經. This was the work of Hsiao Kang 蕭綱, who reigned for one year (+550) as Chien Wên Ti 簡文帝 of the Liang dynasty. A scholar, poet and philosopher, he wrote commentaries on the Taoist classics and many other works.

was by no means restricted to the monks or nuns. At least in the earlier centuries the Confucian prohibition of the mingling of the sexes in bathing was strictly observed.[60] The Taoists also had bathing customs which figured prominently in their purification ceremonies before important festivals.

The emperor and high officials since ancient times had purified themselves as a matter of course before great State liturgies and sacrifices. Bath-houses and bathing-pools were also attached to schools, as we know from a number of stories concerning (for example) poems written on their walls.[61]

Possibly because of a decline in the prosperity of Buddhist foundations after the Thang, the Sung saw the rise and rapid development of public bath-houses. In the Five Dynasties period (about +914), the monk Chih-hui 智暉 became famous for the water-raising machinery he built for the public baths at Loyang, which accommodated thousands of people. Perhaps these were still Buddhist baths, public but not yet commercial. From the +11th century onwards, references to bath-houses which one could enter on payment of a fee become quite frequent. By the time the Sung capital was transferred to Hangchow in +1127, the common name for such public baths was *hsiang shui hang* 香水行, 'perfumed water establishments'. They advertised their presence by hanging up a water-pot or a kettle as their shop sign. These were the baths so much admired by Marco Polo.[62]

In later centuries the public *thermae* became almost as notable a feature of Chinese urban life as of that of ancient Rome, but they often no longer contributed to hygiene. The custom was to use the same water over and over again, heating it in boilers and continuously circulating it by chain-pump to the pool. Possibly for this reason the baths came to be known as *hun thang* 混堂, 'chaos halls', from the revolving motion of primitive chaos, but it is more likely that *hun* here referred simply to the atmosphere of social promiscuity. Pedlars, caravan drivers, butchers and others engaged in the dirtiest of occupations now all used the baths. Access was not forbidden to those suffering from contagious or infectious diseases. What had begun as a laudable measure of hygiene probably sometimes did more harm than good.[63]

[60] See, for example, the *Lo-yang chhieh lan chi* 洛陽伽藍記 (Record of the Buddhist temples and monasteries at Loyang), between +547 and +550, for a story of the jubilation when a well and bathing-pool built in Chin times were discovered at a temple, the Pao Kuang Ssu 寶光寺 (ch. 4, p. 78, tr. Wang Yi-t'ung (1984), p. 177). This was in the north. Broadly speaking, in ancient times and in the early Middle Ages authors in the hot south emphasised bathing customs rather more than those in the north.

[61] See *Kuei-hsin tsa chih* 癸辛雜識 (*Pieh chi* 別集), a work finished by +1308, ch. 1, p. 15.

[62] On Buddhist baths, see Chu Chhi-chhien 朱啟鈐 et al. (1932–6) p. 163. For pertinent descriptions of Hangchow see *Tu chhêng chi shêng* 都城紀勝 (+1235), *Mêng Liang lu* 夢梁錄, ch. 13, p. 3a (+1275) and *Nêng kai chai man lu* 能改齋漫錄, ch. 1, p. 3a, excerpted in Gernet (1959), pp. 123–6. The latter source explains that the sign originated of old from an official, the Hoister of the Water-Pot (*Chhieh hu shih* 挈壺氏), mentioned in the *Chou li* (ch. 7, p. 27a; ch. 30, Biot (1851) tr., vol. 2, p. 201). His duty was to signalise in this way the position of the well or other water-source in an army encampment, and to look after another sort of water-pot, i.e. the time-keeping clepsydra. Cf. Needham et al. (1960), p. 159.

[63] Lang Ying 郎瑛 gave a graphic account of the *hun thang* in *Chhi hsiu lei kao* 七修類稿 (after +1566), ch. 16, pp. 12a–b. Lang tells us that in spite of the decline in hygiene level, poor scholars still used the public baths, attracted by the warmth and facilities which they could not afford at home. Of course, where flowing water was used, as in the hot springs at Fuchow in Fukien, Anning in Yunnan, or Pei-wên-chhüan in Szechuan, hygienic conditions always remained adequate. One of us (J.N.) has the pleasantest memories of visits to them with friends.

A feature of the better public baths, especially those built round hot or medicinal springs, was the presence of 'back-scrubbers', masseurs (*tsha pei jen* 擦背人 or *khai pei jen* 揩背人). Here is part of a +11th-century lyric by Su Shih 蘇軾:

> A word for the masseur:
> You have been working hard all day, elbows in action.
> Take it easy with me!
> A scholar living in retirement isn't dirty.[64]

Lastly, we shall pass over the heady history of the bathing-pools of the imperial palaces, about which much is known from the Han onwards.[65] The emperor and his entourage of beauties could certainly be trusted to make use of the best facilities which the age had to offer. Whether it was a matter of casting into the pools red-hot bronze statues to heat the water, or causing fountains and cooling streams to play in pavilions during summer heat, architects, engineers and other servitors were never lacking to organise whatever was possible.

It is more germane to our purpose to ask a more prosaic question, but one of greater scientific interest. What soap did the emperor, and all the other bathers we have been discussing, use? Although the modern world is now in the 'detergent age', China was always dependent on detergents rather than on true soaps. We can find in any period of Chinese history very little reference to making soap by saponifying fatty acids with caustic alkali.

As is well known, the history of soap-making in the West is strangely obscure. There is some evidence that oil and alkali were boiled together to make soap in ancient Mesopotamia, especially by the Sumerians of Ur. The same has been claimed for the ancient Egyptians, but apparently with less justification. The Greeks and Romans used mainly oil, together with mechanical detergents, and lacked true soap. They employed, however, to some extent, the soap-wort (*Saponaria officinalis*) and the soap-root (*Gypsophylla struthium*), resembling therein, as we shall see, the Chinese. References in later times to the reinvention of soap are quite obscure; some indications point to Gaul and some to the Scythians or Tartars of Western Asia. All one can say is that by the time of Galen (*ca.* +129 to *ca.* +200), true soap was normally in use and continued to be made thereafter. In mediaeval Europe, it was known (if little used) from at least +800 onwards.[66]

Now in China the dependence on the saponins of plants seems to have been complete from early times onward. There were three chief sources of saponins for use as detergents in personal washing and the laundering of clothes. The oldest was the soap-bean tree (*Gleditsia* or *Gleditschia sinensis*) called *tsao chia* 皂莢. We have not succeeded in finding any references to this in the non-medical literature of Chhin and Han. Because it appears in the *Shên nung pên tshao ching* 神農本草經, the first of the pharmacopoeias,

---

[64] [See Su's poetry collected and edited in Lung Yü-shêng 龍榆生 (*1936*), ch. 2, p. 22b. – Editor]

[65] Schafer (1956) deals in particular detail with the baths at Linthung near Sian, associated so romantically with the Thang Emperor Hsüan Tsung 玄宗 and his concubine Yang Kuei-fei 楊貴妃. We had the pleasure of seeing this still delightful place in the summer of 1958.

[66] Cf. Forbes (1954), p. 261; Taylor & Singer (1956), pp. 355–6; Gibbs (1957), p. 703; and Taylor (1957), pp. 129ff.

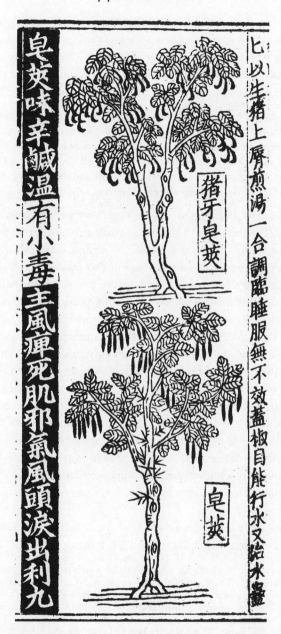

皂莢味辛鹹溫有小毒主風痺死肌邪氣風頭淚出利九

猪牙皂莢

皂莢

七以生猪上屑煎湯一合調臨睡服無不效蓋椒目能行水又殺水盡

Fig. 4. The soap-bean tree (lower illustration), the saponin detergent used earliest in China, and the pig's tooth soap-bean tree (upper illustration), useless for this purpose. From *Chhung hsiu Chêng-ho ching shih chêng lei pei yung pên tshao* 重修政和經史正類備用本草. (Revised materia medica of the Chêng-ho era, classified and verified from the classics and histories, +1249). From 1957 reprint, ch. 14, p. 6a.

we may date its use in the Later Han. The *Pên tshao shih i* 本草拾遺 (Gleanings from the materia medica, +739) recommended a special variety of this tree, the *kuei tsao chia* 鬼皂莢, for bathing the body and washing the hair.[67] The detergent seeds were called *tsao tou* 澡豆, literally 'bath-beans', as we learn from an anecdote of the 'generalissimo' Wang Tun 王敦 (d. +324).

When Wang went to the privy, he noticed a lacquer basket full of dried jujubes [Chinese 'dates']. They were meant for stopping up one's nose (to avoid the smell). Remarking that even in the privy fruit was provided, he began eating them, and proceeded until they were finished. When he returned, the slave girls held out a golden washbasin filled with water and a colored glass bowl filled with soap-beans. Wang emptied them into the water and drank them down, supposing that they were cooked and dried rice. All the slave girls covered their mouths and laughed at him.[68]

The second main source of saponins was the soapberry tree (*Sapindus mukorossi*) which produced the beads ('bodhi seeds') used in the rosaries of Buddhist monks. Its Chinese name was *wu huan tzu* 無患子 (no-trouble seeds) or *kuei chien chhou* 鬼見愁 (worry for demons). Because of its unpleasant smell it was not allowed in public bath-houses. This material seems not to have been much used before it was first mentioned in the *Khai-pao pên tshao* 開寶本草 (Pharmacopoeia of the Khai-pao era) of +973–4. The third and latest saponin source was the soap-pod tree (*Gymnocladus sinensis*, *fei tsao chia* 肥皂莢). It first appears in the *Pên tshao kang mu* (The systematic pharmacopoeia) in +1596. Among other vegetable substances which presumably contained saponins in active form may be mentioned the roots of *Aristolochia recurvilabra* or *Atractylis sinensis* (both called *pai chu* 白朮), and *tou mo* 豆末, a powder prepared from soya bean (*Glycine soja*). The vehicles included the flour of the common pea (*Pisum sativum*, *wan tou* 豌豆 or *pi tou* 華豆), talc, steatite, root powders, beeswax, animal fat and other fatty materials.[69]

Naturally these vegetable products had other uses, for example as emetics, but they were principally used as detergents. They were made up with different kinds of flour, mineral powders, and perfumes into balls analogous to cakes of soap. That saponins poison fish was well known in the +10th century, for Tsan-ning mentions it in the *Wu lei hsiang kan chih*.

Plant biochemists have long recognised that the saponins are especially interesting because they have no deleterious effects upon the most delicate textile fabrics.[70] One of

---

[67] *Shên Nung pên tshao ching*, Mori ed., ch. 3, p. 91; R/387; Stuart (1911), p. 188. See *Chêng lei pên tshao* 證類本草 (+1249), ch. 14, p. 341. Related species useless as sources of detergents were recognised early. Thus Su Ching 蘇敬 in the *Hsin hsiu pên tshao* 新修本草 (+659) noted the uselessness of *chu ya tsao chia* 豬牙皂莢 (pig's tooth soap-bean, *Gleditsia japonica*); Stuart (1911), p. 188. See the whole entry in *Pên tshao kang mu* 本草綱目, ch. 35B, pp. 4aff.

[68] *Shih shuo hsin yü* 世說新語, ch. 34, p. 44a, tr. Mather (1976), p. 479, mod. An almost identical story, which must have involved the same detergent, is told before +863 about Lu Chhang 陸暢 (*Yu yang tsa tsu* 酉陽雜俎, suppl., ch. 4, p. 6b). Jujubes are *Zizyphus sinensis* or *jujuba*; R/292, R/293; Stuart (1911), p. 466.

[69] Soapberry: R/304; Stuart (1911), p. 395. See *Chêng lei pên tshao*, ch. 14, p. 350; and *Pên tshao kang mu*, ch. 35B, pp. 14aff. Soap-pod tree: R/393; Stuart (1911), p. 198; *Pên tshao kang mu*, ch. 35B, pp. 13a–b. *Atractylis* and soy flour: R/585, R/14 and R/388; Stuart (1911), pp. 49, 58 and 189, *G. soja* s.v. *G. hispidia*. Pea flour was used by the time of Li Shih-chen (*Pên tshao kang mu*, ch. 24, p. 20b); R/402; Stuart (1911), p. 335. Li recommended it for removing the traces of smallpox. Pea flour was not, as Schafer supposed, the essential ingredient of *tsao tou*; *ibid.* (1956), p. 64.

[70] Haas & Hill (1928), vol. 1, pp. 261ff. Their detergent power may have been weak in comparison with modern detergents. Comparative estimates would be of some interest.

us (G.D.L.) remembers how her grandmother objected to the introduction of fat soap in modern China, partly because it gave a precipitate in hard water and was harder on fabrics than the old detergents. For example, with the saponins the whiteness of silk was long conserved. Families used to buy the black beans and make their own aqueous extracts from them to launder silk and cotton things; for toilet purposes, people used the dark, moist, scented balls prepared by the druggists, which gave a beautiful and comfortable soft lather. Other preparations of the time included honey and fat from the omentum of pigs.

Late in the +7th century Sun Ssu-mo devoted a fascinating chapter of the *Chhien chin i fang* 千金翼方 (Revised prescriptions worth a thousand) to women's 'facial medicines'; that is, cosmetic, washing and other preparations for personal hygiene. His introduction indicates that these were largely secret formulae:

Facial creams and hand lotions, scents for clothing, and soap-beans for washing and bathing, are things that people of the official class, [not to mention] the highly placed and the powerful, all want. But these days physicians are most reluctant to give out the formulae. They do not permit their disciples to reveal them. Sometimes fathers will not even hand them down to their sons. Yet when the sages established these methods, they intended them to be known to every family, every person. Surely one should not make ignorance of them the norm throughout the realm, so that the perfect Way [of Medicine] does not spread, frustrating the sages' intentions? How odd!

Half a century or so later, Wang Thao's *Wai thai pi yao* quotes these remarks to begin a much more extensive chapter on similar preparations, no longer with specific reference to women. Among the roughly 220 formulae are many kinds of saponin detergent. There are two especially for face washing (*hsi mien fang* 洗面方), five for hair washing (*mu thou chhü fêng fang* 沐頭去風方) and eight for bath soap (*tsao tou fang* 澡豆方).[71] These various detergents, to which various drugs and perfumes are often added, are prepared from *Gleditsia* and other vegetable products.

Wang never mentions caustic alkali. However, there is plenty of evidence from all these texts that sodium carbonate (*chien* 鹼) was used just as in occidental cultures, especially for laundering clothes.

The saponins of plants are mainly of two kinds, those of steroid structure and those of triterpene nature. Both are complex condensed-ring hydrocarbons and in the native state are generally bound as glycosides. All plant saponins contain that double endowment of hydrophilic and hydrophobic groups without which they could not manifest detergent properties. They form monomolecular films between the surface to be washed and the adhering particles of dirt or oil, lifting off the latter as the surface tension is reduced. This brings about the detergent effect.[72] A typical steroid saponin is tigogenin from the digitalis plant, and a typical triterpene saponin is hederagenin.

The modern synthetic detergents are, as is well known, quite different in structure, being largely sulphonic acid derivatives of long-chain non-cyclic hydrocarbons such as

[71] *Chhien chin i fang*, ch. 5, p. 10a (p. 64); quoted in *Wai thai pi yao*, ch. 32, pp. 3a, 51a–65a (pp. 870, 894–901).
[72] See the introductory account of Adam & Stevenson (1953).

lauric acid or the hydrocarbons of castor oil. Some of them, however, are purely organic soaps, ammonia replacing the usual alkali metal. Intermediate between the sulphonated chains or the true soaps and the classical Chinese saponins of highly complex condensed-ring character is the detergent abietic acid, a very surface-active compound obtained from the depolymerisation of pine resin. This has a hydrophenanthrene system of three rings only. It is used in yellow laundry soaps as well as in paper size, varnishes and plastics. It thus forms a link between the new detergents and the old saponins.

Another question which has received too little attention is that of the sanitation of lavatories (*tshê* 廁), both public and private, in China in different periods. Sumptuous lavatories were among the amenities of wealth. Those of the merchant Shih Chhung (+249 to +300), already mentioned (p. 88), were especially ostentatious. In very ancient times, instruments of bamboo, possibly spatulas (*tshê chhou* 廁籌, *tshê pi* 廁篦, or *tshê chien* 廁簡), may have been used with the assistance of water in cleaning the body after defaecation. At other times and places, it seems that pieces of earthenware or pottery were so used. Undoubtedly one material which found employment in this respect was waste silk rag.

After the invention of paper about −100,[73] the increasing availability of this expendable tissue assisted the spread of more hygienic habits among the people. No doubt only the upper class had been able to use silk. The rougher kinds of paper for this purpose were called *tshao chih* 草紙, 'straw paper'. By the time of the Sui dynasty the use of paper in lavatories was probably universal, as we can see from a passage in the *Yen shih chia hsün* 顏氏家訓 (Mr Yen's advice to his clan) written by Yen Chih-thui 顏之推 about +590.[74]

From early times, there must have been some sort of systematic arrangement for the removal of night-soil from urban agglomerations each morning. In the +13th century the workers were called *chhing chiao thou* 傾腳頭 (the 'turners-over of refuse'). This term occurs in *Mêng Liang lu* 夢梁錄 (Dreaming of the idyllic past), Wu Tzu-mu's 吳自牧 memoir of Hangchow under the Sung (+1274).[75]

(8) CARE OF TEETH

A good deal could be said about the care of teeth in China. Chhao Yüan-fang, in the *Chu ping yüan hou lun* (+610), refers to this topic many times. Again quoting the *Yang shêng fang*, he says: 'In the morning, before you get up, rinse and gargle with the saliva that has accumulated in your mouth (*khou chung tho* 口中唾), and then swallow it down. Then gnash your teeth twice seven times. This will make you vigorous and fine in appearance. It will get rid of worms and harden the teeth.'[76] In addition to these semi-magical practices we find much evidence of cleaning agents.

The use of the soap-bean as a tooth-powder or tooth-paste is mentioned in the *Wei shêng chia pao* 衛生家寶 (early +12th century), and in the *Phu chi fang* 普濟方

---

[73] Vol. 5, pt 1, pp. 38–42.   [74] Ch. 5, p. 13b; Fan Hsing-chun (*1953*), pp. 74ff.   [75] Ch. 13, p. 13a.
[76] Ch. 29, p. 9a (p. 156). [Chinese believed that caries were formed by worms eating holes in the teeth. – Editor]

(Prescriptions for universal benefaction) of +1406.[77] This implies the use of some instrument, usually the broken end of a twig or stick but also toothbrushes made with bristles, which originated in the Liao dynasty (+937 to +1125).[78]

Indian influence, entering China with Buddhism, undoubtedly would have brought the use of a simple stick long before the Liao. Indeed, amongst the fresco paintings at the Thousand-Buddha cave-temples at Mo-kao-khu 莫高窟 near Tunhuang, Kansu province, on the border of Central Asia, in a cave which may be dated in the near neighbourhood of +775, it is possible to see a clear picture of a monk cleaning his teeth.[79] A servant holding a towel attends him. As the monk brushes his teeth with his right hand, he holds in his left something at first sight remarkably like a toothpaste tube, in fact perhaps a syringe.

In the early +13th century in the Arabic culture-area, especially Egypt, both tooth powder (*sanūn*) and toothbrushes made from a plant called *siwak* (*Salvadora persica*), were used.[80] As a first approximation, it would seem likely that the use of a roughened stick spread outwards from India and that the toothbrush, in the strict sense, may have developed both in China and in the Arabic lands. The study of the subject, however, has only begun.

(9) SPECIFIC DISEASES: THE EXAMPLE OF RABIES

Although in ancient and mediaeval times it was, of course, impossible to organise effectively the defence of the population against specific disaeases such as hydrophobia and smallpox, preventive measures were taken. Rabies, which was recognised early in Chinese history, will provide examples. We find what is perhaps the earliest reference to it in the *Tso chuan* 左傳 (Master Tso's tradition of the spring and autumn annals, late –4th century), an anecdotal record of events in the feudal States. This record is dated –556: 'In the eleventh month, people [in the capital of Sung 宋 State] were chasing a rabid dog [*chih kou* 瘈狗]. It entered the house of Hua Chhen 華臣. They followed it into Hua Chhen's compound. He was so frightened that he fled to Chhen 陳 State.'[81]

It does not follow that Hua Chhen's fright was due to fear of the disease. The point is rather that this murderous high minister had good cause to fear a mob, and took this one as an omen. The story indicates that an organised effort was being made to get rid of the dangerous animals, so much so that they could not rely on sanctuary even in the private houses and gardens of high officials.

By the +3rd century, we find rabies prominent in Ko Hung's Handy therapy for emergencies (*Chou hou pei chi fang* 肘後備急方) and other medical works. Among the diverse treatments was sucking blood from the bitten area and subsequent moxibustion. Most

---

[77] Cited in *Pên tshao kang mu*, ch. 35B, pp. 13b and 8b. [78] See Chou Tsung-chhi 周宗歧 (*1956a*, *1956b*).
[79] Usually known in the West as Chhien-fo-tung 千佛洞. We had the pleasure of studying this painting in cave no. 159 during the summer of 1958. A similar scene, of slightly later in the Thang, is found in cave no. 196.
[80] Wiedemann (1915).   [81] Duke Hsiang, yr. 17, par. 7; cf. Couvreur (1914) tr., vol. 2, p. 330.

interesting is 'also kill the dog responsible for the bite, remove its brain, and rub it [on the wound]. After this there will be no recurrence.'[82]

China lacked the social organisation that would make it feasible to sequester rabid dogs, but the population was warned at certain times of the year. In the +650s, Sun Ssu-mo tells us in his *Chhien chin fang* that:

[Dogs] mostly go mad at the end of spring and the beginning of summer. It is essential to warn children and the weak to carry sticks for their protection. For those who cannot avoid being bitten in spite of such preventive measures, nothing works better than moxibustion. Only after 100 days, taking care not to skip treatment for a day, can one [be sure of] avoiding suffering. If at first the wound heals and the pain remits, so that [the patient] speaks of recovery, that is most alarming. A great catastrophe may ensue, taking the patient to the brink of death.

When a mad dog bites someone, the patient becomes manic, and his vitalities and spirits (*ching shên* 精神) become abnormal. How do we know this? One need only notice that, during moxibustion, as soon as the fire burns down [to the skin], the patient's mind is roused, fully awake. Thus we realise that a bite means mania. It is deeply necessary to be aware of this. This is a most serious illness, although the uninitiated take it lightly, not giving it a thought. Every year patients die because of [the practitioner's] inaction. . . .

When one is bitten by a rabid dog, the symptoms of the sickness regularly appear in a week. If nothing has happened within three weeks one may escape, but only after 100 days have passed [without signs of the disease] is one safe.[83]

(10) COMPARISONS AND CONCLUSIONS

Looking back over what has been said, one is moved to compare the achievements of ancient and mediaeval Chinese hygienic thought and practice with other civilisations' highlights in the same field. Most of these are, to be sure, common knowledge. For example, in ancient Israel the development of sanitary legislation took the form especially of food prohibitions.[84] Both in Israel and in Islam, great emphasis was laid upon ablutions before prayer, and before and after meals. The Hebrews in ancient times were notable also for their circumcision operation, for their laws on leprosy and for the health-giving Sabbath rest. Ancient India also made such laws, doubtless moved by the heat of the climate, stressing the need for cleansing the body. There perhaps came the earliest realisation of the value of good care of the teeth. To India also much credit is due for the insistence on the cremation of the dead.

In the West everyone knows of the sanitary latrines and water-conduits of Minoan Crete, and the drainage systems of Roman cities, for example the Cloaca Maxima at Rome itself. Fourteen noble aqueducts built between the −4th and the +2nd century supplied its public baths. The love of the Greeks and Romans for gymnastics in the *palaestra* is

---

[82] *Chou hou pei chi fang*, ch. 7, p. 212.

[83] Ch. 25, pp. 18a–b (p. 453). [Sun characteristically quotes the last paragraph from *Chu ping yüan hou lun*, ch. 25, p. 18a (p. 453). – Editor]

[84] [The debate over whether sanitary practice in the scientific sense was a main point of such prohibitions continues unabated. For a typical disagreement see Douglas (1966), pp. 41–57, and *ibid.* (1975), pp. 283–313. – Editor]

part of the European heritage. In Rome, the *quatuor viri viis purgandis*, the officials charged with street-cleaning and the care of the public latrines, recall the similar officials and workers mentioned in the *Chou li* and still found everywhere during the Sung period.

Viewing the general attitude of the ancient and mediaeval Chinese physicians and scholars to hygiene and preventive medicine, one gains the general impression that it compares most favourably with what was done in Greek and Roman civilisations. It is impossible to generalise about the Occident over the last 2,000 years. Roman bathing customs persisting in the *hammam* of Islam strongly influenced the Crusaders, so that bath-houses (in which the sexes were not segregated) became very popular in late medi-aeval Europe. After this there was a marked backsliding in the +16th and +17th cen-turies, paradoxical in view of the spread of classical humanistic culture. Progress was not resumed until the romantic age in the +18th century.

In judging the comparative achievements of European and Chinese preventive medicine through the centuries, personal hygiene and bathing customs should prob-ably be considered independently. Assuredly life in a Buddhist or Taoist abbey of the Thang was far more cleanly than in a Christian one of the same period. Marco Polo noted a similar superiority at the end of the +13th century, yet in the +14th and +15th there may have been less difference. Not until the last half of the 19th century did Europe draw decisively ahead. This brief monopoly of efficient plumbing is rapidly ending.

Once again we meet with an interesting circumstance that we have noted in many other connections. In many branches of science and technology the Chinese of ancient times equalled or anticipated the contributions of the classical Western world. After *ca.* +200 Europe fell behind for some dozen centuries until the renewal of cultural life and sci-entific thought beginning in the Renaissance. Examples can be found in every volume of this book.[85] The evolution of hygiene and preventive medicine furnish another instance. Something very similar may characterise the medical sciences as a whole.

All the civilisations echo certain aphorisms on the preservation of health. When we read the *Regimen sanitatis salernitanum*, for example, those precepts on health which were offered to a king of England, seemingly about +1100, by the doctors of the medical school of Salerno in Italy, we seem to be in the presence of fundamental ideas which the physicians of China and of many other cultures would have endorsed. In Harington's translation:

> The Salerne School doth by these lines impart
> All health to England's King, and doth advise
> From care his head to keep, from wrath his heart,
> Drink not much wine, sup light, and soon arise,
> When meat is gone, long sitting breedeth smart;
> And after-noon still waking keep your eyes.
> When mov'd you find yourself to Nature's needs
> Forbear them not, for that much danger breeds.

---

[85] One of the earliest such discussions is in Vol. 1, p. 242, expanded in Singer *et al.* (1954–8), vol. 2, pp. 770–1. A more comprehensive review will appear in Vol. 7.

Use three Physicians still, first Doctor Quiet,
Next Doctor Merry-man, and Doctor Diet.
Rise early in the morn, and straight remember,
With water cold to wash your hands and eyes
In gentle fashion reaching every member.
And to refresh your brain whenas you rise
In heat, in cold, in July and December
Both comb your head, and rub your teeth likewise.[86]

Words of this kind remind us that men of intelligence in all cultures throughout ancient and mediaeval times would have been at one with Hippocrates and Galen, with Pien Chhüeh and Chang Chi, in their advice on health to Everyman.

[86] Harington (+1607), p. [A6], ed. auct.

(c) QUALIFYING EXAMINATIONS

(1) INTRODUCTION

The origin of examinations in medical and surgical proficiency to protect the public from unskilled practitioners is a remarkably interesting subject. The germ of the idea goes back a very long way, for the beginnings of punishment for malpractice can be found in the famous law-code of Hammurabi (king of Babylon, −2003 to −1961). This continued in Achaemenid Persia. About the −5th century, the Avestan surgeon was bound to practise first on three non-Mazdaeans; then only, if successful, could he perform operations on Zoroastrians.

The science and art of medicine had to advance much further before we find any elaborate system of medical examinations. The evidence available (which we shall mention in due course) indicates clearly that our European system of examinations derived from Arabic culture through the School of Salerno. The question arises, however, whether the physicians of +10th-century Baghdad could have been influenced by still earlier practices in regions farther East.

Chinese civilisation is a milieu of choice for seeking the origins of medical examinations in the modern sense. It is natural to suspect that the bureaucratic feudalism of ancient and mediaeval China, so different from any kind of society known in the Western regions of the Old World, originated examinations for medical qualification. There are striking parallels. When Richard Bentley introduced written examinations in Cambridge in +1702, for the first time in Europe, he was certainly not unaware of the age-old Chinese civil service examinations. Many +17th-century Jesuit writers described them in detail in all the chief European languages. When civil service examinations were introduced in the 19th century in the West, the inspiration came from the mandarinate examinations conducted for 2,000 years previously in China.[1]

As soon as we begin to investigate the examinations in the classical Chinese texts that have survived in great quantity, we find two evolutions beginning early and remaining inextricably combined. First, there was the idea of a State Medical Service, and, secondly, the conception of a National University for educating the Chinese equivalent of 'persons well qualified to serve God in Church and State'. This Eurocentric phrase needs to be taken with a grain of salt, as Chinese cosmic religion was theocratically atheist, so to say, and the State was not separate from the Church. Nevertheless it expresses well enough the institutional aim of State education. We shall see that these two currents combined at a certain point to generate specifically medical colleges, and the introduction of qualifying examinations for medical students naturally followed. The problem is, exactly when?

---

[1] Teng Ssu-yü (1943); Bodde (1946), p. 426.

(2) MEDICAL POSTS

Let us begin with the situation at the end of the Chou period, in the late −3rd century, when the Warring States gave way to the first empires, the brief Chhin and the long-lived Han. We have already discussed the enumeration of medical officials in the *Chou li* 周禮 (Rites of the Chou dynasty, *ca.* −2nd century). This idealised schema does not allude to pupils or examinations.[2]

The Han empire lasted no less than four centuries, the Former Han from −206 to +8, and then the Later Han from +25 to +220, paralleling imperial Rome. Here we see nascent a distinction between two independent medical services, that of the palace and that of the State. In the Former Han we have an Administrator (*Ling chhêng* 令丞) with the title Imperial Physician (*Thai i* 太醫) under the Chamberlain for Ceremonials (*Fêng chhang* 奉常), and another medical administrator under the Chamberlain for the Palace Revenues (*Shao fu* 少府). The title of *Thai i* remains parallel to *Thai shih* 太史 (Astronomer-Royal) over the centuries thenceforward. Special titles for the Emperor's Attending Physicians now appear, such as *Shih i* 侍醫, equivalent to the *Yü i* 御醫 of the Ming and Chhing eras.

In the Later Han period we find listed in the Department of the Imperial Physician (*Thai i ling* 太醫令) 2 Pharmacist Aides (*Yao chhêng* 藥丞), 2 Medical Treatment Aides (*Fang chhêng* 方丞), 293 Official Physicians (*Yüan i* 員醫) and 19 Official Functionaries (*Yüan li* 員吏). At this time also we begin to see provincial medical organisations. Besides the officials at the capital there were Chiefs of Physicians (*I kung chang* 醫工長) on the local staffs of kings of the imperial family. This official system continued into the Three Kingdoms of the San Kuo period.[3]

When we come to the second unification, that of the Chin dynasty (+265 to +420), we must pause to look at a most important parallel development that began much earlier. This was the constitution of the National University. On the one hand the title Professor (*Po-shih*) had appeared in various kingdoms before the Chhin unification. Throughout history it had two distinct meanings. It usually referred to a teacher in an official school, whose responsibility was passing down the authentic text of a classic to carefully chosen pupils, and less commonly to a learned ritualist who served the Chamberlain for Ceremonials.[4]

The Han Emperor Wên brilliantly inaugurated the State examinations.[5] He himself set the questions in −165, probably the earliest occasion of the kind in any culture. In

---

[2] See Section (*b*), p. 70.

[3] *Chhien Han shu* 前漢書, ch. 19A, pp. 726, 731; *Hou Han shu* 後漢書, ch. 26, p. 3592, ch. 28, p. 3629. [The Official Physicians and Official Functionaries are not listed in the official history, but are noted by a commentator. It is unlikely that they were regular members of the civil service. – Editor]

[4] *DOT* 4746. Hucker's literal translation, 'Erudite', is appropriate for both senses. For the title before the Han see Hou Shao-wên 侯紹文 (*1973*), pp. 448–9.

[5] [The Han national examinations did not primarily aim to gauge erudition. Emperor Chang in +83 enumerated the criteria that he was trying to revive. These included personal virtue and incorruptibility, self-cultivation based on study of the classics, knowledge of the legal codes to enable sound judgements, and personal resolution and decisiveness; Hou, *ibid.*, p. 29, note 2. The Han 'filial, incorruptible, and upright' system depended on personal evaluation by high officials, rather than on formal testing. According to Bielenstein (1986), p. 516, the first written examinations for those recommended under this system were imposed in +132. – Editor]

−124 Emperor Wu, at the suggestion of Kungsun Hung (*ca.* −205 to *ca.* −127) and Tung Chung-shu (*ca.* −179 to *ca.* −104), provided a body of Professors with disciples (*Ti tzu* 弟子) supported at government expense, thus establishing for the first time the National University (*Thai hsüeh* 太學). Starting with fifty students, by about −10 it had grown steadily to no less than 3,000. In +4 the regent Wang Mang 王莽, soon to usurp the throne, convoked for it a grand congress of scholars from all over the empire to place the sciences and humanities upon a more definitive basis than before. The first emperor of the Later Han, Kuang-wu, re-established the university in +29 after ending Wang's rule. By the San Kuo period, when Emperor Wên of the Wei moved it to Loyang in +224, the rank of the ritualist *Po-shih* had declined considerably, and this was probably true of the professor's standing as well. In the Western Chin era (which ephemerally united China in the last two decades of the +3rd century), Emperor Wu reformed the university and added to it a School for the Sons of the State (*Kuo tzu hsüeh* 國子學) for the sons of noble families and the highest officials, and the *Thai hsüeh* for promising students of less distinguished, though not common, stock.[6]

In 1958 we had the pleasure of studying a magnificent stele, dated +278, which is still to be seen at the Kuan Kung temple at Loyang. This inscription gives an account of the National University of that time, with its 3,000 students, many from abroad, some from 'East of the Sea' (Liaotung) and some from 'beyond the shifting sands' (*liu sha* 流沙), i.e. Sinkiang. There is a long list of students and professors. A full translation of the inscription would be of much interest.

The student body was divided into Novices (*Mên jen* 門人), Disciples who had passed in one classic (*Ti tzu*), Expectant Instructors who had passed in two (*Pu wên hsüeh* 補文學) and titular Secretaries to the Heir Apparent who had passed in three (*Thai-tzu shê jen* 太子舍人). On graduation, normally at the end of seven years, pupils would be given the title Gentleman of the Interior (*Lang chung* 郎中), as usual for court attendants waiting to be gazetted to a post. In subsequent ages the National University never went altogether out of existence; in times of disunion there were several institutions under competing dynasties. Only in periods of prosperity could the major institution carry out its tasks to the full.

During the Eastern Chin, the National University was strengthened by successive emperors, as for instance by Emperor Yüan in +317. Further chairs for the study of particular classics (including the *Chou li*) were later established.

Towards the end of the Northern Wei, about +490, Emperor Hsiao Wên changed the name of the National University to *Kuo tzu* 國子 (lit. Sons of the State). Under the Sui, about +610, it was reconstituted as an agency responsible for supervising all the schools in the capital, designated the *Kuo tzu chien* 國子監, the standard translation of which is 'Directorate of Education'. This title lasted a very long time, down in fact to the end of the empire in 1911. It is not too much to say that the universities of modern type all over contemporary China descend from the Directorate of the Middle Ages, enlarged,

---

[6] [The discrepancies in accounts of this founding reported by Galt (1951), pp. 280–2, have been resolved by the inscription discussed below. – Editor]

modified and complemented under Western influence in the 19th century. Throughout two millennia the conception of an institution of higher education within the framework of the national bureaucracy remained rooted in Chinese culture.

A decree of +3 issued under the Han Emperor Phing first established (or at any rate first recorded) provincial academies and seminaries. In +466, under the Northern Wei, provincial colleges were set up in every commandery. The larger of these provincial colleges contained 2 Professors, 4 Lecturers and 100 Students (*Hsüeh-shêng* 學生).

(3) Medical teaching

We must now retrace our steps to follow the developments in the national medical administration from the Chin period on. When we look at the Northern Wei records, we find little significant difference until we suddenly come upon two striking new posts, *Thai i po-shih* 太醫博士, which we can translate only as Regius Professor of Medicine, and *Thai i chu-chiao* 太醫助教, Regius Lecturer in Medicine, of the lower 7th rank and lower 9th rank, respectively (9B was the lowest regular rank in the civil service). These appear as part of a great reorganisation and settlement of the official hierarchy carried out by Kao Tsu (the Emperor Hsiao Wên) in +492.[7] The posts parallel a number of other didactic posts at the National University not only in general education but also in astrology, mathematical astronomy and geographical communications. The picture is of such interest that it is worth tabulating. In Table 3, (NU) stands for National University, and (OIP) for Office of the Imperial Physician.

Thus the medical men held low positions in dignified company. How many medical teaching posts there were in each category we do not know. Similar posts spread rapidly beyond the borders of China.[8]

(4) Medical examinations

By the +3rd century, examinations for the position of Professor had existed for nearly 500 years. In the +4th, many new professorships were established, and the university was enlarged in the +5th. When, as we have just seen, the Wei dynastic history enumerates medical teaching officers, the implication is quite clear that there were not only lectures but also examinations for competence in medical knowledge. We do not yet have documentary evidence. But it comes soon.

[7] [Tabulated in *Wei shu* 魏書, ch. 113, pp. 2980–93. It is difficult to agree that the translations of these titles require the word 'regius' in a way that others do not. All official titles originated in principle with the emperor, and these did not so originate except in principle. Hucker's standard translation for *Thai i po-shih* (*DOT* 6182), for instance, is simply 'Medical Erudite'. A Thang source (*Ta Thang liu tien* 大唐六典, ch. 14, p. 23b), notes that a Liu Sung official proposed a Medical School (*I hsüeh*) in +443, but the history of that era does not record it. A documented Sung proposal of +469, meant to combat popular religious curing by training physicians in the Academy, was not accepted; *Sung shu* 宋書, ch. 82, p. 2100. It is at least possible that the Wei, in medical education as in other respects, followed the lead of the south. – Editor]

[8] In +553 the Korean kingdom of Paekche sent a Medical Professor named Wangyu Rungtha to Japan to reorganise medical education there; his mission included two Masters of Drug Production. This embassy bore full fruit in +702, when the Japanese Emperor Mommu established an Imperial Medical College with five departments, and monthly and annual examinations.

Table 3. *Technical administration, +492*

Rank	Post
5 ii A	Regius Professor of Classics (NU)
	Libationer (NU)
6 i B	Senior Regius Professor of Astronomy (NU)
	Regius Professor of Classics (NU)
6 i C	Regius Professor of Music, Imperial Music Office
7 i B	Student (NU)
7 ii C	Manager (NU)
	Junior Regius Professor of Astronomy (NU)
	Regius Professor, Imperial Divination Office
	Manager, Imperial Music Office
	Regius Professor (OIP)
8 i B	Regius Lecturer in Classics (NU)
9 i B	Regius Lecturer in Astronomy (NU)
	Regius Lecturer (OIP)
9 ii B	Regius Professor, Provincial Post-Station Service

The third great unification of China occurred in +581 when a single House, that of the Sui, took over again the dominance of the entire country. Apart from the usual administrative staff of the Department of the Imperial Physician we read that there were 2 Pharmacists (*Chu yao* 主藥) and 200 Master Physicians (*I shih* 醫師), as well as 2 Curators of Physick Gardens (*Yao-yüan shih* 藥園師) where medicinal plants were systematically cultivated. On the educational side, around +585 we find that the Imperial Medical Office has been expanded to include 2 Medical Professors (*I Po-shih*), 2 Lecturers (*Chu-chiao*), 2 Professors for Physiotherapy (*An mo po-shih* 按摩博士) and 2 Professors for Apotropaics (*Chou chin po-shih* 咒禁博士). The Professors for Physiotherapy must have taught medical gymnastics and massage, among other things, while the Professors for Apotropaics drew on the various incantatory and talismanic methods of driving away diseases which had existed among the Chinese people, as among all ancient nations, from high antiquity downwards. Among the staff of the Chamberlain of the Imperial Stud (*Thai phu* 太僕) were 120 Teachers of Veterinary Science (*Shou i po-shih yüan* 獸醫博士員).[9]

For the great Thang dynasty (+618 to +906) we have truly abundant information. It is in the middle of the +8th century that we shall see the full development of medical examinations.

We might first take a look at the National University. The Directorate of Education, as it was now called, seems to have consisted of two parts, one more socially exalted than the other (as indeed had been the case in the Northern Wei), perhaps something like 'Collegians' and 'Oppidans'. The 'Noblemen's' group consisted partly of relatives of the Imperial House itself, and partly of sons of officials of the 2nd rank and higher. These,

[9] *Sui shu*, ch. 28, pp. 774, 776.

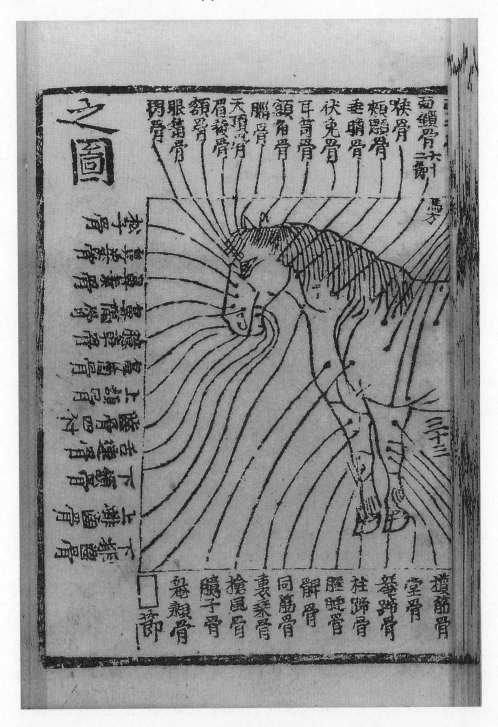

Fig. 5. Veterinarian equine bone anatomy. From *Ma niu i fang* 馬牛醫方 (Therapeutic methods for horses and cattle, +1399). From Needham (1970), Fig. 94, p. 386.

numbering 300, were accommodated in the School for the Sons of the State. In addition there were the 500 ordinary students of the University, sons or nephews of officials of the 5th rank and above; and 1,300 from all over the country enrolled in the School of the Four Gates (*Ssu men hsüeh* 四門學). Five hundred of the latter came from the families of officials of the 7th rank and above, but 800 were essentially 'Scholars' or 'Sizars' drawn from the most intelligent and best families who, although hardly plebeians, were not sons, grandsons, or great-grandsons of civil servants. Fifty places were reserved for students specialising in law, and thirty each for students of calligraphy and mathematics, all from families of low-ranking officials or commoners 'who understood these subjects'. Lastly quotas were specially reserved for students from the capital and the provinces, something rather like the scholarships tied to localities which existed until recently in Cambridge and Oxford colleges. The full complement of students was thus 2,100.[10]

Turning to the Imperial Medical Office (*Thai i shu* 太醫署), we find that it was one of the eight administrations under the Office of Imperial Sacrifices (*Thai chhang ssu* 太常寺). Below the two Imperial Physicians, Senior and Junior (*Thai i ling*) with their Aides (*Thai i chhêng* 太醫丞), there were 4 Medical Supervisors (*I chien* 醫監) and 8 Principal Practitioners (I chêng 醫正). Among the unranked staff of the Office we hear of 2 Stores Officials (*Fu* 府), 4 Scribes (*Shih* 史), 8 Pharmacists (Chu Yao), 24 Apprentices in Materia Medica (*Yao thung* 藥童), 2 Gardeners (*Yao-yüan shih*), their 8 Apprentices (*Yao-yüan shêng* 藥園生) and lastly 4 Clerks (*Chang ku* 掌固). It is rather remarkable to find that administrators still evaluated the rank and file of state physicians according to the proportion of patients who recovered after they had treated them. This procedure, as we know, goes back to the end of the Chou period.

The New history of the Thang dynasty is informative about teaching:

The Imperial Physicians were in charge of all methods used in therapy. Their subordinates included the Master Physicians (*I shih*), the Master Acupuncturists (*Chen shih* 鍼師), the Master Physiotherapists (*An mo shih* 按摩師) and the Master Apotropaists (*Chou chin shih* 咒禁師). In all of these, teaching was carried out by Professors. Testing and appointment were similar to [the practice] of the Directorate of Education. As the Master Physicians, the Principal Practitioners (*I chêng*) and the Medical Apprentices (*I kung* 醫工) treated patients, the number of those who recovered were recorded in order to test them. . . . In the capital region, good fields were set apart for [medicinal plant] gardens. Sixteen or more subofficial functionaries were assigned to be Apprentice Gardeners. When their training was finished, they were made Gardeners. The duty of these men was to recognise from where every item [i.e. plants submitted as tax payments] came, and to choose the best to be the portion of the palace.[11]

Here we have a quite specific statement about examinations and practical tests which must refer to the middle decades of the +7th century. Each of the teaching divisions included 1 Medical Professor and 1 Medical Lecturer. The Thang dynastic histories give us a great deal more information about this Imperial Medical Office. Because many

---

[10] *Hsin Thang shu* 新唐書 (New history of the Thang dynasty), ch. 44, p. 1159.
[11] *Hsin Thang shu*, ch. 48, pp. 1244–5; see also *Ta Thang liu tien* 大唐六典, ch. 14, p. 23a (p. 331).

of its therapeutic services had been taken over in +605 by the Palace Medical Service (*Shang yao chü* 尚藥局), it became largely a teaching and certifying agency. It was not formally subdivided into departments, but each of the professors had his own staff.

The Professor of General Medicine was supported by a Lecturer, 20 Master Physicians, 100 Medical Apprentices, 40 Students and 2 Pharmacists (*Tien yao* 典藥). Only the Professor and Lecturer had rank in the official hierarchy. They taught all the students and practitioners from the already sizeable medical literature, including the funda-mental classics, e.g. the *Su wên* 素問 (The plain questions [and answers]) and *Ling shu* 靈樞 (The vital axis) of the *Nei ching* 內經; the *Shên nung pên tshao ching* 神農本草經 (Pharmacopoeia of the Divine Husbandman, +2nd or +1st century) and its descendants; and the sphygmological classics such as the *Mo ching* 脈經 (The pulse manual, *ca*. +280).

The courses into which the subjects of study were divided have also come down to us. They were five in number: general medicine (*thi liao* 體療); external medicine, mainly on the treatment of surface lesions (*chhuang chung* 瘡腫); paediatrics (*shao hsiao* 少小); otology, ophthalmology, stomatology and dentistry (*êrh mu khou chhih* 耳目口齒) and cupping (*chiao fa* 角法). This last therapy, literally 'the horn method', originally used an animal horn (in which a vacuum had been produced by burning leaf pulp or some other material) to draw the blood to the surface of the skin. In some instances cupping involved phlebotomy. This widely practised folk therapy is generally considered by mod-ern physicians to be a form of counter-irritation. Its use in China has been documented since the −2nd century.[12]

Next, the subordinates of the Professor of Acupuncture included a Lecturer, 10 Mas-ter Acupuncturists, 20 Acupuncture Practitioners (*Chen kung* 鍼工) and 20 Students (*Chen shêng* 鍼生). The students and young practitioners specialised in sphygmology and acupuncture, learning the system of small areas on the surface of the body at which to needle and burn moxa as the pulse and other diagnostic aids indicated. The mysteries of many kinds of instruments, too, were opened to them.

The other two departments were somewhat less important. The Professor of Physio-therapy had no Lecturer to second him; his staff included 4 Master Physiotherapists. They were in charge not only of medical gymnastics and massage but also dealt with traumatic injuries and fractures, having under them 16 Physiotherapeutic Practitioners (*An mo kung* 按摩工) and 15 Students (*An mo shêng* 按摩生). Lastly the Professor of Apotropaics was seconded by 2 Master Apotropaists, 8 Apotropaic Practitioners (*Chou chin kung* 咒禁工) and 10 Students (*Chou chin shêng* 咒禁生). This gives a grand total of 271 established posts and studentships in the Imperial Medical Office: 162 in General Medicine, 52 in Acupuncture, 36 in Massage and Gymnastics and 21 in Apotropaics.

[12] [Formula 144 in the *Wu shih êrh ping fang* 五十二病方, for haemorrhoids, specifies 'cup it with a small horn, for as long as it takes to cook seven litres of rice, and then remove the horn'. See the Chinese text in Harper (1982), pp. 415–16. Harper interprets this not as cupping but as 'goring' the haemorrhoid with a horn, based on an erroneous statement in Vol. 4, pt 1, p. 38, that cupping was not characteristic of Chinese medicine. The practice is documented in many early sources, from *Chou hou chiu tsu fang* 肘後備急方 on, under such names as *huo kuan* 火罐. Yü Wên-chung 于文中 (*1981*) studies its history. There is no evidence in the med-ical literature that patients were gored for therapeutic reasons. On phlebotomy see Epler (1980). – Editor]

The administrative statutes afford us a vivid glimpse of policy regarding examinations:[13]

On 18 March +758, an announcement (*chih* 制) specified that 'henceforth those to be given official appointments on account of their medical skill should be treated like [those appointed on the basis of examinations for] the degree in classics (*ming ching* 明經)'.

On 1 February +760, the Aide of the Right Imperial Insignia Guard, Wang Shu 王淑, memorialised to the throne, requesting that 'when personnel are selected for medical skill the procedure be like that for the degree in law (*ming fa* 明法). Henceforth let each be tested with 10 papers on the application of medical classics and techniques; 2 papers on [knowledge of the] *Shên nung pên tshao ching*; 2 papers on the *Mo ching*; 10 papers on the *Su wên*; 2 papers on Chang Chi's *Shang han tsa ping lun* 傷寒雜病論 (Treatise on febrile and miscellaneous diseases, *ca.* +200); and 2 papers on the interpretation of medicinal formulae in various manuals. Those who pass in seven or more papers are to be kept on, and those who pass fewer dismissed.'

Wang also asked that [the procedure for examining personnel for] the Food Service (*Shang shih chü* 尚食局, under the Chamberlain for Palace Revenues) and the Pharmacy (*Yao tsang chü* 藥藏局, in the Secretariat of the Heir Apparent) be like those of the Heir Apparent's Foods Service (*Tien shan chü* 典膳局); and that those under the Imperial Medical Office should be similar to those under the Imperial Music Office (*Thai yüeh shu* 太樂署).

This interesting passage requires a little commentary. The Thang dynasty gave six special degrees in qualifying talented scholars for the administration. The first of these, Cultivated Talent (*Hsiu-tshai* 秀才), was given to a very few broadly learned scholars examined on 'modes of action to be taken in certain situations' between +595 and its abolition in +651.[14] Five more specialised degrees were important: Presented Scholar (*Chin-shih* 進士), primarily literary, the most prestigious and competitive; Classicist (*Ming-ching* 明經) for textual interpretation; and, until +1060, Law Graduate (*Ming-fa* 明法), to do with the dynastic legal code; Calligraphy Graduate (*Ming-tzu* 明字); and Mathematics Graduate (*Ming-suan* 明算). Because there was no medical degree, the standards for appointment in that discipline were, as Wang proposed, assimilated to those for Law Graduates. The final remark about the Imperial Music Office explains itself when one peruses its regulations. There it appears that the professors themselves were periodically examined, as well as the executive musicians and acoustics experts, so that Wang Shu was presumably asking for a periodical test of the learning and efficiency of the medical teachers.[15]

It does not look as though the trials of the embryo medical officers ended with the passing of their written examinations, for elsewhere in the *Hsin Thang shu* we read that among the duties of the Director of the Bureau of Sacrifices, 'he tests the skill of the sons and disciples of famous physicians at treating medical disorders. Their superiors supervise and report on them. He sends forward the names of those who have succeeded in this probationary practice for three years.' The recruitment of doctors' sons was not new: 'In the Chin, it was ordered that Instructors train the hereditary successors of outstanding doctors.'[16]

[13] The quotations that follow are drawn from the section on medical examinations in *Thang hui yao* 唐會要, ch. 82, pp. 1522–5.
[14] Wright (1979), p. 86. On degrees offered and texts set in the early Thang see *Hsin Thang shu*, ch. 44, pp. 1159, 1161–2.
[15] *Hsin Thang shu*, ch. 48, p. 1243.   [16] *Hsin Thang shu*, ch. 46, p. 1195; *Ta Thang liu tien*, ch. 14, p. 51b.

(5) Provincial medical education

We gain further insight into the Thang system of medical education by following the institution of provincial colleges of medicine (*I hsüeh* 醫學). As early as +629, medical schools were established in prefectural capitals. As often happened, orders of the central government about local administration had no lasting effect. This we can see from an edict of +723 which observed that

in remote administrative circuits and prefectures no medical skills at all are found. When the lowly suffer, on whom can they rely? It would be appropriate to have every provincial capital appoint one functioning Medical College Professor, with rank equivalent to that of an Office Manager. Each prefectural government office should stock a pharmacopoeia and a collection of 101 proven medicinal formulae (*Pai i chi yen fang* 百一集驗方) along with its classics and histories.

Although the purpose of provincial schools over most of history was to serve the needs of local officials, the Thang history states clearly that the duty of the professor was 'to be in charge of treating the diseases of the people' (*chang liao min chi* 掌療民疾).[17] We learn that 'before long the positions of Professor and Student were abolished, and the remote prefectures, as before, lacked medical care'.[18] The failure of this policy is not surprising, since each school was in charge of a physician whose rank equalled that of an Office Manager, at the lowest level of official service, or more likely below it.

Following these institutions into the Sung, we find that at the beginning of the +12th century the examination regulations stated clearly what may have been a long-established policy of awarding local posts to the worst-qualified graduates. The students 'who receive high scores will be given appointments as Master Physician and below in the Palace Medical Service. The rest will fill vacancies, according to their scores, as Professor, Director, or Office Manager in the College, [and then] as Instructor (*Chiao-shou* 教授) in provincial colleges.'[19]

To inaugurate the reformation of the provincial medical schools in +723, Emperor Hsüan Tsung compiled, published and sent to each prefecture 'formulae for widespread benefaction' (*kuang chi fang* 廣濟方). This no doubt meant the collection cited in the edict above, for the Chinese phrase is not the title of a single book. In +746 this imperial pharmacist announced in a policy statement issued for the guidance of officials (*chhih* 敕) that local authorities were to choose suitable formulae and post them on public notice boards so that the masses could take advantage of them.[20]

In +739 another policy statement had asserted that 'every prefecture with more than 100,000 families is to have 20 Medical Students (*I shêng* 醫生) and those with fewer than 100,000 are to have 12. Each is to make rounds within its borders providing therapy.'[21]

---

[17] See above, note 13. [This was not a command but a suggestion, and it is ambiguous about whether a Medical College was to be established to house the single professor. – Editor]

[18] Liu Po-chi (*1974*), vol. 1, p. 194.

[19] [*Sung shih* 宋史, ch. 157, p. 3689. The great discrepancy between expectations of the central government and actualities in the provinces is typical of imperial China, as anyone who has studied the problematic evolution of hospitals knows. – Editor]

[20] The Arabic traveller Sulaimān al-Tājir, who was in China in +851, remarked upon this.

[21] [In other words, people were given this title to practise rather than for academic study. The term applied not only to those enrolled as students in a State school, but to practitioners who had finished their educations (as it does today). – Editor]

In subsequent years the number of professors and students was increased. In +796 the Emperor Tê Tsung personally compiled the *Chen-yüan kuang li fang* 貞元廣利方 (Medical formulae of the Chen-yüan reign-period for widespread benefit) and had it distributed to every prefecture. On 27 April a policy statement said:

Early during the Chen-kuan reign-period (+629) the prefectures established Medical Professors. In the Khai-yüan reign-period (+723 to +741) Lecturers were added. [These officials] were charged to evaluate and examine gentlemen with medical skills, and make sure that they understood the proper organisation of rounds of medical visits. From time to time the authorities made further enactments but, although the appointed [Professors and Lecturers] remained in place, their skills were not at all honed. Few were of any use.

I contemplate that the provincial governors share my concern. Now that they are aware of my intention, I shall entrust the selection to them. Henceforward, when the prefectures are filling vacancies for Medical Professors, it is appropriate that executive Aides be ordered personally to seek out and examine applicants, choosing those of superior skill and standing, so that [appointees] will be useful. Their names are to be reported to the court. Those already holding appointments and formally qualified will not be subject to confirmation by the Ministry of Personnel.[22]

All this has a considerable bearing on when the Imperial Medical College was founded. Since the provincial medical colleges were initiated as early as +629, it is clear that the central medical college must have been functioning within the first decade of the foundation of the dynasty, i.e. from +618. As we have seen (note 8), the institution was duplicated in Japan from +702 onwards.

A final document on medical examinations suggests that the system did not remain vital throughout the Thang era. In January +802 a policy advisory noted that candidates for Medical Officer and Apprentice Pharmacist of the Medical Institute (*Han lin* 翰林) 'henceforth, even if they pass the examinations, may not be selected by the relevant authorities. The appointments of those selected are to be terminated.'[23]

We do not propose to follow further the development of the medical administration. We have already said enough to establish the point. The mature system of medical education and examination in the Sung, more ambitious in conception but less so in practice, is worth examination.

(6) SUNG MEDICAL EDUCATION

The reformer Fan Chung-yen 范仲淹 reported in +1044 that centrally regulated medical instruction had died out by the beginning of the Sung. Even in the capital at that

---

[22] [Two points in this document are significant. First, it does not mention Medical Colleges, and refers to Medical Professors rather than Regius Professors (i.e. Medical College Professors), which leaves open the question of whether they and the lecturers were appointed to staff an institution. Secondly, it does not mention a curriculum, but merely specifies that the appointees are responsible for evaluating staff practitioners and organising their therapeutic rounds. We have seen (p. 104) the Professors' duties described as clinical rather than instructional. The point of the emperor's final instruction is apparently that Professors and Lecturers already holding posts need not be discharged.

These omissions do not necessarily invalidate the authors' interpretation. It is possible, however, to conclude that, despite a policy more than a century old, at the end of the +8th century no government medical schools existed, and that provincial appointees were responsible for managing rather than educating low-level staff physicians. – Editor]

[23] *Thang hui yao*, ch. 82, p. 1525.

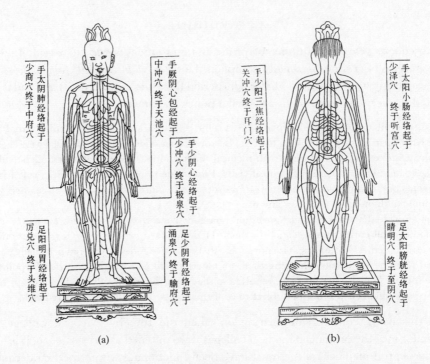

手太阴肺经络起于　少商穴终于中府穴　　中冲穴　终于天池穴　手厥阴心包经络起于　　手少阴心经络起于　少冲穴　终于极泉穴　　厉兑穴　终于头维穴　足阳明胃经络起于　　涌泉穴　终于腧府穴　足少阴肾经络起于

手少阳三焦经络起于　关冲穴终于耳门穴　　手太阳小肠经络起于　少泽穴　终于听宫穴　　足太阳膀胱经络起于　睛明穴　终于至阴穴

(a)　　　　　　　　　　(b)

Fig. 6. Hollow bronze statue used *ca.* +1026 to examine students of acupuncture in the palace. The holes that represented the loci were stopped with wax and the statue filled with water. When a student inserted the needle correctly the water spurted out. From *Thung jen shu-hsüeh chen-chiu thu ching* 銅人俞穴針灸圖經 (Illustrated canon of loci for acupuncture and moxibustion for use with the bronze instructional statue, +1026).

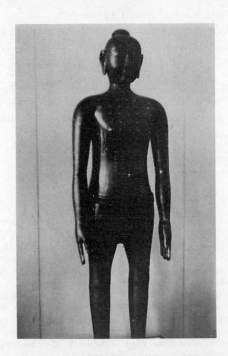

Fig. 7. Bronze acupuncture statue; reproduction of a Sung figure (Peking Historical Museum, original photo, 1964). From Needham (1970), Fig. 86, facing p. 441.

time, its population over a million, there were only a few thousand physicians. Fan proposed a comprehensive plan for improving medical care through education and testing. Like most of the other reforms of his brief ascendancy, the examinations were not carried out.[24]

Between +1068 and +1085, the enlightened premier Wang An-Shih 王安石 and his allies carried out a great series of reforms that reconfigured the institutions and aims of government. Their Three Halls (*san shê* 三舍) policy structured education by dividing the National University into three grades, with stringent requirements for promotion through steps meant to become the only route to success in the examinations. In +1101 to +1104, Wang's successor Tshai Ching 蔡京 expanded this system of graded, government-sponsored education down to the local elementary schools. This system was meant simultaneously to feed outstanding students to the National University, to abolish all but the classics degree (*chin shih* 進士) and to standardise the curriculum to the end that all officials would think uniformly and act virtuously. This system rose and fell, grew and shrank, in time with the furious battles between the reformers and their enemies. It was abolished in 1121, but left a heritage of closely linked education and examinations.[25]

In +1076, as part of this process, Wang gave the Imperial Medical Service (*Thai i chü* 太醫局) its definitive form as a primarily educational institution. Under the Director and his Aide there were a number of Professors and Lecturers, teaching what by +1085 became no less than 300 Students in nine specialities.[26]

In +1103, Tshai Ching moved the Imperial Medical Service into the Directorate of Education. The Emperor Hui Tsung, who was knowledgeable and enthusiastic about medicine, backed him in this innovation. The aim was to raise the status of medicine by training the sons of families with scholarly traditions (*shih lei* 士類), who would be social equals of the Directorate's students of classics. Those who survived rigorous pruning on the basis of frequent examinations would be qualified for high civil posts.

The number of students fluctuated greatly over the remainder of the Sung. In fact, as Hucker remarks, over the course of the Sung the Service 'had a particularly erratic existence as one of several medical agencies in the palace and central government, repeatedly abolished and re-established'. From the late +12th century on, both authorisations and incumbents shrank. By the end of the Southern Sung, the system of training and examination seems to have shrivelled up, to be revived in part by the Yüan. This was in the first instance due to the legacy of the Wang reforms, but also depended on finances, the general vigour of government, official perceptions of the importance of medicine, and the favour of individual monarchs. Despite these vicissitudes, the Service influenced the less ambitious medical examinations of later dynasties.[27]

---

[24] *Fan Wên-chêng kung chi* 范文正公集 (Collected writings of Fan Chung-yen), 'Tsou-i' 奏議, ch. 2; for a related edict see *Sung hui yao*, 'Chih-kuan' 宋會要, 職官, ch. 22, p. 35a (p. 2877).

[25] Chaffee (1993), pp. 77–80, 184; Lee (1985), pp. 64–6, 102.

[26] [Although the Service kept its educational character, its separation from the Office of Imperial Sacrifices was rescinded in +1078. *Sung hui yao*, ch. 22, p. 38a (p. 2879). – Editor]

[27] *DOT* 6179; Kung Shun 龔純 (*1955, 1981*); Liang Chün 梁峻 (*1995*), p. 99; Lee (1985), pp. 96–8. On the Ming and Chhing systems see Kao Yeh-thao 高也陶 (*1991*) and Liang Chün (*1995*).

The Service chose each advanced class from low-ranking medical officials, advanced students, and 'medical personnel outside [the government] who are of established reputation and achievement'. The record adds that 'those who agree to accept the status of students are to be informally tested, and added to the rolls without quota. After the spring examinations every year, three hundred passing [students] are to be retained.'[28] The instructional facilities of the Imperial Medical Service were housed in spacious buildings.[29]

After the fall of Khaifêng to the Chin Tartars in +1126, and the setting up of the new Southern Sung capital at Hangchow, the residential aspect of the organisation comes into greater prominence. The *Mêng Liang lu* 夢梁錄 (Dreaming of the idyllic past, +1274) has much to say of the elaborate buildings provided in that city for the Imperial Colleges.[30] The Medical College had lecture-halls for the teaching of the 4 Professors, and a temple for the worship of the tutelary deities of medicine. The 250 students, who got excellent food in the refectory, were accommodated in eight halls (*chai shê* 齋舍), the very names of which have come down to us. They were granted a special cap and belt which distinguished them from ordinary citizens, but they had to face examinations every month and every quarter. Such was the state of the Imperial Medical College about +1275, when Marco Polo was on his way to China.

(7) Sung medical examinations

Ho Ta-jen 何大任, as the official in charge of the Service in +1212, compiled for posterity a book entitled *Thai i chü chu kho chhêng wên* 太醫局諸科程文 (Model examination papers for diverse courses given by the Imperial Medical Service). He explains that the system instituted by Wang An-shih had been revived after the move to the south. Although the court wished to draw the best medical talent from throughout the empire, it is clear that the ideal of regional medical schools had not been realised.

The pupils in the Service's school number roughly 300, but they are, generally speaking, all from the capital region. It can hardly be true that there is no [talent] elsewhere. Is it not that the examination subjects and the texts on which the papers are based have never been promulgated, so that scholars in distant regions have no compass to guide them? Even if they want to follow the curriculum, they have no way to obtain it.

With this in mind, Ho and his colleagues compiled the best papers from recent examinations.

Although the official history of the Sung gives prominence to three main divisions in medical education, sources closer to the changing curricula reflect 6, 8, 9, 13 and 14.[31] The 13 of Ho's time became the usual number in later eras:

(1)   Internal and general medicine (*ta fang mo* 大方脈)

(2)   Miscellaneous medicine (*tsa i* 雜醫)

---

[28] *Hsü tzu chih thung chien chhang pien chi shih pên-mo* 續資治通鑑長編紀事本末, ch. 81, p. 14a (p. 2605).
[29] *Sung shih*, ch. 157, p. 3689.   [30] Ch. 15, pp. 4a–4b.
[31] See, for instance, the frequent changes reflected in *Sung hui yao* 宋會要.

(3) Paediatrics (*hsiao fang mo* 小方脈)

(4) Convulsive and paralytic diseases (*fêng* 風)

(5) Gynaecology and obstetrics (*chhan* 產)

(6) Ophthalmology (*yen* 眼)

(7) Stomatology and dentistry (*khou chhih* 口齒)

(8) Laryngology (*yen hou* 咽喉)

(9) Orthopaedics (*chêng ku* 正骨)

(10) External medicine (*chin chhuang chung* 金瘡腫)

(11) Acupuncture and moxibustion (*chen chiu* 針灸)

(12) Apotropaics (*chu yu* 祝由)

(13) Interdiction (*chin* 禁)[32]

At other times these courses were combined to make a smaller number. As late as the 20th century, examinations in traditional medicine followed a somewhat similar division into categories, with the addition of physiotherapy and history of medicine, and of course minus the last two categories.[33]

The examinations which Ho and his colleagues administered were based on a plan established in +1104. They presupposed a three-year course, with monthly and quarterly written exercises to determine the standing of the pupils. The questions in the final examination took six forms, which reflect the content of medical education:

(1) The so-called 'black-ink purport' (*mo i* 墨義), in which the examinee was given a sentence such as 'those who treat illness must have a clear understanding of the way of Heaven and the patterns of the Earth'. The object was not to explain or interpret it, but to quote exactly from memory the long passage from the *Nei ching* in which it occurs.

(2) The 'large purport', also called 'purport of the classic' (*ta i* 大義, *ching i* 經義). The examinee was expected to explain a canonical passage chosen to test mastery of the foundations of medicine, such as 'the *chhi* of the liver flows to the eyes; when there is concord in the eye [functions], one can distinguish black and white'. To answer this question satisfactorily required ability to apply yin–yang and Five Elements reasoning.

(3) The 'pulse purport' (*mo i* 脈義), which called for comment on a passage about pulse diagnosis from the *Nei ching*.

(4) The 'discussion of a medicinal formula' (*fang lun* 方論), taken from a canonical source such as the *Shang han tsa ping lun* or the *Thai-phing shêng hui fang* 太平聖惠方 (Imperial grace formulary of the Thai-phing-hsing-kuo era, +982). The object was to explain the physiological activity of each ingredient, as well as the results of their synergy.

---

[32] [In ordinary usage these last two kinds of practice greatly overlapped. The first had to do with the use of charms and spells in curing. The second was concerned with both preventing and undoing spirit possession. The institutional distinction is unclear. – Editor]

[33] [This old dispensation was drastically changed after *ca.* 1980, with the introduction of substantial content taken directly from biomedicine, and less attention to certain aspects of the traditional art. See the summary of an actual curriculum in Sivin (1987), pp. 481–2. – Editor]

(5) The 'hypothetical instance' (*chia ling* 假令) of diagnosis or therapy, meant to test the ability to apply learning to patients. One instance asked the student to explain a formula recommended for given symptoms by a standard source, containing ingredients that according to standard principles ought not to be combined.

(6) The question on 'phase energetics' (*yün chhi* 運氣), which required the student to determine what effect the cosmic dispensation of a given year would have on therapy. This practice was not altogether unlike the astrology that was a routine part of medical practice in early modern Europe. In China no astronomical computation was involved, merely simple arithmetic.[34]

Although it was essential to memorise many volumes of medical classics before one could cope with an examination of this sort, it also evaluated a command of abstract concepts and the ability to apply this learning to diagnosis and therapy. On the other hand, it did not presuppose clinical experience.

One may not conclude that practical knowledge was never relevant to the evaluation of medical students. The short-lived proposals of Wang An-shih and his supporters, at least, newly applied the Thang government's pragmatic evaluation of State physicians (p. 101) to students in the Imperial Medical Service.

When pupils in the National University, the Law School, and the Military School, as well as officers in the military garrison, fall ill, the students are to be sent in rotation to treat them. A printed form (*yin chih* 印紙) is to be issued to the student for each instance. The officials of the school or garrison are to write down the diagnosis and manifestations, and whether the illness ended or [the patient] died. The report is to be certified by officials of the Service.

If after examining the patient [the student declares that the illness] is incurable, someone else is to be dispatched to treat it, and determine whether [the patient] dies or recovers. The details are to be written down as a basis for adding points to, or deducting them from, the student's score. At the end of the year [the scores] are compared, and the [100 most successful] students divided amongst three levels.[35]

The ranking determined not only the student's standing but his salary for the next year. The document adds that those who rank below the top 100 might be punished or expelled.

After the Sung period, there were many minor modifications but little in the way of significant innovation. The Yüan and its successors tended to recruit students from medical families, although the Chhing relaxed this policy. Local practitioners could apply to fill teaching vacancies in the Yüan provincial system. The Imperial Directorate of Medicine (*Shang i chien* 尚醫監) demanded that applications contain the details of the doctor's social standing as well as of the medical cases he had treated. From 1285 on the State periodically tested provincial medical teachers. The Directorate graded their examination papers as well as those of their students.

---

[34] [The questions cited are from: ch. 1, p. 1a; ch. 3, p. 1a; and ch. 3, p. 9b. See also *Sung shih*, ch. 157, p. 3689. 'Purport' for *i* is due to Sun (1961). On phase energetics see Porkert (1974), pp. 55–106. The extant version of Ho's book is apparently incomplete; it does not, as the preface claims, contain examination questions for all thirteen departments. – Editor]

[35] See note 28, above. The text is not clear about whether 'someone else' was another pupil or an examiner.

The Ming–Chhing system relied upon quarterly local examinations and a triennial central examination. Both were oral as well as written. The government used grade rankings to promote and demote incumbents, and to appoint neophytes Physician or, if the grade were lower, Student of General Medicine.[36]

(8) ISLAMIC INFLUENCE ON EUROPE

We have run over the centuries which corresponded to the rise and fall of the School of Salerno. This famous seed-bed of Western European medicine began in the +9th century when the Thang dynasty was drawing to its close. It reached its apogee in the +12th century during the highly cultured period of the Southern Sung after the fall of the capital. It continued until the end of the +14th, which would take us to the close of the Yüan period in China. It appears that Arabic influence in Salernitan medicine is not to be detected much before +1050, but after that it becomes extremely strong, with the *Antidotarium* of about +1080. That was the time of Constantine the African, after which a massive transmission of Arabic knowledge and practice occurred, culminating in the *De aegritudinum curatione* of the +12th century.[37]

We now approach the dénouement of this paper. It is almost certain that the examinations and the licensing of Western Europe were borrowed from Arabic practice. The later European pattern of general education without vocational trend, followed by a course of theoretical medical study and a year or two of practice under supervision, was foreshadowed as far back as +1224 in an edict of the Emperor Frederick II, which held good for Sicily, South Italy and Germany. Medical students at the School of Salerno were required first to study the logical treatises of Aristotle for three years and then to learn medicine from the books of Hippocrates, Galen and Avicenna for five years, finally carrying on clinical practice for one year under an experienced physician. The candidate was eventually subjected to a searching examination on the works of the Greeks and the Arabs. On passing he obtained a licence and graduated as *Doctor medicinae*, a term which originated in Salerno. The edict of Frederick II was apparently not quite the first of this kind, for in +1140 Roger of Sicily had decreed laws concerning State examinations for physicians. Examinations for surgeons appeared first in Paris at the Collège de St Côme in +1210. After the +13th century the monopoly of Salerno declined with the rise of the Schools of Montpellier, Paris and Bologna. We hear also of medical examinations in Cairo in +1283.

The year +931 constitutes a focal point in this matter. As a result of a death occurring through the mistake of a medical practitioner, the Caliph al-Muqtadir charged an eminent physician, Sinān ibn Thābit ibn Qurrāh (*ca.* +880 to *ca.* +942), to examine all those practising and studying medicine at the time. Sinān was the son of the great

---

[36] See Hsüeh I-ming 薛益明 (*1997*), valuable for post-Sung dynasties but unreliable for Sung, Kung Shun (*1955*) for Yüan, and reports of the College's last phase in Dudgeon (1870) and Cowdry (1921). Kung also has a monographic study of the Southern Sung (*ibid.* (*1981*)).

[37] Discussed in detail in Garcia-Ballester *et al.* (1994), pp. 13–29. See for orientation Kristeller (1945).

astronomer and mathematician Thābit ibn Qurrāh (+825 to +901). The decree which he received was written out by the Caliph himself. Sinān proceeded therefore to examine all those who came before him, passing 860 practitioners, both old and new, in the first year. The examinations were continued by his son Abū Ishāq Ibrāhim ibn Qurrāh (+908 to +947). There is reason to think that similar tests continued regularly in Egypt, if not in Persia, until the end of the Caliphate and afterwards. By +980 the Buwayhid emir, 'Adud al-Dawlah, founded a large new hospital in Baghdad.[38] Twenty-five physicians taught there, examining pupils and attesting to their proficiency.

(9) CHINESE INFLUENCE ON ISLAM

In view of all that we have found concerning the long history of Chinese medical examinations, we cannot but ask ourselves whether some influence from further east could have initiated the important development in Baghdad in +931.

There is indeed evidence that the Muslims of Iraq and the Chinese were in contact from at least the beginning of the +8th century. After the Battle of the Talas River in +751, which marked the farthest limit of Islamic expansion eastwards, many Chinese artisans settled in Baghdad, including paper-makers and metal-workers. It would be surprising if there were not physicians among them. Some prisoners returned home from the Gulf in Chinese ships in +762, but many remained and founded families in Baghdad. From that time on we find a mass of other material on Chinese–Arab contacts.

At just the moment our enquiry calls for it, there is evidence of medical contacts between the two civilisations. The *Fihrist al-'ulūm* (Index of the sciences), by the famous Abū'l-Faraj al-Nadīm in +988, contains a story told by perhaps the greatest physician and alchemist of his time, the great Rhazes (Abū Bakr Muḥammad ibn Zakarīyā' al-Rāzī, +865 to +923, or even +932). This story concerns his friendship with a Chinese physician, who asked him to read Galen aloud as fast as he could. The Chinese translated at equal speed, taking down notes or whole passages verbatim in the shorthand known as 'grass-writing' or, more accurately, as 'draft script'.[39] In view of this bridge between Islamic and Chinese cultures, it is easy to imagine that one or another of al-Rāzī's Chinese medical friends suggested to him that periodical examinations should be held, as they were in China, to test the fitness of young physicians.

It would seem that the Arabs took up this suggestion with energy and enthusiasm, handing on the torch of public safety and medical self-respect to the Western world. They may have been prepared to accept it by the Galenic tract 'On identifying the best physicians' (*De optimo medico cognoscendo*), quoted by al-Rāzī himself in a tractate with the same title, and still extant in an Arabic MS in Alexandria. It does not deal

---

[38] The background of all the Islamic hospitals and medical education was the great Sassanid medical school of Jundī-Shāpūr, founded in the +5th century. It continued the Greek tradition by absorbing the medical school of the University of Edessa (the foundation of which can be traced to Seleucus Nicator in −304) when the Emperor Zeno closed it in +489. It also drew abundantly from Indian, and later Chinese, sources. Nestorian Christianity, its dominant faith, tied it to both civilisations.

[39] The anecdote is translated in Vol. 1, p. 219.

with qualifying examinations, but rather with how 'rich men and heroes' can choose true physicians over charlatans.[40]

(10) Conclusion

To summarise: examinations of proficiency were held in China from −165 onwards; the National University was founded in −124; Regius Professorships and Lecturerships in medicine, implying examinations for qualification, date from +493; an Imperial Medical Office (in effect a medical college) and provincial medical colleges were established by +629; and medical appointments were awarded to local graduates from then onwards. In the light of all that we now know, we are able to estimate at its true worth the opinion of John Barrow, who wrote in 1804: 'The Chinese . . . pay little respect to the therapeutick art. They have established no public schools for the study of medicine, nor does the pursuit of it lead to honours, rank or fortune.'

[40] [See Iskander (1988) and Nutton (1988). Sir Geoffrey Lloyd agrees with Iskander that this is an authentic writing of Galen (personal communication). – Editor]

(d) THE ORIGINS OF IMMUNOLOGY

(1) INTRODUCTION

It will be generally allowed that inoculation for smallpox was the beginning of all immunology, one of the greatest and most beneficent departments of modern medical science. In what follows we propose to show that the practice can be documented a good deal earlier in China than in any other civilisation (i.e. from about +1500), with a weighty tradition taking it back much earlier still (to about +1000). Its numerous appearances in less developed societies spread widely over the Old World may not unreasonably be interpreted as emanating from the Chinese focus.

Since the beginning of the +18th century every Western historian of epidemiology and public health has known that something important happened in this connection long ago against an East Asian background, but almost no scholars have made it their business to dig the facts out of the Chinese literature. We often meet with situations of this kind. For example, modern historians of rocketry and firearms acknowledge that the Chinese over half a dozen centuries developed the first chemical explosive known to man, but no one before 1986 drew out for the benefit of the whole world the treasures which Chinese scholars have uncovered in the texts.[1]

It is good to begin with an interesting paradox. In Section (b), we drew attention to a basic dichotomy in the history of medical thought: aid to the healing and resisting power of the body, as opposed to direct attacks on invading influences. Traditional-Chinese and modern-Western medicine differ fundamentally in the relative value given to these conceptions. In Europe, especially since the time of Louis Pasteur and the beginnings of bacteriology, the notion of the direct attack on the pathogen has tended to dominate, culminating in the sulpha drugs and the antibiotics. Reliance on the *vis medicatrix naturae*, the natural healing power of the body, was the mainstay of occidental therapy before modern times. It could act not only against invasive micro-organisms or other parasites entering by infection or contagion, but also against malfunctions of the body's own organic machinery. This spontaneous resistance became less focal in medicine as time has passed.

Chinese medicine shared both therapeutic strategies, which it called *kung* 攻, 'attack', and *pu* 補, 'replenishment'. A physician might choose to repulse malign or sinister *chhi* (*hsieh chhi* 邪氣) from the environment, infective or meteorological.[2] Some of these were the essential pneumata (*ching chhi* 精氣) of harmful and venomous animals left behind on food afterwards eaten by man.[3] Therapeutic measures which counteracted these

---

[1] See Vol. 5, pt 7.   [2] *HTNC/LS*, ch. 79.

[3] *Wai thai pi yao* 外臺秘要, ch. 23, p. 11b (p. 641). [A given *chhi* is not a distinct substance. It is a kind of the universal *chhi* – *ching chhi*, *hsieh chhi*, etc. – according to its function with respect to the material or creature of which it is a part or on which it acts. See Sivin (1987), pp. 46–53. – Editor]

invaders were termed 'expelling the bane' (*chhü hsieh* 去邪) or 'dispersing the poison' (*chieh tu* 解毒). On the other hand, the Taoist conception of *yang shêng* 養生 or 'nourishing the vitality' amounted to strengthening the natural healing power of the body. Just as care of the patient, adequate nursing through the crisis, and administration of drugs in innumerable combinations attacked the pathogenic *chhi*, acupuncture and moxa gave the body more defensive strength to resist and throw off pathological agents.

Another pertinent idea was expressed in a common phrase of medical authors, *i tu kung tu* 以毒攻毒, 'using poison to combat poison'. *Tu* meant 'active principle' as well as 'poison'.[4] The thought must have been compelling for those Taoist medical adepts, whoever they were, who first thought of conferring permanent immunity from small-pox on a young person by a small inoculation of the poison itself. They must have consciously conceived of their technique as aggressive or combative, though unconsciously (as we now know) what they were doing was protective, immeasurably increasing the individual's resistance by building up a store of antibodies.

To be sure, their aggression was against an illness which had not yet appeared. That again was entirely in line with a medical conviction which started very early in Chinese history, the conviction that preventive medicine was the best (Section (*b*)). With that in mind, it is hardly surprising that China should have been the culture where we find the earliest evidences of preventive inoculation.

The search for immunity must have arisen from the ancient folk observation that nobody suffered from smallpox twice in a lifetime. In regions where it was endemic, however, everybody was due to get it once. It was one of the gates of life that children, or sometimes adults, had to pass through. One could advisedly pray for a mild attack and not too much scarring.

On one visit to the cave-temples of the Thousand Buddhas (Chhien-fo-tung 千佛洞), near Tunhuang 敦煌, I well remember finding a cave where the country people had pasted up pieces of yellow paper along the processional circumambulatory way round the central group of statues, where of old the monks would pass chanting their sūtras. Each paper bore the character *kuan* 關, 'gate',[5] and there were the names of the diseases as well, for example one for cholera, one for chickenpox, one for whooping-cough, and of course one for smallpox. Each disease that might be expected had its gate. No doubt the children were taken there to make the rounds, with a station at each flag where the resident Taoist would say the appropriate prayers.

Accordingly, with the background of preventive medicine in mind, it would have occurred naturally enough to some Taoist physician that if one could instil or 'engraft' the disease artificially in a very mild form, somehow gently, ensuring a lenient attack, then the patient would have 'got it over'. That gate at least would have been success-fully traversed.[6] He or she could not have had the remotest conception of all that was

---

[4] See below, p. 129.

[5] Cf. the expressions *kuan thou* 關頭, a critical juncture, or *kuan khou* 關口, a critical point in a given situation.

[6] Originally the heart of the technique must have been to keep the infective dose as small as possible, and to apply it when the patient was in the best of health and vigour. But later on, as we shall see – and much earlier in China than in Europe – it was realised that the virus dose could be 'attenuated' (pp. 143ff.).

being set in motion thereby, for the concepts of antibody formation and active immunity were as yet far ahead in the womb of time.

It may occur to the historian to wonder why this happened first with smallpox and not with any other of the exanthematous diseases. The answer lies near to hand. Smallpox produces pustules with an abundant content of infectious lymph readily available for transfer. These when they die down are covered with a scab which is also rich in variola virus particles. Centuries later immunologists would make 'vaccines', killed or living, sera and antisera, for many other human and animal diseases. All would require a much more sophisticated methodology than sufficed for the first smallpox inoculations.

Continuing this introductory *tour d'horizon*, we may proceed from the known to the unknown, and take a look at the coming of smallpox inoculation to Europe.[7] The first that anybody heard of it was in letters from China to the Royal Society just before +1700 (by which time it had reached Russia). But no one paid much attention to them, nor to letters from Jesuits in China later in the century.

It seems likely that at some time during the +17th century the technique spread from China to the Turkish regions, and shortly after to Europe. It is well known that Lady Mary Wortley Montagu (+1689 to +1762), the wife of the British ambassador at Constantinople, allowed the technique to be used on her own family in +1718 (p. 146). Details of the process were published in Western memoirs by +1714. Variolation or 'engrafting', as it was often called in the West, was introduced to Europe between that date and about +1721.[8] We shall see that the technique originated in China in the +15th or +16th century, or earlier still.

This transmission set the stage for a whole century of inoculation, first in England and America, then more slowly in France, Germany and the other countries of the continent. The appalling ravages of smallpox – no other description is adequate – were thus for the first time checked. At the end of the century, in +1796, came the discovery of Edward Jenner that cowpox lymph, which was not dangerous for man, would give considerable protection against smallpox itself. Thus the familiar vaccination came into being.[9]

Jenner's discovery retains, and always will retain, a high importance in the history of medicine. It is not quite the isolated peak of scientific achievement that many writers have thought it to be, for a number of reasons.

(1)   Smallpox inoculation was not really so dangerous as has sometimes been supposed. It is true that if patients were not isolated it could perpetuate disastrous foci of infection in the population. But mortality as a result of variolation was not nearly as great as some have represented, and the technique became a great deal safer in the course of the +18th century.

---

[7]   The best narrative is the classical monograph of Miller (1957), a work to which we shall often have occasion to refer. It is briefly and clearly summarised in Langer (1976). On Russia, see p. 149, below.

[8]   Cf. the papers of Blake (1953) and of Stearns & Pasti (1950).

[9]   Razzell (1977a), p. 8. Jenner actually had predecessors, and knew about some of them, such as Benjamin Jesty (see p. 150, below). The term 'vaccine' came later to be used, by extension, for many 'biologicals' which had nothing at all to do with cows. Kahn (1963) has reckoned that the world death-rate from smallpox for the +18th century was no less than sixty million people; see also Crosby (1993). At the most modest estimate, early vaccination reduced mortality from the disease by a factor of at least ten; Henderson (1976).

(2) Inoculation powerfully reduced mortality from smallpox during the +18th century. Historians claim that its demographic effects can be demonstrated.

(3) Inoculation permanently protected against the disease, since the active immunity gained was so strong, but vaccination, contrary to Jenner's original belief, has to be repeated every few years.

(4) Jenner's cowpox lymph was quite soon 'contaminated' with variola, so that mixed inocula were given in many thousands of cases.

(5) His intervention led to something which he could never have anticipated, namely the creation of vaccinia virus. This virus has no known natural host. It survives only in vaccine institutes, where it is bred in animals or cultivated on the chorioallantoic membranes of hen's eggs. Serologically the three viruses are quite distinguishable. The most likely of several contending views is that vaccinia was a genetic hybrid between the cowpox virus and the human smallpox virus, variola.[10]

The general upshot of all this is that inoculation was not the crude and dangerous 'folkloristic' predecessor of vaccination, but the first step towards that vast armamentarium of vaccines and sera, antitoxins and toxoids, which mankind now possesses, with all its accompanying developments in the new science of immunology. Today it is based on, but it did not originally stem from, virology and bacteriology.

Reconstructing how early inoculation and vaccination evolved is difficult. The physicians of those days were not able to write with such precision as became usual afterwards, their practices were often not meticulously recorded, and there is no possibility today of examining the strains of virus which they used. Statistical information is also uncertain and imperfect, often available only in local records fitfully kept, so that it is generally not possible to be sure of the effects of the various procedures. But all this must not deter us from fitting together as carefully as possible a picture of the events which took place at the birth of immunology.

The history of smallpox is obviously indispensable. Only too many medical historians have airily said that 'it has been known for countless centuries', but in fact a disease can be identified for certain only when it has been specifically described. Galen did not accomplish this. The great Baghdad physician and alchemist al-Rāzī about +900 left a good account. Again significantly, perhaps, he had been anticipated in China by Ko Hung 葛洪 (ca. +340), whose statement was subsequently amplified about +500, as we shall duly see.[11] Not long after +500 we find detailed descriptions of the papules. For instance, in the first systematic treatise on diseases and their aetiology, *Chu ping yüan hou lun* 諸病源候論 (+610), under the heading of 'Shang han têng-tou chhuang hou 傷寒登豆瘡候' (Symptoms of 'chalice papules' in febrile disorders), Chhao Yüan-fang 巢元方 wrote:

When hot poisonous *chhi* is preponderant in febrile diseases, chalice-shaped papules, white or red in colour, appear on the skin. If they are raised and contain whitish pus, the toxin is relatively weak. If they carry a purple or black coloration and form a root, indistinctly visible within the

---

[10] Crosby (1993), p. 1013, sums up the debate. In culture the viruses are very labile, readily exchanging DNA with each other. Cowpox itself may have been a rodent virus passing through cows.
[11] See p. 125.

flesh, the toxin is more serious. In the worst cases, lesions occur in the viscera, as well as round the seven orifices of the human face. Since the pustules (with their 'roots') are shaped like footed, covered serving dishes, one of the names of the disease is 'chalice lesions' (*têng-tou chhuang*).[12]

The first known monograph on therapy for childhood eruptive diseases, the *Hsiao êrh pan-chen pei chi fang lun* 小兒斑疹備急方論 (+1093), does not make a point of distinguishing smallpox from other macropapular rashes, but describes its development in great detail. Only a little over a century later came the first book entirely devoted to smallpox, the *Tou chen lun* 痘疹論 (Treatise on smallpox, +1223) of Wên-jen Kuei 聞人規. Often reprinted until the Yüan period, it opened a floodgate of publication. From that time until the end of the empire, 441 books on the same subject are extant.

Many medical historians have claimed that inoculation 'was practised for untold centuries as a folk custom'. This assertion rests upon what we may call modern ethnological evidence from Central and Western Asia and many parts of Africa, as well as European information supposedly pre-dating the introduction of inoculation in China. These data need to be tested against what the Chinese texts reveal.

Documentation in China from the beginning of the +16th century onwards, incorporating a tradition that it had first been practised towards the end of the +10th century, has to be taken seriously. From the earliest days of medicine in China, there were 'forbidden prescriptions' (*chin fang* 禁方), confidential remedies and techniques handed down from master to apprentice, among the physicians as well as the alchemists, and sometimes sealed with oaths of blood.[13] Books were also passed down in the same way, as in the case of Pien Chhüeh 扁鵲 (−6th century), whose master Chhang Sang Chün 長桑君 conferred upon him private scrolls with warnings (*chieh* 戒) that their contents should not be revealed to uninitiated practitioners.[14] In early times there had been a strong element of taboo about these 'forbidden prescriptions', together with the conviction that injudicious disclosure would make the medicine ineffective.[15] Of course this social situation lent itself to abuse by mystagogues and quacks chiefly interested in making money, but of the existence of secret traditions there can be no doubt. Particularly where a technique was somewhat dangerous or daring, they would have applied with particular force.

From the early +16th century onwards there grew up in China a specialist literature easily identifiable because its titles usually begin with the words *chung tou*, 'Transplanting the smallpox', instead of *tou chen*, 'Smallpox, measles and chickenpox'. The secrecy was breaking down, the technique was becoming widespread, even entering royal and imperial families, just about two centuries before the spread of smallpox inoculation in Europe.

---

[12] Ch. 7, p. 44.1. A *têng tou* is a long-stemmed serving dish, rather resembling a chalice, with a convex cover. Tamba no Motohiro 丹波元簡 (*1809*), ch. 2, p. 8a, argues that *têng* is a misprint for *wan* 豌 (garden pea). Chhao's treatise still borrows the old *tou* 豆 for 'soybean' to write 'smallpox', rather than the specialised word *tou* 痘 that eventually took its place. Note that generally *tou* and *tou chen* do not strictly correspond to smallpox, but include other eruptive diseases from which it was not distinguished, such as scarlet fever and chickenpox. Our translations of titles are therefore best understood as referring to 'smallpox and related diseases'.

[13] *Pao Phu Tzu nei phien* 抱樸子內篇, ch. 4, p. 5b., tr. Ware (1966), p. 75.

[14] *Shih chi* 史記, ch. 105, tr. Bridgman (1955), pp. 17–18; Nguyên Trân-huán (1957), p. 60.

[15] *I hsüeh yüan liu lun* 醫學源流論, ch. 1, p. 120.

Besides, if we accept the tradition going back to the Sung, there had been eight or nine centuries for this bold exercise in preventive medicine to spread out in all directions over the Old World and Africa. This is just what we think it did.[16]

An interesting problem arises with regard to the method used. In China it generally involved implanting the contents of the pustule, or more often the scab extract, wrapped in a pledget of cotton-wool, into the nose, so that the nasal mucous membrane was the point of entry. It shows great acumen on the part of the Chinese physicians to have guessed that the respiratory tract was the normal route of infection. In the cultures between China and the West, as also in Africa, scarification and introduction of the lymph into the epidermis was the commoner method. We shall return to this subject.

Another matter which we must take up is the various theories developed to explain the nature of smallpox – and indeed many other epidemic diseases as well. One finds such an extraordinary similarity between the Chinese and the European ideas that it is hard to believe there was no intellectual contact or interchange. Broadly speaking, there were two possibilities. The 'morbific agent' may be internal to the patient, a matter of intrinsic predisposition. It may also be external, a matter either of the airs or seasons, at times unhealthy or even mortally poisonous, or of invisible malign animalcules in the surroundings of human beings, liable to break out from their hiding-places when the conditions were just right. These three possibilities could be called the genetic, the meteorological and the contagional, respectively. Let us examine them in turn, first in Europe and then in China.

Many +18th-century Western medical writers supported warmly the theory of the 'innate seed' of smallpox. They supposed that some inherited contagion from the maternal blood, some virus, venom or ferment latent there, was destined to burst into the flower of smallpox whenever the conditions were favourable. Sooner or later every individual would have to go through it. It was as if something sinister, something almost like 'original sin' inside each person, was struggling to get out, or needed to be expelled. Many physicians opined that luxurious living and too rich a diet exacerbated this tendency.[17]

Chinese medical writers concurred, without ever having the slightest idea, so far as we know, of what the Western doctors were thinking. The Chinese theory involved what was called *thai tu* 胎毒, literally 'womb poison', within the child and due to come out sooner or later. The very late metaphor of a flowering plant was also telling, because among the names of the disease was *thien hua* 天花, 'flowers of Heaven',[18] a phrase mirrored etymologically in the term 'exanthematous', from a Greek root for 'flowering'.

On the other hand, in Europe many other authors supported a meteorological explanation, believing that unseasonable weather released 'morbific seeds' or 'putrefactive effluvia' into the human environment so that smallpox resulted.[19] A perfect balance

[16] We take up this topic on pp. 154ff.

[17] For Western views of the aetiology of smallpox see Miller (1957), pp. 241ff., and Ranger & Slack (1996).

[18] 'Heaven' in China meant the whole order of Nature, not only the visible heavens.

[19] Hence the intriguing name of 'seminarists'. The papers of Singer (1913) and Singer & Singer (1913, 1917) and the books of Rosen (1958) and Bulloch (1930/1938) discuss the connection with seed doctrines such as that of the Stoics, as well as the connection between epidemiology and the knowledge of fermentation and putrefaction.

in the elements of the circumambient air, those *eukrasias aeras* which the Liturgy of St John Chrysostom (*ca.* +6th century) prays for, was needed for health. An imbalance gave rise to epidemic diseases such as smallpox.

The idea went back, of course, to Hippocrates, with his 'Airs waters places'. Its most prominent Renaissance advocate was Guillaume de Baillou (+1538 to +1616). This French physician was the first to describe whooping-cough, and introduced the idea of rheumatism, in his book *Epidemiorum et ephemeridum*, published posthumously in +1640. Later Thomas Sydenham (+1624 to +1689) supported the same conception, introducing a long-lived phrase, 'the epidemic constitution'. It became the watchword of the atmospheric-miasmatic school as it battled with the contagionists in the 19th century.[20]

One finds in China a somewhat similar idea, that of a miasmal *chhi* (*chang* 瘴), 'bad air' in the original sense of the Italian word 'malaria'. This deadly pneuma was responsible for malaria and other infectious diseases of the south. As Liu Hsün 劉恂 put it *ca.* +900, 'The mountains and rivers of the far south are twisted and jungly. [The vaporous *chhi*] knots up and concentrates, and is not easily dispersed or diffused. Thus there is much mist and fog to cause pestilence.' As this passage indicates, Chinese explained miasmata by a stasis in the macrocosmic circulation of *chhi*. It corresponded to the stoppage that, in the bodily microcosm, was the most frequent cause of disease. This does not resemble the usual European belief that miasmata were caused by decay.[21]

The third aetiological theory, that of the *contagium vivum* or *contagium animatum*, 'atoms, corpuscles, bodikins in the ayre', decisively alive, had no parallel in China. Out of this indeed, after many vicissitudes, arose the germ theory of disease.

The turning point for the idea of contagion was the posthumous publication in +1546 of the treatise of Girolamo Fracastoro (+1478 to +1533) entitled *De sympathia et antipathia rerum, liber unus; de contagione et contagiosis morbis et eorum curatione, libri tres*. It was a landmark in the history of pathology. Fracastoro was a 'seminarist' because he believed in the existence of widely dispersed seeds of disease. He also believed that each was specific, and above all he believed that they were alive. He distinguished between a poison which cannot multiply itself and an infection which can do so. The seeds were transmissible and self-propagating. Infection was the cause, epidemic disease the result. Fracastoro also distinguished between three kinds of infection, by direct contact from person to person, by carriage through the air at some distance, and through intermediate objects contaminated by the patient. In the following century, with the work of Athanasius Kircher, *Scrutinium physico-medicum . . . pestis* (+1658), and the immortal letters of Antoni van Leeuwenhoek to the Royal Society (+1673 to +1724), the ground was firmly prepared on which modern bacteriology would later build.[22]

---

[20] See Rosen (1958), pp. 103ff. On the idea of *krasis*, or perfect balance in the constituents of things, with its successor terms in Arabic, see Vol. 5, pt 4 *passim*.

[21] *Ling piao lu i* 嶺表錄異, ch. 1, p. 1a, tr. Schafer (1967), p. 130, mod. Schafer quotes Hoeppli (1959), p. 274, to the effect that the agent of malaria was said to be 'produced locally by decaying plants and animals'. This is a modern datum from Yunnan or Kweichow, and has no counterpart in Thang sources.

[22] Wright (1930) has translated Fracastoro's *De contagione*. Singer & Singer (1913, 1917), Goodall (1937), Bulloch (1930) and Colombero (1979) have examined the book closely. The ancient idea of sympathies and antipathies was of great importance for the history of the notion of chemical affinity; Vol. 5, pt 4, pp. 305ff.

So far as we can see at present, there was nothing quite like *contagium vivum* in China.[23] The classical term for epidemic disease was *i li* 疫癘. Either of these two characters could be combined with that for the omnipresent pneuma as *i chhi* 疫氣 and *li chhi* 癘氣. The graph *i* 疫, characterised by its 'disease' radical, is cognate to *i* 役, to serve involuntarily. *Li* 癘 combines the same radical with the character denoting 10,000, again perhaps a reference to the number of patients contracting or succumbing. *Tou* 痘, the term for smallpox itself, was obviously derived from *tou* 豆, a bean, because of the pustules. *Jan* 染 means primarily 'to dye', secondarily 'infection'. The common modern phrase *chhuan-jan ping* 傳染病 does not occur in the classical Chinese literature.[24] *Jan* itself refers to infection, as in the following passage from Ko Hung's *Pao phu tzu nei phien* 抱樸子內篇 (Inner chapters of the Preservation-of-Solidarity Master, *ca.* +320):

Man exists in the midst of *chhi*, and *chhi* is within him as well. From Heaven and Earth to all things within them, none can live without *chhi*. He who knows how to circulate the *chhi* internally can nourish his body and repel all external evils; ordinary people rely on this [circulation] daily without being aware of it.

Among the people of Wu and Yüeh there is a clearly effective method of exorcism (*chin chou* 禁咒) which renders those who practise it more vigorous. He who knows it can pass safely through the worst epidemics, and even share a bed with a sick person, without himself being infected. And several dozen of his followers similarly can be rendered free from fear. That is what mastery of the *chhi* can do to exorcise (*jang* 禳) natural disasters.[25]

This passage displays not only Ko's strong belief in the efficacy of Taoist respiratory techniques, but a clear understanding of person-to-person infection. Ancient and mediaeval Chinese writings recognise infectivity. That is evident from one of their methods of 'inoculation', namely enveloping a child in cloths or clothes which had been worn by a smallpox patient. Here is how Chang Lu 張璐 (+1695) explains it:

If you are unable to take [lit. 'steal'] lymph from the pustules, you can use scabs to culture the inoculum. If there are no scabs to be taken, you can obtain clothing from a child who has just developed smallpox and give it to another child to wear; it too will develop smallpox. The point is to employ a similar pneuma (*chhi*); inchoate though it be, it can serve to guide out the womb poison.[26]

What seems to be missing is the idea of living particles. It is essential to recall that Chinese natural philosophy and science were perennially averse to the idea of particles at all. Atomism must have been introduced many times, as by Buddhist monastic

---

Contaminated objects were called fomites, from Latin *fomes* (tinder). The idea of indirect contagion via fomites was first clearly stated in the writings of the School of Salerno (+10th or +11th century); see Klebs (1913a), p. 70. Kircher, the polymathic Jesuit, made some microscopical observations. He claimed to have seen animalcules in the blood of plague patients, but they were most probably just erythrocytes. On Leeuwenhoek, see the classical monograph of Dobell (1932), and Hall (1989). On believers in 'invisible animalculae', 'venomous corpuscles', 'noxious animated atoms', etc. during the +18th century, see Nutton (1990).

[23] But see p. 130, and Hsieh Hsueh-an 謝學安 (*1983*).   [24] *Thien hsing ping* 天行病 occurs instead.

[25] Ch. 5, p. 5b. On respiratory techniques see Vol. 5, pt 5. [The sentence *to chhi êrh* 多氣耳 is evidently missing at least one graph. By analogy to the only similar passage in the book, ch. 15, p. 1a, the lacuna is perhaps a *li* 力 after *chhi*. – Editor]

[26] *I thung* 醫通 (+1695), ch. 12, p. 28. Much earlier, Tsan-ning gave directions for disinfecting clothing with steam in his *Ko wu tshu than* 格物麤談 (+980); Vol. 5, pt 4, p. 315. See also p. 142, below. [Note that no ancient or mediaeval source is cited for immunisation by clothing. – Editor]

philosophers from India, but it never gained a footing. Chinese thought remained invariably faithful to a prototypic wave theory, the rises and falls of yin and yang, with a conviction of the reality of action at a distance in a continuous medium.[27] In Europe, the Stoic seeds, grounded in ancient atomist thought, made the idea of infective particles, and then living infective particles, natural. In China there was no analogue. Perhaps the intellectual turmoil of the Renaissance, an upheaval not paralleled in China, had something to do with the new perspectives which Fracastoro set forth.

The miasmatic anti-contagionist theories survived for a very long time, perhaps in part because they provided a rationale for public hygiene.[28] In spite of Spallanzani and all the animalculists of the +18th century, the 'epidemic constitution' dominated during the first half of the 19th.[29]

Ackerknecht (1948) has well said that 'it was shortly before their final and overwhelming victory that the theories of contagion and the *contagium vivum* experienced the deepest depression and devaluation in their long and stormy career; and it was shortly before its disappearance that anticontagionism reached its highest peak of elaboration, acceptance and scientific respectability'. The miasmatic meteorological theory inspired many of the early sanitary reformers such as Edwin Chadwick and Southwood Smith, because it justified their campaigns for healthy air, pure water supplies, and adequate drains and sewerage. Anti-contagionism coincided interestingly with the rise of liberalism and the idea of individual freedom as it was understood in developing capitalist society. What was needed was to clean up the community, not to quarantine and otherwise restrict the movement of people and goods; to abolish pathogenic effluvia, not to support bureaucratic controls.

The triumphant rise of modern bacteriology after 1860 put an end to effluvia. But Hermann Boerhaave had got it right when he said about +1720 that after people have smallpox something must remain in their bodies which overcomes subsequent contagious infection. Nobody paid any attention to this except Kirkpatrick, who agreed in +1754 that the disease 'left some positive and material Quality in the Constitution'.[30] More than a century had yet to pass before antibodies were recognised, showing that this brilliant guess had hit the mark.

From those first simple steps taken in mediaeval China flowed developments of incalculable importance, formerly unheard-of protection against the terrifying activities of human micro-parasites. The essential feature in Jenner's work was the inoculation of an animal virus related to the one that endangered man. Cowpox was relatively harmless, but conferred substantial protection.

---

[27] Vol. 4, pt 1, pp. 3–14. [The yin–yang notion was one of temporal and spatial alternation, but calling it a wave theory, even a prototypic one, does not reflect early Chinese ideas of it. One may well doubt that the wave-particle distinction is pertinent. More suggestive is that of continuity *vs.* discontinuity, which Sambursky (1956, 1959) gainfully applied to early Greek physics in ways that invite comparisons. It would, for instance, range Chinese natural philosophers with Aristotle and against Democritus. – Editor]

[28] An example could be drawn from John Caius' college in Cambridge, where we worked for so many years. *Ca.* +1566 he provided that the south side of Caius Court should forever be kept open and not built upon, for the greater health of the residents.

[29] Rosen (1958), pp. 278, 287ff.

[30] Boerhaave (+1716), vol. 5, p. 508, noted by Miller (1957), p. 263; Kirkpatrick (+1754), pp. 29, 30.

Curiously enough, this was not destined to be the main theme in subsequent immunology. Only in a few cases did heterologous vaccines come to have interest. For example, a killed vaccine of endemic murine typhus caused by *Rickettsia mooseri* has been used to give protection against epidemic louse-borne human typhus caused by *R. prowazeki* (Castaneda).

What came to dominate was attenuation, the artificial reduction in the virulence of virus particles or bacteria destined for the inoculum. Attenuated micro-organisms were implanted at least as early as the +17th century in China, and during the +18th in Europe. The next step was injecting suspensions of the killed organisms, since it was found that the body would react by forming antibodies just as well in such conditions, and the procedure was generally safer. Since these protein antibodies were just what was most needed to protect human beings against many disease agents, researchers learned to obtain them in bulk from the blood serum of hyperimmunised animals, mostly horses, and thus developed the many antitoxins in use today.[31] Since Ehrlich in 1892, the outcome of their administration has been called passive immunity, as distinguished from the active immunity which results from the patient's own response to the pathogenic organism introduced. Still later it was found that some pathogenic bacteria produce soluble toxins, as in the case of diphtheria and tetanus. These proteins could be used in the same way as the killed organisms, duly producing protective antitoxins. In closely parallel fashion the toxins, if chemically inactivated with alum and other precipitants, would still produce the desired antibodies. These toxoids, like the toxins themselves, could be used either directly in the human subject or indirectly to produce antisera in large mammals. Such an inactivation logically corresponds to the use of killed organisms as against live ones for injection, and hence derives from the old attenuation procedures already mentioned.

These immunological developments began in the 19th century *pari passu* with the growth of bacteriology itself. Perhaps the greatest landmark in post-smallpox prophylactic inoculation was the work of Louis Pasteur (1822 to 1895) on sheep and goat anthrax, in which he attenuated the virulence of *B. anthracis* by heat treatment and carbolic acid (1880). He supposed that cowpox was a form of smallpox which had been weakened or modified by passage through cattle. Therefore one might search for other methods to decrease virulence further. This led him to the attenuation concept.

Pasteur is said to have asked himself whether vaccination could not become a general technique. Inoculation would have been the better word to use, for the pathogens were generally to be the same, not related ones that attacked other mammals. About the same time, Pasteur attenuated the organism of chicken cholera (*Pasteurella septica*) by re-culturing it for some time or growing it anaerobically, and found that it remained fully antigenic. Then came the heroic story of hydrophobia, and Pasteur's drying of the spinal cords of experimental animals so that the rabies virus was attenuated and could be used for inoculation (1885).[32]

---

[31] Parish (1965), pp. 325, 136ff. Careful proteolytic treatment was needed to prevent serum allergic reactions.

[32] *Ibid.*, pp. 2, 43ff., 53. To this day we do not know whether Pasteur's supposition was right, because natural cowpox seems to have died out (as has horsepox), at least in Western Europe. It is possible, though rather difficult, to infect cattle with variola.

After that the development of immunology was explosive, as all the main possibilities obvious today were explored and used. Pasteur's live attenuated virus vaccine for rabies has had parallels in that for poliomyelitis (Sabin) and yellow fever (Sawyer). His live attenuated bacterial vaccines for sheep anthrax and chicken cholera have similarly had their successors in the combat against human cholera (Ferran and Haffkine), tuberculosis (Calmette) and plague (Strong). Killed bacterial vaccines were first successful for hog cholera (Salmon and Smith), for human cholera (Kolle), for plague (Haffkine) and for typhoid (Wright). Next came the use of cell-free and virus-free toxins to induce the formation of antitoxins. The latter were then injected, as in the cases of diphtheria (Behring and Kitasato), tetanus (Behring and Kitasato) and gas gangrene (Bull and Pritchett). Finally the use of toxoids or 'detoxicated' toxins joined the other methods for protective inoculation, as for tetanus (Ramon and Glenny) and diphtheria (Glenny).[33]

It would be impossible to compress a history of immunology into the previous three paragraphs. This brief survey merely aims to disengage the principal lines of thought and logic which led, from the first inoculations in mediaeval China, through the Jennerian phase, to the vast expansion and flowering of research on antigens and antibodies during the last and the present century. No field has produced more intrepid investigators and martyrs of medical exploration; no sequence of ideas and experiments has more greatly benefited the human race.[34]

(2) SMALLPOX IN HISTORY

The origins of the variola virus are lost in the mists of the past. The famous elevated pustules on the mummy of the Egyptian King Rameses V (d. −1157) cannot be certainly identified; evidence from early Mesopotamian, Egyptian, Hebrew and Indian writings is ambiguous. Smallpox was not known to Hippocrates, and probably not to Galen or his successors either. The earliest Western references which seem to describe it are those of early mediaeval chroniclers, notably Gregory of Tours (*ca.* +540 to +594). In his *Historia francorum* he tells of a severe epidemic with pustular rashes on the skin which ravaged southern Gaul in +580. But not until the Arabic centuries do we find the first clear accounts. There are passing references to something that looks like smallpox in the *Kunnāsh* or *Pandectae medicinae* written by Aaron the Priest (Ahrūn al-Qass), a mid +7th-century Alexandrian, and in the *Firdaws al-ḥikma* (Paradise of wisdom), a medical encyclopaedia due to Alī ibn Sahl Rabban al-Ṭabarī (fl. +850).[35]

The turning point comes with the great physician and alchemist Abū Bakr Muḥammad ibn Zakarīyā' al-Rāzī (+865 to *ca.* +923), whose *Kitāb al-jadarī wa'l-ḥasba*

---

[33] *Ibid.* A curious parallel for non-attenuated smallpox inoculation is the work of Sewall in 1887, who succeeded in making pigeons immune to rattlesnake venom by increasing sublethal doses of the venom itself. The account by Sherrington (1948) of the earliest inoculations with diphtheria antitoxin in England will long remain a classic of medical literature.

[34] In addition to the books cited above and Parish (1968), see the wide-ranging survey of McNeill (1977).

[35] On Rameses see Ruffer & Ferguson (1911), but cf. Crosby (1993), p. 1009. McNeill (1977), p. 116, argues that the disease accounted for the Roman plagues of +65, +165, +251 and so on, but this is pure guesswork. *Historia francorum*, V. 8. 14, tr. Dalton (1927). On Aaron see Sarton (1927–48), vol. 1, p. 479, and, on the *Firdaws al-ḥikma*, Mieli (1938), pp. 71–2 and Browne (1921), pp. 37ff.

(On the variola and the measles) is still regarded rightly as a landmark in the history of medical literature.[36] It gave the first clear account of these diseases and their differentiation in the Western world. Smallpox, al-Rāzī believed, arose essentially from an internal cause. Bad menstrual blood not discharged during pregnancy remained within the child, and boiled up in an evil fermentation as the body became less moist during growth and puberty. Thereafter all medical treatises among Arabs, Greeks and Latins alike, had something to say about smallpox.[37] Our understanding of its history remains much as Marchamont Needham in his *Medela medicinae* of +1665 stated it:

In the dayes of Hippocrates and Galen, the Small-Pox and Measels were either altogether unknown, or else so light and easie, that they were never reckoned as particular diseases among the Grecians, but look't on as Accidents only, and Critical Eruptions supervening putrid and Malignant Fevers: And in after-time we hear no such News of them, till the Arabians began to describe the Small-Pox as a Disease distinct from others; but then they were very gentle, and thus continued till about 40 years agoe, and less. Just so, Riverinus tells us, it was in the West-Indies till the Spaniards arrived there, and then it hapned that a Black Slave sickning of the Small-Pox with Pestilential Symptoms, the maligne and venomous quality thereof being communicated by contagion to others, the Distemper began to rage at such a height, that a great part of the Indians were destroyed by it, though so milde before, that it was not reputed worth the Trouble of Physick or Physicians.

In China smallpox was accurately described several centuries earlier than in Islam. The key passage occurs in the *Chou hou pei chi fang* 肘後備急方 (Handy therapies for emergencies) finished by the great physician and alchemist Ko Hung 葛洪 about +340, and revised by Thao Hung-ching 陶弘景 (also a great physician and alchemist), in the close neighbourhood of +500. What it says is this:

Recently some people have suffered from seasonal epidemic eruptions which attack the head, face and trunk. *In a short time they spread all over the body. They look like fiery [red] boils* (huo chhuang 火瘡), *all containing a white fluid.* The pustules arise all together, and later dry up about the same time. If they are not treated immediately, many of the more severely afflicted patients will die in a few days. Those who recover are left with purplish or blackish scars, the colour of which takes years to fade. *This is due to a severely poisonous chhi. People say that it first appeared from the West in the fourth year of the Yung-hui* 永徽 *reign-period, and passed eastwards, spreading all over the country . . .*[38] In the mid Chien-wu 建武 era [our soldiers] caught it when attacking marauders at Nan-yang 南陽; for this reason one of its names is still 'marauders' pox' (lu chhuang 虜瘡).[39]

The datings in this passage are unfortunately not so simple as they look. Ko Hung's book was lost, probably in Sui times. Only Thao's expanded version of it survived. There

[36] Major (1955), p. 196; Vol. 5, pt 4, p. 398 and *passim*.
[37] About its treatment, that is to say, for until the +18th century none of them could prevent it. On al-Rāzī and his book see Ullmann (1978), pp. 83–4, and Browne (1921), p. 47. The work was translated into Latin in +1565 and +1766; there is an English translation, together with relevant chapters in other books of al-Rāzī, by Greenhill (1848). The ferment was an Aristotelian version of the 'innate seed' and the 'womb poison' of which we have already spoken (p. 119) and to which we shall return (p. 129). Al-Rāzī distinguished, for example, between variola major and minor; Greenhill, *ibid.*, pp. 71ff.
[38] We omit several lines on treatments.
[39] Ch. 2, p. 35.2. The text is reproduced, with slight variations, in *Chu ping yüan hou lun* (+610), ch. 7, p. 44a. Cf. Wang Thao's *Wai thai pi yao* (+752), ch. 3, p. 119.1, and Li Thao 李濤 (1954), p. 189. 'Seasonal' is *shih-hsing* 時行 in Ko's text, *thien-hsing* 天行 in Wang's version; there is no essential difference in meaning.

was only one Yung-hui reign-period in Chinese history, in the Thang (+650 to +655), but that is too late. For the Chien-wu reign-period the case is even trickier, for there were seven eras literally 'established by force of arms' in Chinese history, four prior to the time of Ko Hung's text if it was written *ca.* +340, and two more before that of Thao's recension.[40]

Scholars have taken two tacks in dealing with the first problem, doubtless due to a copyist's error. Fan Hsing-chun 范行準 has suggested that the era should be Yüan-hui 元徽 (+473 to +477) of the Sung, which would make the 4th year +476, during Thao's lifetime.[41] There was fighting in Hunan and around Nanyang at this time. On this basis Fan reasoned that Thao wrote the whole passage.

Ma Po-ying 馬伯英 has more recently pointed out, however, that the version of the passage in Wang Thao's 王燾 *Wai thai pi yao* 外臺秘要 (Arcane essentials from the Imperial Library, +752) quotes separately the accounts of Ko and Thao. This was still possible in the +8th century because Thao's additions were customarily copied in red ink. With the advent of printing, which was ordinarily monochrome, such discriminations were lost. We echo it by italicising in the above quotation the parts that Wang credits to Thao.[42]

Ma has proposed a different emendation for the date, to Yung-chia 永嘉. That opens two options earlier than both Ko and Wang, namely in the Eastern Han (+145) and Western Chin (+307 to +313). Only the second of these eras could have had a fourth year, and the name of the first is debatable.[43] The emendations of Fan and Ma are equally ingenious and equally unsupported by ancillary evidence. It is not possible to choose between the years +310 and +476.

Since Ko recorded the Chien-wu date, we must next consider which of the possibilities up to his time would have been long enough to have a middle. Of the pertinent Chien-wu reign-periods, only those of +25 to +56 and +335 to +348 meet this condition. The former, but not the latter, is supported by known historical circumstances. Ma Yüan 馬援 (d. +49) conquered part of present-day Vietnam in a campaign of +42 to +44, in the middle of the reign-period. His biography notes that by the time the army returned to the capital, 'four or five of every ten soldiers and officers had died of miasmal diseases and epidemics (*chang i* 瘴疫)'.

Ma's astute argument tallies with the opinion of Wan Chhüan 萬全, author of the *Tou chen hsin fa* 痘疹心法 (Heart method for smallpox) of +1549. Nevertheless it remains inconclusive. There is no record of a Nanyang in the area where Ma fought, either in the south or in the far northwest, where he spent the years shortly before the southern

---

[40] These are Later Han, +25 to +56, Western Chin, +304, Eastern Chin, +317 to +318, and Later Chao, +335 to +348. Ko Hung could have meant any of these. Then came Western Yen, +386, Southern Chhi, +494 to +498 and a brief period in the Northern Wei, +529. Ma Po-ying 馬伯英 (*1994*), p. 805, implausibly dates Ko's book *ca.* +303, which would rule out all but the Han reign-period.

[41] Fan Hsing-chun (*1953*), pp. 107ff.

[42] *Wai thai pi yao*, ch. 3, p. 119. Wang attributes part of the description of the papules, the first italicised section as well as the first sentence of the second italicised section, to both Ko and Thao, but this does not compromise Ma's argument.

[43] Li Chhung-chih 李崇智 (*1981*), p. 10.

campaign.[44] We can only sum up by saying that what may have been the first recorded epidemic of smallpox, *ca.* +45, was recorded by Ko Hung three centuries later, and the second such record, of an outbreak in either +310 or +476, was set down by Thao about +500. The priority for the first definite description of the terrible illness remains Chinese.[45]

We might further ask who was the first person said to be pock-marked. The answer may be Tshui Shan 崔瞻, a scholar-poet. His biography says that 'he had had a febrile disorder, and there were many scars (*pan hên* 瘢痕) on his face. But poised, dignified, he was pleasant to look at, and his poetry was gentle and refined. The southerners greatly respected him.' *Pan hên* is a very general term, not a specific one for pockmarks. Tshui Shan lived under the Northern Chhi dynasty and died between +567 and +570. In the late +9th century, we learn from the *Tsa tsuan* 雜纂 of Li Shang-yin 李商隱 that 'girls who had had the smallpox were ashamed of the marks, and did not go out into society'.[46] Perhaps a similarly concrete mention before the time of Thao Hung-ching will yet come to light.

(3) Aetiology and theory in China

The moment has now come to take a closer look at the traditional Chinese theories about the causation of smallpox and its prevention. The two are so closely bound up that they have to be considered together. In classical medical thought, as it developed out of Han themes, illnesses were due to one or other of three causes, the *san yin* 三因. First there was *wai yin* 外因, environmental or external abnormality. It involved a *wai hsieh* 外邪, malign *chhi* in the surrounding world, or a *wai yin* 外淫, *chhi* in the circumambient air characterised by some excess, attacking human beings and inducing all kinds of diseases.[47]

Before the Han the emphasis was primarily on the environmental factor, imagined meteorologically,[48] as in the six *chhi* of pathogenesis. As time went on, environmental was gradually enlarged so as to include everything external, notably agents of specific infection which, though of unknown nature, were a good deal more concrete than the effects of weather. When Chhao Yüan-fang, the great pathological systematist, was writing *Chu ping yüan hou lun* in +610, he spoke of extraordinary infectious agents. What classical medicine calls Warm Disorders

are caused by disharmony in the seasonal phenomena, warm and cool weather out of order. [When this happens] people respond to perverse and violent pneumata (*kuai li chih chhi* 乖戾之氣) by

---

[44] Ma Po-ying (*1994*), pp. 803–6. Ma deals with the latter problem by emending 'Nanyang' to *nan chiang* 南疆, 'the southern border region'. This is ingenious, but can hardly settle the question. Neither Fan's nor Ma's emended characters are graphically similar to the originals, nor are they homonyms. For the southern campaign see *Hou Han shu* 後漢書, ch. 24, pp. 838–40.

[45] Without wishing to detract from the glory of al-Rāzī, we must mention his Chinese connections. He entertained a physician from China for a long while in his home (Vol. 1, p. 219); and Chinese alchemy influenced his work and that of other Islamic alchemists (Vol. 5, pt 4, pp. 388ff.).

[46] On Tshui, *Pei shih* 北史, ch. 24, p. 876. *Pei Chhi shu* 北齊書, ch. 23, p. 336, does not contain the phrase about scars. The term commonly used later was *ma phi* 麻皮. *Tsa tsuan* 雜纂, ch. 5, p. 23a.

[47] The theory of the three causes can be studied at leisure in two books, *Chu chieh Shang han lun* 註解傷寒論 of +1144 and *San yin chi i ping yüan lun tshui* 三因極一病源論粹 of +1174.

[48] The parallel with the European concept of 'miasma' is quite close.

(a)   (b)

Fig. 8. Variations in smallpox symptoms owing to internal abnormality: (a) 'empty' (non-suppurative) pustules in smallpox, owing to an abnormally high or low level of somatic *chhi*. (b) pustules that resemble small nail-shaped boils, a sign of toxic *chhi* that has lain dormant beneath the skin. From *I tsung chin chien* 醫宗金鑑 (Golden mirror of the medical tradition, +1743).

becoming ill. The pathological *chhi* moves on to infect others, even to the point of wiping out a family and spreading outside it. Thus it is advisable to take drugs or have religious rites performed to prevent it.

This implies contagion, even if it does not describe epidemics.[49]

In contrast to environmental causes, there was internal abnormality, *nei yin* 內因, caused by various forms of *nei hsieh* 內邪, internal malignity, or *nei yin* 內淫, internal excess. This was just a way of talking about what modern physicians call organic malfunction.[50] It included all imbalances of yin and yang, all shortcomings in the perfect *krasis* or equilibrated mixture which meant health.

The third division was *pu nei pu wai yin* 不內不外因, causes neither internal nor external; this included all traumatic accidents, such as falls, snake bites or war wounds. We might naturally think of them as external, and the specialists in external therapy (*wai kho* 外科) certainly dealt with them, but 'external' in this classification confined itself to the miasmatic and the infective.

In ancient times there was another expression, *mien i* 免疫, avoidance of epidemic disease. Though it later came to imply immunity, it originally included everything from flight to avoid the epidemic to apotropaic propitiation of disease deities in the celestial bureaucracy to inoculation itself.

Now we must take a closer look at the term *thai tu* 胎毒, literally 'womb poison'.[51] This

[49] Ch. 10, p. 8b. The idea that seasonal abnormalities open the way for Warm Disorders is much older; see *Shang han lun* 傷寒論, Introduction, p. 7. Chhao introduced the idea of specific infectious agents that take advantage of the body's abnormal relation to its environment.

[50] [This is debatable, since a large proportion of *nei yin* were matters of excessive, undisciplined emotion. – Editor]

[51] On womb poison see *Hsiao êrh yao chêng chih chüeh* 小兒藥證直訣 (+1119), ch. 1, p. 12. Fu Fang 傅芳 (*1982*) briefly discusses the concept of *mien i*, and Liu Chêng-tshai 劉正才 & Yu Huan-wên 尤煥文 (*1983*) take up its application in Chinese medicine today.

word referred to congenital pathogens that cause eruptive disorders. As has often been pointed out, in Chinese therapeutics throughout the ages it has implied 'active principle'. One of the basic approaches to therapy was to 'attack' the toxicity of the disease with that of an agent that would just overcome it.[52] The oldest compilation of materia medica, the *Shên nung pên tshao ching* 神農本草經 (Pharmacopoeia of the Divine Husbandman, late +1st or +2nd century), for instance, directs that the dosage of toxic drugs be regulated by systematic trials with small doses. Under the *Chou li's* 周禮 entry for the Imperial Physician, the commentator Chêng Hsüan 鄭玄 wrote about +190, 'medicinal substances generally contain a large amount of toxicity' (*yao chih wu, hêng to tu* 藥之物恆多毒), implying that they are rich in active principles.[53] The pharmaceutical natural histories universally characterise their materia medica as *ta tu* 大毒, very powerful because highly toxic, *hsiao tu* 小毒, mildly toxic and thus active against disease, or *wu tu* 無毒, non-toxic whatever the dose.

The propensity to what Chinese considered congenital eruptive childhood diseases was probably not susceptibility in the European sense. The oldest extant monograph on paediatric disorders explains it without using a technical term for fetal poison: 'As the foetus is within the womb for ten months, it ingests the pollution of the yin pneuma (*hsüeh* 血) of the [mother's] five viscera. After the child is born the poison must come out. The form of the eruptive disorder depends on the fluid associated with the [predominant] viscus.' That is, if the toxicity was associated with the mother's cardiac functions, the eruptions will take the form of measles; if the spleen functions, smallpox, and so on.[54]

The term 'womb poison' (*thai tu*) is not a very old one. Perhaps its first appearance was in the *Hsiao êrh ping yüan fang lun* 小兒病源方論 (Discourse on the aetiology of children's disorders, with prescriptions, *ca.* +1180).[55] The word has been current only since the Yüan period, as in the *Chêng chih yao chüeh* 證治要訣 (Essential oral instructions on diagnosis and treatment), written by Tai Yüan-li 戴元禮 in +1380. Chang Yen 張琰, in the author's preface to his *Chung tou hsin shu* 種痘新書 (New book on smallpox inoculation, +1741) wrote:[56]

Smallpox is a congenital poison (*hsien thien chih tu* 先天之毒) rooted in the conjugation of yin and yang at the very beginning of conception. Once the temporal cycles of susceptibility to illness (*wu yün* 五運) and the seasonal epidemic *chhi* stimulate it, [the disease] will unfailingly break out. If one waits to deal with it until it has broken out, because the epidemic *chhi* is already rampant the symptoms will generally be unmanageable.

---

[52] Here we may recall the formulation of Paracelsus that 'it is the dose alone which decides whether a thing is a poison or not' (*Sieben Defensiones*, +1537). Cf. Vol. 5, pt 3, p. 135. [Note, for example, the statement in *Shêng chi tsung lu* 聖濟總錄, ch. 4, p. 179. As Obringer puts it in the earliest of his several studies of this duality, 'We prefer to keep the translation "toxic" for all occurrences of *tu*, as a reminder that therapeutic activity and toxicity are inextricable . . . , of which it seems Chinese medicine was always intuitively aware'; see Obringer (1983), p. 151, and esp. *ibid.* (1997). – Editor]

[53] Ch. 1, pp. 33–4; ch. 2, p. 1a.

[54] *Hsiao êrh yao chêng chih chüeh*, ch. 1, p. 12. [Chinese often used inclusive reckoning for elapsed time. – Editor]

[55] Ch. 1, p. 2.   [56] P. 2a. Chang inoculated the imperial princes in +1681.

Chu I-liang's 朱奕梁 *Chung tou hsin fa* 種痘心法 (Heart method for smallpox inocula-tion, 1808), written several decades later, expressed much the same opinions.

Authors often claimed that the poison was due to excessive pleasure in sexual inter-course, with its associated ideas of vitiated meconium or menstrual blood, but this belief was not universal.[57] Chinese aetiologists also attributed womb poison to a blood clot or lump of meconium not properly removed from the mouth of the foetus at birth. The sexological interpretation by no means went unchallenged. All physicians agreed on was that the hidden flaw was congenital in every boy and girl. This folk belief has persisted, with much modification and recurrent dissension, to the present day.[58]

Next we must refine the statements about the *contagium vivum* that were made on p. 120. Particles were not characteristic of Chinese natural philosophy, but on the other hand the extended, protean *chhi* 氣 was living matter.[59] It included many sorts of *chhi* that were alive in some sense, but not particulate living 'animalcules' or virulent 'atomies'. There was also the ancient term *chi* 幾, which in one of the commentaries to the Book of Changes means the ultimate minute beginnings of things, out of which both good and evil come.[60] So far as we are aware, the physicians did not adopt this word as a technical term, but one might say that they used *miao* 苗 in its place. The dictionary meaning of *miao* is 'sprout', but our words 'seed' and 'germ' also do it justice. Tradi-tional Chinese medicine knew nothing, naturally, of the germ theory of disease as it appeared in Europe in the late 19th century. Still, *miao* corresponds to the older sense of 'germ' as the form from which a new organism develops. The implication of life is never far off.

Ku miao 穀苗 and *tshao miao* 草苗 were common terms for the sprouts or cotyledons of plants and grasses, but *miao* was also used in animal contexts, as for *yü miao* 魚苗, tiny fish fry just hatched from eggs. But the word came into its own when it began to be used, at least from the +15th century onwards, in the sense of inoculum.

As we already know (p. 118), inoculation was termed *chung tou* 種痘, implanting, or transplanting, the sprouts or germs (*chung miao* 種苗). This linguistic usage is easier to understand when one recalls the procedures of rice farming, particularly the planting out of the seedlings at much larger intervals from each other than they had when sprout-ing in the seed-bed paddy-field (*miao thien* 苗田). An +18th-century pharmacopoeia says while discussing the effects of incense smoke that 'the scabs (*yen* 靨 or *chia* 痂) of the smallpox pustules are called "sprouts" (*miao* 苗), and the outbreaking of the smallpox is called "blossoming (*hua* 花)" '.[61] So also Chêng Wang-i 鄭望頤, in his *Chung tou fang* 種痘方 (Method for smallpox inoculation, *ca.* +1725), now lost, says that those who

---

[57] Ch. 1, p. 1. The connection with carnal pleasure might seem Confucian in its restraint and sobriety, but one senses the presence perhaps of some ancient Manichaean influence.

[58] For twentieth-century field studies, see Topley (1974) and Kleinman (1980), pp. 87–8, 93–4.

[59] Think only of the many kinds of *chhi* in the human body, described by Porkert (1974). He likes to inter-pret them all as different kinds of energy, but we remain uneasy about this; see Needham & Lu (1975). [The last sentence greatly oversimplifies Porkert's interpretation. – Editor]

[60] On the history of this term see Vol. 2, p. 80.

[61] *Pên tshao kang mu shih i* 本草綱目拾遺, ch. 2, p. 31a. The citations that follow are from Fan Hsing-chun (1953), p. 118. Other names for material from natural scabs were *yeh miao* 野苗 (wild transplant) and *huo miao* 禍苗 (disastrous transplant).

choose the *miao* are careful to take the scabs from children who have been inoculated (lit. 'smallpox which has been planted and has emerged', *chung chhu chih tou* 種出之痘). These are the true transplanted sprouts (*chung miao*), in contradistinction from scabs taken from natural or epidemic smallpox cases (*shih miao* 時苗).

Read in the opposite sequence, and with *chung* pronounced in a different tone, the words *miao chung* 苗種 meant various 'species' of sprouts. Indeed a great part of the skill and expertise of the earliest inoculators consisted in selecting or choosing the scabs (*tsê miao* 擇苗, *hsüan miao* 選苗). Yü Thien-chhih 俞天池 (+1727) favours hard thick scabs, with the form of a snail; thin damp irregularly shaped ones are to be avoided. The *I tsung chin chien* 醫宗金鑑 (Golden mirror of the medical tradition, +1743) recommends those which are large and thick, waxy and slightly bluish in colour. Chu I-liang (1808) says that the size matters little, but they should be thick, rounded and of a clear purplish colour.[62]

Authors often warned against using the natural *shih miao* scabs. Chu I-liang specifies mature inoculum (*shu miao* 熟苗) that has been refined (*lien* 鍊) by seven passages through inoculated persons. The best kind of inoculum was also termed pure, *shun miao* 純苗, elixir, *tan miao* 丹苗, numinous, *shên miao* 神苗, or immortals' transplant, *hsien miao* 仙苗.[63] Judging from the later descriptions, these must have meant an attenuated virus, either obtained from the scabs of patients who had already been inoculated, or artificially weakened in virulence by special methods shortly to be described (p. 143). Finally, after Jenner, there was *niu tou miao* 牛痘苗, cowpox sprouts. This appellation shows the continuity between inoculation and vaccination in the minds of the Chinese physicians.

There is something to be said for the suggestion that in their careful choice of the 'best' scabs the Chinese inoculators were selecting for the micro-organisms of variola minor or alastrim, a less serious, rarely fatal disease, as against the major form, a distinction that al-Rāzī may have already noted. The Chinese physicians were aware that there were two types of the disease. *Chu ping yüan hou lun*, the great aetiological handbook of +610, distinguishes between the light (*chhing* 輕) and the heavy (*chung* 重) affliction.[64] The selection of alastrim may not have been conscious, but it was a classic example of using poison to combat poison (see p. 115). As we have seen, the beneficent 'poison' had to be most carefully chosen. All the books give strict instructions not to inoculate when natural smallpox is already within the house, only under suitable conditions of relative isolation some time beforehand.

The theorising in traditional Chinese medicine about the nature of inoculation is surprisingly elaborate. To elucidate this, we have to recall a conception of urinary endocrinology, that of *yin tao* 引導, 'leading something out by the same way that the

---

[62] *Sha tou chi chieh* 痧痘集解, +1727; *I tsung chin chien*, ch. 60, p. 118; *Chung tou hsin fu*, ch. 1, p. 6.

[63] P. 6. Cf. *Chung tou chih chang* 種痘指掌, before +1796. The primary meaning of *tan* was of course elixir; cf. Vol. 5, pts 2–5 *passim*. Its secondary meaning, red, hardly applies here. The term in itself clearly points to the Taoist origins of inoculation (see p. 156). [This interpretation assumes that *tan* implies alchemy, and that alchemy implies Taoism. Most scholars today would find these links unconvincing. See the Introduction. – Editor]

[64] Ch. 9, p. 58b.

guiding agent previously came itself'.[65] Urine or the protein, steroid and other hormones isolated from it, administered as medicine, can 'lead the undue heat (lit. "fire", *yin huo* 引火) which is the cause of the illness downwards to be excreted and got rid of' because the urinary substances had already passed that way themselves. If it was their very nature to follow the path of urinary excretion, they could perhaps combine with another substance or *chhi* which was causing the trouble, and lead it forth. These ideas originated rather early. The *I shu* 遺書 (Posthumous writings) attributed to the physician Chhu Chhêng 褚澄, who died in +501, state them clearly. The works of Sun Ssu-mo 孫思邈 (fl. +673) and Chu Chen-hêng 朱震亨 (+1231 to +1358) express them again.[66]

One of the most explicit inoculators of Chhing times, Chang Yen 張琰, emphasised the importance of leading substances. In his author's preface to the *Chung tou hsin shu* (+1741) he writes:

If one has used excellent inoculum to lead out the yin womb poison, it will be unable to spread freely [through the body], and the symptoms will be manageable without further interference. One may dare call [inoculation] a fine method . . . by which man's designs can wrest from the shaping powers of Nature their control of disease![67] He can hasten good fortune, avoid evil, preserve peace and safety, and escape danger. This technique is a wonderful way in which people can show their love and compassion for children.[68]

In another passage Chang writes more exactly of the mechanisms which he and his contemporaries visualised.

The individual's smallpox poison is congenital, a veritable original taint of disaster (*nieh* 孽). This toxin is stored in the heimartopyle.

What is the heimartopyle?[69] *Ming mên* 命門 is a term with many denotations in traditional Chinese medicine. Earlier it was one of the acu-points (TM 4), but that is certainly not the meaning here. Then it was one of the kidneys, the right one, so named because it was in early times believed to be the repository of vitalities gathered from throughout the body to form the sperm (*ching* 精).[70] Other descriptions make it something heart-shaped and membranous between the two kidneys. Others yet again say that the top was connected to the kidney and the bottom with the bladder; perhaps the ancient anatomists were seeing the suprarenal glands and the ureter. Li Shih-chen 李時珍 in the late +16th century identified it with the pancreas (*i* 胰), of which he gave a very clear description.[71] In sum, like many other things in antique medical thought, both Western and Eastern, it does not have any exact correlate in modern anatomy. We must be content to accept it simply as the site where, according to Chang, the interaction between

[65] Vol. 5, pt 5, p. 311.   [66] Cit. *PTKM*, ch. 52, pp. 16a–19a.

[67] In Vol. 5, pt 5, we collected many examples of this optimistic view of the relations between man and Nature, at first sight unexpected in the days before modern science. 'Shaping powers' usually refers to yin and yang.

[68] Pp. 2b–3a. This work was copied into the imperial medical compendium, *I tsung chin chien* (+1743; see p. 141).

[69] [This is Lu & Needham's coinage for what is usually translated as 'Gate of Life' or 'Gate of Destiny'. – Editor]

[70] *ICK*, p. 1485, citing the *Nan ching* 難經, ch. 36.   [71] *PTKM*, ch. 50, p. 35.

the external and the internal germs or seeds took place. The translation which we use is derived from the Greek roots of *pyle* (πύλη, a gate) and *heimarmenê* (έίμαρμένη, fate or destiny).[72] Chang continues:

Transplanting the smallpox is nothing more nor less than using an external seedling (*wai miao* 外苗) to lead out the internal poison. After the inoculation by blowing, the germ goes to the abdominal region, travelling gradually deeper, each day traversing one of the [six] tracts, and taking six days to reach the heimartopyle.[73] Stimulated by the germ from outside, the internal poison reacts (*tung* 動). After you have counted seven days, the poison manifests itself [in the reaction]. If this happens after the eighth or ninth day, that is because the *chhi* of the inoculated germ is weak, so that its transmission is slow. Hence the smallpox reaction is also a bit retarded. But in normal circumstances it is bound to happen on the seventh or eighth day.[74]

This remarkable passage seems to envisage that the external germ links up in some way with the internal germ and, after producing the mild reaction expected, leads it out through the surface of the body to trouble the individual nevermore. But there is more to it than that.

How now did he think of this interaction? Was it not another form of procreation, engendering the immunity? In the *ming mên* the external seedling encountered the internal poison. The words used of their meeting are very evocative. 'Stimulation' might also apply to the mutual arousal in sexual congress. The seedling from outside would naturally be yang, and the womb poison buried inside is specifically yin.[75] Male and female spontaneously interact to produce permanent immunity as the used poisons are excreted through the skin. The transplanting of the seedling, if of the right species, was a male or yang operation. Introducing the inoculum through the nose might subconsciously have corresponded to sexual ejaculation.

All in all, these accounts show that the medical men practising inoculation in China towards the end of the +17th century had detailed, but not at all modern, ideas about the mechanism of the process. And one may remember that the passage just quoted came from the pen of a man who had himself inoculated nearly 10,000 healthy patients, with (by his account) only twenty or thirty failures.

The last conception which ought to be noticed here is that of *thung lei* 同類, 'of the same category as . . .'.[76] It played a great part in mediaeval Chinese alchemy as well as in pharmaceutics. The inoculation idea was an eminent example of the Chinese maxim that like responds to like. Where smallpox was concerned, categorical affinity affected

[72] The word was cognate with *moira* (μοῖρα, allotted portion), similar to *fên* 分, as we saw in Vol. 2, p. 107.

[73] As we shall see, an aqueous extract of the most suitable pustule scabs was normally inserted into the nostril on a piece of cotton-wool (p. 141), but the dry powdered scabs were often blown in, as in this instance. We use 'germ' in the sense discussed on p. 130 above. The tracts are not those of acupuncture, but rather those of the *Shang han lun* 傷寒論.

[74] Ch. 3, p. 9b.

[75] It is true that Kuo Tzu-chang 郭子章 said in +1577 that the womb poison was fiery yang, but everything would depend on the degree of yang concerned. See Kuo's *Po chi hsi tou fang lun* 博集稀痘方論. [Yin and yang are not absolute qualities. Each is always relative to the other member of the particular pair. Unless the latter is specified, to say that something is yang or yin has no definite meaning. – Editor]

[76] Vol. 5, pt 4, pp. 307ff. discusses the great importance of category theory in relation to the earliest developments in the theory of chemical affinity.

the therapy more than the prevention. It came into play once the disease had established itself and needed palliative treatment until it had run its course.[77] It was of venerable antiquity, for Chêng Hsüan comments with respect to the Royal Surgeons in the *Chou li*: 'fortify the patient [with medicines] appropriate to the category. To fortify means to impose order (*chih* 治), and to manage means to heal and cure (*liao* 療).'[78]

(4) THE EARLIEST MENTIONS OF INOCULATION

Smallpox inoculation came out into the open, as it were, some time during the first half of the +16th century. To see clearly what happened afterwards, the story has to be pieced together from flashbacks and hindsight, traditions reported by medical writers, and statements about the practice of families in which the calling of inoculator had been passed down through several generations. The earliest reference (apart from the Taoist tradition which we discuss separately below, p. 154) seems to be in the book of Wan Chhüan 萬全 on smallpox and measles, *Tou chen hsin fa*, first published in +1549 and reprinted half a dozen times in the Chhing dynasty. Speaking of treatments, he casually mentions that smallpox inoculation is liable to bring on menstruation unexpectedly in women.[79] His book gives no information on the technique. His remark suggests that inoculation was common in his time, even though no one else was writing about it.

By +1727 much was being written on smallpox, including Yü Thien-chhih's 俞天池 *Sha tou chi chieh* 痧痘集解 (Collected commentaries on smallpox) of that year. This was based on the *Tou chen chin ching lu* 痘疹金鏡錄 (Golden mirror of smallpox, +1579) of Weng Chung-jen 翁仲仁. Yü writes:

Smallpox inoculation arose in the Lung-chhing 隆慶 reign-period [+1567 to +1572], especially at Thai-phing 太平 district in Ning-kuo 寧國 prefecture [modern Anhwei]. We do not know the names of the inoculators, but they got it from an eccentric and extraordinary man who had himself derived it from alchemical adepts (*tan chia* 丹家). Since then it has spread all over the country. Even now the inoculating physicians come mostly from Ning-kuo, but not a few people from Li-yang 溧陽 [Kiangsu] have learnt and appropriated it. The strain of inoculum obtained from the strange and eccentric man has been kept and used to this day, but you have to pay two or three pieces of gold for enough to inoculate one person.[80] Physicians who want to make some profit pass it through the children of their own relatives in winter and summer, and have no mishaps. Others eager for money steal away the scabs from [severe] smallpox cases and use the material directly. It is called *pai miao* (敗苗, ruined inoculum). In such cases there will be 15 deaths in 100 patients.[81]

Thus we can say with fair confidence that inoculation for smallpox was a general practice in China in the time of Thomas Linacre, John Caius and Henry VIII, long before the days of Lady Mary Wortley Montagu at the turn of the +18th century.

---

[77] *Tung i pao chien* 東醫寶鑑, ch. 11, p. 655.
[78] Ch. 2, pp. 2a, 3a. Cf. Needham & Lu (1962), p. 436.   [79] Ch. 10, p. 3b.
[80] Here we first encounter the idea of payment for inocula (cf. p. 138, below). It is important because it links with similar practices in proto-inoculation Europe, as we shall see.
[81] Ch. 2, pp. 23aff.

Similarly, the *Chêng tzu thung* 正字通 (Comprehensive guide to correct use of characters), a dictionary published in +1627, has this to say about smallpox:

Smallpox (*tou chhuang* 痘瘡). The formularies attribute it to an innate flaw or womb poison. Some people never get the disease [in spite of this]. A remarkably effective way of dealing with smallpox (*shên tou fa* 神痘法) is to take the liquid pustule contents (*tou chih* 痘汁) and instil it into the nose, so that simply by breathing the patient will get infected with a light eruption [and be protected].[82]

Apart from its early date, this passage is interesting because it explicitly recognises the role of the respiratory passages in the infection.

The next step concerns the Chu family, who practised medicine through several generations. A book of Chu Shun-ku 朱純嘏 (*ca.* +1634 to +1718) entitled *Tou chen ting lun* 痘疹定論 (Definitive discussion of smallpox, +1713) describes inoculation. It was printed in +1713 as an appendix to the *Tou chen chhuan hsin lu* 痘疹傳心錄 (Smallpox: record of direct transmission, +1594) of an older Chu, Chu Hui-ming 朱惠明.[83] It is highly probable that the elder Chu knew and practised inoculation, though living at a time when it was not generally written about.[84] Again, a novel by Chou Hui 周暉 entitled *Chin-ling so shih* 金陵瑣事 (Trifling affairs in Nanking, written +1610) appeared in +1621. This mentions two cases of inoculation during the Wan-li reign-period (+1573 to +1619) in which children were badly infected.

Just a year before the fall of the Ming, Yü Chhang 喻昌 related an experience of inoculation in Peking:

The young sons of Ku Shih-ming 顧諟明 were inoculated for smallpox, and he invited me to watch. The pustule lymph (*tou miao* 豆苗) was faintly pink and clear, with an attractive watery, lustrous appearance. The solid particles in it were scattered like stars in a night sky. The family physician was positive that it was an 'optimum inoculum' [*chuang-yüan tou* 狀元痘, i.e. the best lymph that could be had].[85]

Yü disagreed with the family physician, concluding from his examination that the inoculum had been taken from a donor who had not recovered from a seasonal epidemic disorder. We can glimpse the care which physicians took to choose only the best inoculum (*chia miao* 佳苗) infected with the mildest forms of the disease.

This was about the time, too, when Nieh Shang-hêng 聶尚恆 (b. +1572) was in full activity as physician and inoculator. His book *Huo yu hsin fa* 活幼心法 was printed in +1616. This 'Heart method for saving children's lives' did not mention the inoculation technique.

---

[82] Cit. Fan Hsing-chun (*1953*), p. 117. We also learn from Fan that *chen* 疹, which normally now means measles, was then used with *tou*, especially by the common people, to mean smallpox.

[83] [I can find no record of a +1713 ed. of *Tou chen chhuan hsin lu*, e.g. in *LHML*, item 7720. Lu & Needham believed that *Tou chen ting lun* was written *ca.* +1680, but the author states in his preface of +1713 that, as an Imperial Physician, he wrote the book in the same year for the Emperor Khang-hsi's auspicious sixtieth birthday, auspicious because it completed a sexagenary year cycle. All the prefaces are reprinted in *CIT*, pp. 4353–5. – Editor]

[84] [Since the elder Chu's book does not mention inoculation, the likelihood that he practised it is not high. – Editor]

[85] *Yü i tshao* 寓意草, p. 52b. It looks as if a little blood had been drawn as well as the pustule contents. Yü Chhang's book also contains dietary precepts to help the inoculation to take.

Our next quotation comes from Chang Yen's preface to his *Chung tou hsin shu* of +1741. Tung Han 董含 said in +1653 of Chang's family that its members had practised inoculation for at least three successive generations, which tallies with Chang's claim below that an ancestor had studied with Nieh, presumably early in the +17th century.[86] The whole of Chang's preface, challenging conventional physicians' rejection of inoculation, is worth quoting. It contains many interesting and instructive sidelights on the story.

The works of those who devise ingenious novelties (*chhi* 奇) unknown to antiquity, and set out their unique knowledge to reopen the ears and eyes of physicians, may justly be called new (*hsin* 新). Among the experts in treating smallpox are many eminent and worthy persons. The therapies for smallpox fill whole libraries. [Physicians] shut away in their studies, steadfast in their narrow views, are unable to understand what is written in [such books]. Those who move along the old ruts, bogged down in conventional therapies, know nothing of the change. They simply conform to the old rules.

If we proceed according to the old rules and consider the medical formularies that are abroad, they ignore the doctrine of inoculation. Can it be that although formulae for treating smallpox are to be freely transmitted in writing, techniques of inoculation cannot be spoken of to anyone? It seems that the practitioners are keeping the instructions secret by transmitting them orally, and are unwilling to record them in books. They keep the techniques as private property, and do not wish to make them available for the public good.

I now divulge (*hsieh* 洩) what no one divulges, and transmit what others do not transmit. If this book does not deserve to be called new, what can it be called?

Someone objected: 'Smallpox is a most dangerous disease. When an epidemic strikes, people are terrified, and worry about being unable to flee far away. Apropos of nothing you appear, and want to collect infants and children and inoculate them. This amounts to gathering people without illness and for no good reason infecting them. What is a humane person to think of that?'

I replied that that is incorrect. Smallpox is a congenital poison . . . People may put their cases in the hands of quacks who cannot distinguish disease manifestations and do not understand the functions of drugs. Not only are they unable to restore the health of their dying patients, they blame their failures on mischief wrought by heavenly visitations. This situation has brought pain to my heart and anxiety to my mind. I cannot refrain from expressing admiration for those who can disperse the trouble before it germinates, and make sure of safety before danger arrives. Their merit is enormous. . . .

For several generations, since an ancestor received instruction from Master Nieh Shang-hêng, my family has practised inoculation for smallpox as a hereditary occupation (*chhi-chhiu* 箕裘). I [in turn] studied the writings of my father. I have inoculated [children] everywhere, perhaps 10,000 of them. I have spent many years of hard thinking and sedulous practice on the methods of treating smallpox. I have made the old practices my teacher without being mired in them. I have read the books without putting my faith in books. I have distinguished the manifestations of illness types and prescribed drugs accordingly, making the formula fit the disorder. I have lightened the serious case, made contrary phenomena compliant, changed dangerous conditions to safe, sometimes curing even the incurable.

Alas, I am now old and tired, with no further ambitions in the world. I have therefore set down in a book my lifetime's effort at study. I dare not claim that I have devised ingenious novelties unknown to antiquity. I merely lay my unique knowledge before the public to enrich the arts that

---

[86] *San kang chih lüeh* 三岡識略. Tung was a high official in the Chhing Ministry of Personnel.

express compassion for children, and to refresh the ears and eyes of physicians. For these reasons I call my book 'New treatise on smallpox inoculation'.[87]

From this it is clear that during the latter half of the +17th century a new spirit in Chu Shun-ku and others tore off the veils under which inoculation had been concealed, and published and propagated the method. What a strange coincidence it was that this was the same spirit as that of the contemporary Royal Society, far away in the Western world and quite unknown to Chang Yen and his generation.[88]

Another preface to the book came from the brush of Chang Yü-cho 章玉琢, who tells us that Chang Yen's work was so much appreciated that grateful parents and patients called him 'the Second Heaven' (êrh thien 二天).

When Chang Yen was asked about the original source of his expert arts, this was his reply: 'In the old times when Nieh Shang-hêng was an official in Ning-yang, my grandfather studied under him, and had access to his personal notes on the pox diseases (tou kho 痘科), besides receiving his oral advice. Thus our family has been in practice as pox physicians for three generations.'

His friend Mr Wu again asked: 'As your family has inherited such secrets, why not propagate them for the benefit of all living beings?'

Dr Chang answered with a sigh: 'There exist libraries full of medical writings, yet never a word has been said on the technique of inoculation. This is because of the selfishness of the experts (shu chia 術家) who practise it. They have maintained exclusive knowledge of the art in order to enrich themselves. But I myself have used it for several decades, and accumulated much clinical experience, attending many smallpox cases and seeing all the dangers. I have given the subject much thought, and have administered many prescriptions that proved effective. Now in my declining years I do in fact intend to publish and propagate this technique, so I have prepared this manuscript which includes the details of all that I inherited from my grandfather.'

There may have been other constructions of the universal silence of the early Ming inoculators, as we shall see. So far as this field was concerned, Chinese made the same break with the Middle Ages that occurred at the scientific revolution in Europe. No secrecy, but plain words for all men to read, was the watchword there as it was at Gresham College.

At the end of the +17th century, the I thung 醫通 (Compendium of medicine) gave details of several methods, later included in an official compilation which we shall look at carefully (see p. 140, below). In +1707 the Tou kho ta chhüan 痘科大全 (Comprehensive treatise on smallpox) was printed. The author of this rare work, Shih Hsi-chieh 史錫節, mentions that in +1673 he did a particularly large number of successful inoculations.[89]

We may now examine two sections of Yü Thien-chhih's Sha tou chi chieh (+1727), entitled 'Explaining inoculation' (Chung tou shuo 種痘說) and 'Method of inoculation' (Chung tou fa 種痘法). Yü opens the first by saying that inoculation is now more or less universal. 'Although it is an act [seemingly] going against Nature, it has great merits

---

[87] The two omitted passages are separately quoted above, on pp. 132 and 129.
[88] As we shall see, the Royal Society was greatly involved in the first stages of inoculation in England (p. 146).
[89] Preface, p. 3a.

for the people in the world.' He is not very enamoured of the 'innate seed' theory. 'If it is all put down to the congenital womb poison, why should some cases be so severe and others so mild?' Many things may go wrong, but only one or two out of thousands of inoculated children die. These exceptions are perhaps due to lack of care, not some fault of the inoculator. The latter must be an expert (*miao shou* 妙手), and must use mature inoculum (*lao miao* 老苗) that has undergone several passages through those inoculated. Yü directs the practitioner to insert in the nose overnight a plug of cotton-wool imbued with an aqueous extract of scabs (*yen* 靨) from mild cases. After three days there will be a rash of red patches which will turn to a few pustules, ultimately with white contents, just like naturally occurring smallpox but very mild. If the *chhi* of the inoculum is not absorbed or does not take, as may happen if a child's inborn *chhi* is adequate to resist it (*yüan chhi su tsu* 元氣素足), then after an interval the inoculation may be repeated.[90]

In the second section, Yü enjoins inoculators not to press families who do not really want their services, and not to treat weakly children who already have boils or lesions, or who have had convulsions. Since one cannot always get elixir inoculum or numinous inoculum, one must buy the pure variety from colleagues who happen to have it. Thus the inoculum was a commodity like other medicaments.

Yü specifies snail-like pustule scabs, hard and thick, not flat mis-shapen pustules. Seven scabs should be used for one inoculation, comminuted in milk with a trace of musk; then steep the cotton pledget in the liquid and place the latter in a nostril. It need not be left in more than six hours or so. A week later the fever appears, with lumps on the neck called 'pox-mother' (*tou mu* 痘母).[91] The red rash changes to maculae and finally to a few pustules, which in time give the scabs. These may be used again for other patients. If the congenital womb poison is slight, there will be very few pustules and no need for medicine. If it is severe there will be many, and the child will need treating as for ordinary smallpox.

Yü relates that in +1725 he watched a Li-yang man inoculate. The technique was rough and careless, without precautions. All the children contracted a high fever, with many pustules.

Shih Hsi-chieh [according to Yü] says that one should use only the numinous transplant; this seems absurd, but is actually very acute.[92] It leads out the womb poison and does not permit even a speck of contamination by the epidemic type of the disease to mix with it. Therefore it is absolutely safe. For this reason I say that what kills people is not the womb poison but the epidemic *chhi* [*i li chih chhi* 疫癘之氣].

[90] Ch. 2, pp. 23aff. Normally after successful inoculation there is a little fever, the breathing is slightly affected, and the maxillary lymph glands are enlarged. *Lao miao*, here referring to a 'mature', good transplant, elsewhere could denote 'old' scabs which had lost their activity by too long keeping or other unfavourable conditions.

[91] Ch. 2, p. 25a. The lumps are swellings of the cervical, submaxillary and mastoid lymph glands. The plug is inserted in the left nostril for boys, right for girls, as many of the other books also say. It is to be tied to an ear with red thread. If it is lost, another one may be put in as substitute (*thi miao* 替苗). There is also talk of choosing a lucky day. Yü repeats the theory (pp. 131–2, above) that the inoculum meets the womb poison and leads it out, in this instance via the stomach tract.

[92] Here again the reference is to choosing lymph or scabs from the lightest cases, probably variola minor.

As for the accompaniments of the technique, some people swallow water in which written charms (*fu* 符) have been [burnt and the ashes] mixed, or use incantations (*chou* 咒).[93] These are, no doubt, unorthodox arts of spirit-worship (*shên tao hsieh shu* 神道邪術). Nevertheless, if one examines carefully what these [methods] are to accomplish, it is only to expel the evil influences and to avoid infection. They do no positive harm. There is an unintelligent group of people who simply perpetuate what their teachers said, understanding nothing about medical treatment and not following the changes of the disease. Many children have suffered from them. Above all one needs doctors who know how to choose the best inoculum . . .

During the +18th century there were still many books that discussed smallpox without mentioning inoculation, for example the *Yu yu chi chhêng* 幼幼集成 (Compendium for the proper care of children) by Chhen Fu-chêng 陳復正 in +1750, and the *Fu yu pien* 福幼遍 (For the welfare of children) by Chuang I-khuei 莊一夔 in +1777. These men were, it seems, mainly interested in the treatment of severe cases. Chhen may have typified the conservatives of whom Chang Yen complained; his chapters 5–6 abridge the *Tou chen hsin fa* of two centuries before![94]

There is an interesting work by Chu I-liang 朱奕梁, the *Chung tou hsin fa*. It is difficult to date, for although it was not printed until 1808, internal evidence would put it nearly a century earlier.[95] Chu goes over much the same ground as the other inoculation books, emphasising the choice of material. Indicating in his term *chu miao* 貯苗 (stored inoculum) some knowledge of attenuation (cf. p. 143, below), he advocated many passages through successive patients, as many as seven times. He tells how in former days the technique was 'handed down by oral transmission and memorisation from master to disciple', and expresses the usual admiration at the prolonged immunity that it confers. 'How can one deny that the works of man can exceed those of Nature if only we put our trust in the temporal cycles!'[96]

We learn from Chu that during the +18th century two schools of inoculation evolved,[97] that of Hu-chou 湖州 (Chekiang), who believed in using scabs from epidemic patients, and that of Sung-chiang 松江 (Kiangsu), who were convinced (in the orthodox phraseology which we have already given) that this was most dangerous. They called this *shih miao* by the name of *hsiung miao* 凶苗, ominous germ (*inoculatio nefas*). They used only material derived from the mildest cases. Unfortunately, the Hu-chou views prevailed among the learned editors of the imperially commissioned *I tsung chin chien*.

To end this Subsection we must say something about the Khang-hsi emperor (r. +1661 to +1722). When his father, Shun-chih, the first emperor of the dynasty, lay dying of smallpox, Hsüan-yeh 玄曄 (for that was his personal name), only eight years old, was designated heir-apparent. He was chosen in part, it is said, because, having already survived an attack of the disease, he would be immune to it and therefore likely to have a

---

[93] Note the association with Taoist religious practices. [As the next sentence makes clear, the author is thinking not of Taoism but of popular religion (*shên tao*). – Editor]

[94] Sung Hsiang-yüan 宋向元 (*1958*) pointed this out.

[95] [No evidence is apparent in the source that the book was written long before it was printed. Chu's preface indicates that he was familiar with *I tsung chin chien* of +1741, an imperial compilation not widely accessible outside official circles until the second half of the +18th century. – Editor]

[96] Preface, p. 2; text, p. 4.   [97] P. 5. Chang Yen and Yü Thien-chhih are also aware of the two schools.

long reign, as indeed happened. In later life he used to reproach himself for not having
gone to visit his father on his deathbed, but about inoculation he had good things to
say, as the following passage from his 'Family instructions' shows:

At the beginning of our dynasty, when anyone was attacked by smallpox, people made use of a
thousand precautions, and this disease was very greatly feared. The method of inoculation having been brought to light during my reign, I had it used upon you, my sons and daughters, and
my descendants, and you all passed through the smallpox in the happiest possible manner. The
Forty-nine Banners of the Mongolians, and even the Chiefs of the Khalka people, have made use
of inoculation, and all of them were perfectly healed. In the beginning, when I had it tested on
one or two people, some old women taxed me with extravagance, and spoke very strongly against
inoculation. The courage which I summoned up to insist on its practice has saved the lives and
the health of millions of men. This is an extremely important thing, of which I am very proud.

Old men used to consider it a necessary precaution never to go into houses where the smallpox was. They did not even dare to pronounce its name, just as they abstained from many expressions which they regarded as of bad augury. But nowadays people take no notice at all of that sort
of thing.[98]

These memorable words record another 'first'. The Khang-hsi emperor had the experiments carried out in 1681 or slightly later, anticipating by some considerable time the
famous Newgate Prison experiments in London in +1721 (p. 146). We learn from another
of his writings that he had the regular troops of his army inoculated 'as I did my own
children'.[99]

(5) Methods of inoculation

The official 'Golden mirror of the medical tradition' (*I tsung chin chien* 醫宗金鑑, +1743)
incorporated a chapter on inoculation. One might say that the first of all immunisation
techniques had, after a number of centuries, achieved official recognition. This was strikingly contemporaneous with the public decision of the Royal College of Physicians in
London in +1755 to set the seal of their authority on variolation.[100]

The 'Golden mirror' was superintended by the Manchu official O-êrh-thai 鄂爾泰,
and prepared by a large committee of medical men under the chairmanship of Wu Chhien
吳謙.[101] They drew on earlier works, notably the *I thung* 醫通 (Comprehensive medicine) of +1695.

Four chapters of the work are devoted to 'Essentials of the heart method for treating
smallpox and other eruptive disorders'. Chapter 56 considers in detail the aetiology,
symptomatology and therapeutics of these diseases. Chapters 57 to 58 take up the
complications which are liable to accompany smallpox or appear as sequelae. These

---

[98] Tr. de Poirot (+1783), p. 111, eng. auct. Fan Hsing-chun (*1953*), pp. 125–6, gives the Chinese text. The
last paragraph occurs only in the Manchu text. Four books in the Manchu language on smallpox and inoculation are preserved with twelve on other medical topics in the Library of the Oriental Institute in Moscow.
For this information we are indebted to Dr Hartmut Walravens of Hamburg.

[99] Spence (1975), p. 18. The Manchus, very susceptible to smallpox after they conquered China, instituted many measures to control the disease, as Chang (1996), pp. 171–6, has noted.

[100] Miller (1957), pp. 24, 169–70. Probably neither group of physicians knew of the other's work.

[101] Cf. Hummel (1943–4), pp. 601ff. O-êrh-thai (+1680 to +1745) took charge of many famous government
publications, e.g. the agricultural compendium *Shou shih thung khao* 授時通考 (+1742).

chapters contain more than forty drawings (cf. Fig. 8, p. 128) showing the distribution of the pustules on different parts of the body. These are meant to aid in interpreting the syndromes in terms of the tract systems, and discussing damage to the internal organs.[102] They are particularly attentive to aftereffects of ulceration in the eye, mouth and throat. Chapter 59 gives accounts of causes, symptoms and treatments in other sicknesses with fever and rashes, such as measles and chickenpox, especially in adults, including pregnant women.

Chapter 60, 'Essentials of the heart method for smallpox inoculation', interests us most. In spite of its title, it was based, not on the book of Chu I-liang, but on the 'New treatise' of Chang Yen, at that time the latest publication on the subject. In a moment we must look at the four inoculation methods which the 'Golden mirror' described. But first it is interesting to read the comments on it in Huang Thing-chien's 黃廷鑑 preface to a valuable but anonymous work, finished before +1796, *Chung tou chih chang*.

Inoculation was not mentioned at all in the old formularies. The imperially commissioned 'Golden mirror of the medical tradition' was the first to add this topic to its contents. It revealed the essential points, so that scholars now can certainly get a general idea of the subject.[103] But masters of the art still insisted that there were secret teachings on the techniques of diagnosing, selecting the inoculum, and inoculating – as a pretext to maintain it as a specialty and preserve their livelihoods. Those who studied this art, if they did not receive the secret teachings, dared not try the methods, so people did not trust them. This is truly regrettable.

In the spring of +1796 I unexpectedly got hold of this *Chung tou chih chang*, in one chapter. It does not record the name of its author. It discusses the methods of using the inoculum and of selecting it, and relevant prohibitions. It and the 'Golden mirror' clarify each other. The so-called specialist teachings do not go beyond it. The method of inoculation is a matter of human intervention manipulating the shaping forces of Nature. Those who carry it out following [this book] truly will not lose one patient in ten thousand.

Even veteran practitioners are unable to avoid errors, for two reasons. They make their mistakes, if not at the height of an epidemic, then on children who should not be inoculated. When physicians do not respect the prohibitions, and experiment for no good reason, leading to disaster, how can [the technique of] inoculation be held responsible?

This passage squarely confirms the view that the inoculators, at least following publication of the 'Golden mirror' before the mid +18th century, were discreet and clandestine because they wanted to maintain the value of a trade secret.

What then were the four methods described in the *I tsung chin chien*? We summarise the instructions in the order given.

Aqueous inoculum method (*shui miao fa* 水苗法). Allow a moistened plug of cotton-wool to imbibe an aqueous extract of a number of pulverised scabs (*chia* 痂), and insert it into a nostril of the child to be inoculated. It should stay there at least six hours, and could be re-inserted if it should be sneezed out.[104]

---

[102] [This does not refer to organic lesions, but to impairments of the vital functions with which the organs are associated. – Editor]

[103] I.e. the prestige of this encyclopaedia eventually made the technique generally accessible to non-specialists.

[104] If possible the scabs were to be taken from the pustules of persons undergoing inoculation; in other words, passages of mild smallpox should be used.

Dry inoculum method (*han miao fa* 旱苗法). Use slowly dried scabs, grind them into a fine powder, and blow it into the child's nostrils by a suitable tube of silver.

Smallpox-garment method (*tou i fa* 痘衣法). Wrap the child or the patient in a garment which has been worn by a smallpox sufferer during the illness.

Smallpox lymph method (*tou chiang fa* 痘漿法). Impregnate a plug of cotton-wool with lymph from the perfectly matured pustules of a smallpox patient, and insert this into the nostril of the child to be inoculated.

Of these methods the first was preferred.[105] The order of enumeration is not the historical order. Fan Hsing-chun argues plausibly that the third must have come first, then the fourth, and finally the second and the first successively.[106] One could easily imagine that the garment-impregnation method, the most primitive, was used perhaps in Thang or Sung, while by the time that inoculation found its way into medical writings in the Ming, the other three methods would all have been practised. But there is no pre-Ming evidence for any of them.

The clothing method was still used in the +17th century, for a curious account by Tung Han in +1653 refers to it.[107] Lymph from a mild case with few pustules, conserved some time in a small porcelain container, was smeared on the child's night-clothes, thus infecting them (*jan*).

For this, and no doubt other methods as well, a soybean was coated with medicinal substances and buried in the ground. Its subsequent sprouting and withering marked the rise and decline of the inoculation fever. Incense was also burnt on an altar, the child's name written on a supplication paper, etc. As Yü Thien-chhih would have said, this sympathetic magic did no positive harm.

Yet another Chinese method was employing a wet-nurse who had just nursed a child with smallpox. We are assured that only rich families could afford this expedient.

A problem remains unsolved. It was very acute of the Chinese physicians to have guessed that the normal route of infection in smallpox is by the respiratory tract, hence their use of the nose. In all other parts of the world inoculation by scarification was universal, reaching the lower layers of the epidermis or the capillary circulation and the blood stream. Was this a Turkic variant that originated after the technique passed from China through Mongolian and Manchurian lands to Sinkiang and all points West?

Did the Chinese inoculators practise scarification from time to time without its being mentioned in contemporary medical books? Much may be said for this supposition, for late foreign accounts of variolation in China describe as alternative methods introducing lymph or scab material into sores on the body, or else into three or four scarified patches on the upper arm.[108]

We note below (p. 146) that Richard Mead tried out the Chinese nasal method as part of the Newgate Prison experiments in London in +1721. This is so early that it suggests Mead was using an unaltered technique from China.

---

[105] Ch. 60, pp. 122–4; Lü Shang-chih 呂尚志 (*1973*), p. 58.   [106] Fan Hsing-chun (*1953*), pp. 118ff.
[107] *San kang chih lüeh*, pt 2, sect. 2. The full text is to be found in Fan's discussion.
[108] Besenbruch (1912), Seiffert & Tu Chêng-hsing (1937) and Manson (1879), cit. Wong & Wu (1936), pp. 275, 293.

(6) ATTENUATION

The fact that Chinese discovered the attenuation principle is remarkable. From the scientific point of view, perhaps most remarkable are the measures taken to attenuate the virulence of the inoculum. Today we know that the attenuation phenomenon includes doing two distinct things, on the one hand reducing the population of fully active virus particles or bacteria and, on the other, inducing the appearance of genetically distinct strains or clones of organisms with intrinsically diminished virulence. The old Chinese methods probably involved mainly the former. Whether the virulence of individual virus particles can be diminished artificially is still a moot point among virologists, but the number of living particles in a population can certainly be reduced. Here is the 'Method of storing the inoculum' (*tshang miao fa* 藏苗法) from the *Chung tou hsin shu* of Chang Yen (+1741).

After collecting the scabs, wrap them carefully in paper and put them into a small bottle. Cork it tightly so that the *chhi* does not leak out. The container must not be exposed to sunlight nor warmed beside a fire. It is best to carry it for some time on the person so that the scabs dry naturally. The container should be marked with the date on which the contents was collected.

In winter the material has yang *chhi* within it, so it remains mostly active even after being kept from thirty to forty days. In summer the yang *chhi* leaks away, but even after more than twenty days it remains slightly active. What matters most is that the inoculum be new, for then the yang *chhi* is abundant. It will take regularly in nine persons out of ten. As it gets a little older its activity lessens, taking in perhaps half the cases. When still older, there is no *chhi* in it; when it is implanted it will not work at all. In situations where new scabs are rare and patients are many, it will be necessary to mix in older inoculum. In this case a little more of the powder should be blown into the nostril. After inoculation it will still be able to take.[109]

Other +18th-century books give substantially the same directions. In the *Chung tou chih chang*, for example, we read that the inoculum, whether lymph or scabs, should be put in a small bamboo tube, between two septa, and carefully corked up.[110] It should then be carried round regularly in one's pocket so that it can absorb adequate human *chhi*. If the weather is very hot, the container should be stored in a cool place. There follow the usual directions about grinding and extracting the material if dry, and implanting into the nostril on the cotton-wool plug. The writer was clear that lifelong protection would be afforded by inoculation. That was more than could be said, in spite of Jenner's initial optimism, about vaccination.

Thus the general system was to keep the inoculum sample for a month or more at body temperature (37°C) or rather less. This mild heat would have inactivated some 80 per cent of the living virus particles. Since their dead protein would have been present, inoculation would have strongly stimulated interferon production as well as antibody formation.[111]

There is no way of telling how far back these attenuation procedures went. Many treatises of the +18th century ignored them, among them the *I tsung chin chien*. It would not be unfair to guess that they evolved as clinical experience accumulated from the mid

[109] Ch. 3, pp. 4b–5a. [110] Pp. 25aff. [111] On interferon see Burke (1977).

+16th century, when inoculation began to come out into the open and take its place in medical writings.

As we shall see, Europeans may have got their information about inoculation from the reports of Westerners in China from about +1700 onwards (p. 145). They paid little attention to these in comparison with the powerful influence exerted by the Turkish version after +1720. Attenuation, not a conscious part of the Ottoman technique, remained unknown.

In +1723, when inoculation was rapidly gaining ground in the West, de la Coste published a review in French on the practice 'as it is commonly done in Turkey and England'. A review of this book in +1724 stimulated the Jesuit missionary d'Entrecolles in Peking to write an account in +1726 (published +1731). It included news of attenuation, yet no one took it seriously. He told how Chinese inoculators had been sent to Tartary at urgent need in +1724, and how he himself had got an account of the secret techniques from certain lesser court physicians for communication to Europe only. D'Entrecolles knew of the long keeping of the scabs in the sealed tubes, and remarked that 'it suffices to moderate them by the gentle perspiration of a healthy man who carries them on his person for some time before they are to be used'. He also knew of other methods of interfering with the virulence of the particles, saying 'new scabs need a preparation to moderate this sharpness'. For example, in one method the scabs were steamed with black salsify (*Scorzonera austriaca* or *glabra, yeh tshung chü* 野蔥菊) and 'liquorice' (presumably *Glycyrrhiza uralensis, kan tshao* 甘草).[112] This heating doubtless also did a good deal 'to dissipate the malignity of the poison'. D'Entrecolles also reported that waste silkworm cocoons were used as the plug for the nostril. He translated a lost MS, 'Chung tou kan fa 種痘乾法' (Dry inoculation method), which described sealing and storing the scabs for as much as a year, then treating them with realgar before implanting them in the nose.[113] Finally he gave his opinion that the nostril method was much better than the deep incisions which inoculators were then using in Europe. Many years later, in +1779, another Jesuit, P. M. Cibot, also wrote on smallpox and inoculation, summarising the five chapters of the *I tsung chin chien* but saying practically nothing about attenuation.

Historians have not generally realised that if only the precautions of Shih Hsi-chieh and Yü Thien-chhih had been transmitted to Europe early in the +18th century, thousands of lives could have been saved there. Europeans had to find out about attenuation by themselves, the hard way.[114] In the middle of the century Kirkpatrick knew that some attenuation occurred when pustule lymph from one inoculated person was used upon another ('arm-to-arm' transfer), but he did not believe that this was in any way necessary or desirable (+1743, +1754). Offray de la Mettrie first clearly stated the point in +1740, but he did not positively recommend attenuation. Angelo Gatti did, however (+1763, +1764, +1767). He used the lymph, dried in air and pulverised, as well as aqueous

---

[112] *Scorzonera*: Steward (1930), p. 416; *ECMM* 2025; *CYTT* 3386, s.v. *ya tshung* 鴉蔥. *Glycyrrhiza*: *ECMM* 617.
[113] Here is another hint of the connection with Taoist alchemy. [This comment implies that realgar is peculiarly associated with Taoist practice, which is not the case. – Editor]
[114] The history of this has been conveniently written by Miller (1956).

extracts of scabs in the Chinese manner, for several successive inoculations.[115] He was followed by the Dane, Christen Friis Rottbol (+1727 to +1797), and others. By the nineties the era of Jenner was approaching, and the attenuation of variola did not have much future. What past variolation itself had had in Europe now merits brief consideration.

(7) VARIOLATION IN THE WEST

It is not generally known that the first information Europeans ever obtained about the inoculation process came from China. This 'engrafting' from one civilisation to another did not 'take' until the Turkish influence was about to be strongly felt in England.[116] In +1698 Joseph Lister, who was in the service of the Honourable East India Company at Amoy, began to correspond with his namesake Martin Lister, FRS, in London about botany, conchology and the like. His letter of 5 January 1699 contained the following:

I have here in China been informed by Credible People of a thing I believe only practiceable in these parts, and therefore shall make bold to relate it to you, Viz., a Method of Communicating the Small Pox when and to whom they please, which they perform by opening the pustules of one who has the Small Pox ripe upon them, and drying up the Matter with a little Cotton, which they preserve in a close box, & afterwards put it up the nostrils of those they would infect. The benefit they pretend by it is that they can prepare the body of the Patient & administer it at what Season of the year or Age of the Person they think most proper.

He did not say that the object of the exercise was to give lifelong protection against the dread disease. Martin Lister, perhaps because he doubted whether that effect was possible, did not disseminate the report. But only six weeks later (14 February 1699/1700), Clopton Havers, FRS, who must have got his information from another source, reported it at a Royal Society meeting. Thus, it has been said, by +1700 a considerable number of Englishmen, including reputable physicians, were aware of the inoculatory art.[117] None had tried it. They were probably uncertain about its credibility, its efficacy, its admissibility under law, and its status in theology. They did not know how climatic, racial and even pathological differences might affect the result.

No detailed account of the Chinese procedure appeared until Walter Harris' tractate of +1721, soon after the first inoculations had been done in London. In this same year Richard Mead, as part of the royal experiments at Newgate Prison, inoculated a girl by the nasal technique. As it happened, her resulting attack of smallpox was more severe than that of the other subjects, but the procedure was effective. She was afterwards sent to Hertford where an epidemic was raging, and remained unscathed. Not surprisingly, the method did not spread.[118] We have already seen that neither d'Entrecolles' classical account (published +1731) of the Chinese procedure, nor Cibot's translation from the *I tsung chin chien* (+1779), was influential in the Western world.

[115] This was reminiscent of the seven passages specified in the older Chinese books. Cf. p. 131, above.
[116] The main sources for our account are the book of Miller (1957) and the papers of Klebs (1913a, 1913b, 1914), Stearns & Pasti (1950) and Langer (1976); and very interesting reading they are.
[117] Stearns & Pasti (1950), p. 108.
[118] Miller (1957), pp. 75, 80ff., 86. For Mead's own account see Mead (+1748), pp. 88–9.

What made a great mark on Western Europe was the experience of Westerners living in the Turkish culture-area centred on Istanbul. For several generations before +1700, Turkish practitioners had widely performed inoculation by scarification and lymph. Every author attributed its origin to a different people. Some inoculators were Greek women from the Peloponnese. More often they came from the Tcherkess people of Circassia, north of the Caucasus mountains beside the Black Sea, or else from Georgia, nearer to the Caspian. The technique was probably transmitted from the Turks of Sinkiang by way of Kirghizia, Uzbekistan and Kazakhstan. It would be of the greatest interest to know from Turkish sources how old the practice was, but almost no evidence is available. De la Mottraye, writing in +1727, described the operation as he had seen it performed during his travels in those parts. Edward Tarry in +1721, describing a great smallpox epidemic at Aleppo fifteen years earlier, had much to say about inoculation in the Ottoman lands.[119]

Meanwhile, the Royal Society had been redoubling its efforts to engage in 'philosophical correspondence' with scientific and medical men in foreign and colonial climes. On 22 October 1713, at the Royal Society, 'Mr Taylor said that Mr Williams, who had long lived in Turkey, told him that the Small Pox, when it was of a favourable kind, was often propagated there by inoculating a Scab of Persons Sound, which communicated it to them, and that they had it after a favourable manner . . .'[120]

Two months later Emanuel Timone, a Greek doctor with degrees from Oxford and Padua, wrote a letter to the Society with a full account of the practice as it was done in Turkey; it was published in the *Philosophical Transactions* (and in Germany) in +1714. One curious echo of China appears: 'a convenient quantity of the Matter being thus collected, it is to be stop'd close, and kept warm in the Bosom of the Person that carries it, and as soon as may be, brought to the place of the expecting future Patient'.

This suggests that the 37°C method of attenuation known from Chinese practice (see p. 143) had been observed but not understood, since the time of holding should have been substantial. In +1715 came a further, even more judicious, account, from the pen of another Greek physician (the former Venetian consul at Smyrna), Jacob Pylarini, also an alumnus of Padua, who had already printed a longer version at Venice. A Scottish surgeon, Peter Kennedy, who had studied the technique himself in Istanbul, supplemented these publications in the same year.

Then came the first practice upon Europeans, beginning with the families of diplomats. Two sons of the secretary to the British ambassador at the Sublime Porte, Sir Robert Sutton, returned to England in +1716 bearing the marks of the operation. There followed the celebrated intervention of Lady Mary Wortley Montague, an aristocratic Englishwoman married to Sutton's successor, who first had her own son done in +1718. Upon her return to England in the following year, she set about propagating inoculation. She had written to a friend that she intended to 'bring the useful invention of ingrafting

---

[119] Rifat Osman (1932) states that the first mention in a Turkish text says that a man from Anatolia inoculated a number of children in Istanbul in +1679, but gives no bibliographical reference. Nor is anything to be found in Adnan Adivar's (1943) discussion of variolisation, p. 194n.

[120] *Journal-book*, vol. 10, pp. 512–13.

the smallpox into fashion in England'. Mary Montague was extremely critical of the medical profession, anticipating that their revenues from therapy would make them dislike any form of prevention. She had to some extent the ear of the court, and found a powerful, if discreet, ally in Sir Hans Sloane, the president of the Royal Society.[121]

Charles Maitland (+1723), who had been physician to the embassy and who had also returned to England, performed the first inoculations in Europe in +1721. Not only this, but the Newgate Prison experiments were made before the year was out. In +1722 two of the royal princesses were inoculated, just as had been done at the Peking imperial court forty years previously. A number of tractates in favour of inoculation appeared, for example à Castro, Harris & le Duc. They knew that it had originated in oriental parts but they could not say where.

But opposition was brewing, as in the book of William Clinch (+1724). It continued strongly until mid century. Some of it came from the theologians, such as Edward Massey, who in a famous sermon of +1722 called inoculation dangerous and sinful. Disease, he said, was present in the world by holy ordinance as a trial of our faith, or for the punishment of our sins. Job's affliction might have been smallpox, 'perhaps conveyed to him by some such way as Inoculation'. This idea became so famous as to generate long afterwards a popular rhyme:[122]

> We're told by one of the black robe
> That the Devil inoculated Job.
> Suppose 'tis true what he doth tell,
> Pray, neighbours, did not Job do well?

It is fascinating to find here a belief closely similar to that of the Brahmins in India long before, and to the philosophy of the early Taoist Church in China (pp. 163ff.). The latter was not sufficient to inhibit the invention of inoculation.[123] In England as in China theological opposition could not prevent the progress of the technique. By +1752 the wheel had come full circle. The Bishop of Worcester, Isaac Maddox, preached from the very same pulpit a sermon which became one of the most influential pro-inoculation pamphlets in the literature.

Now began the first efforts to decide the question by statistical evidence. John Arbuthnot in +1722, using a twelve-year run of the London bills of mortality, calculated that if one caught the smallpox the chance of dying was 1 in 9, but for those inoculated it was much better. The major influence was James Jurin (+1722, +1724), one of the Secretaries of the Royal Society. He embarked on a voluminous correspondence and organised systematic enquiries in towns and parishes to ascertain what the true facts of the case were. This was a characteristically European gambit, appropriate to the developing scientific revolution. Jurin amply proved his point. For example, one set of figures,

---

[121] Halsband (1953, 1956).

[122] This was in connection with the epidemic of +1752, though not printed until +1774.

[123] [The practice of public confession in the Celestial Masters' communities of the +3rd century indeed equated sin and illness. This idea (which can hardly be called philosophical) and indeed the communities were too short lived to affect ideas about inoculation. – Editor]

sent from a New England minister to Henry Newman in England, gave particulars for the city of Boston, Massachusetts:[124]

Circumstance	Sick	Died	Proportion	Per cent
Natural smallpox	5,742	841	1 in 6.8	14.6
Inoculated persons	282	7	1 in 40.3	2.5

Jurin's figures argued strongly in favour of inoculation.[125] They did not convert all his opponents, but by +1730 the practice was substantially vindicated. Nevertheless, the debate continued. Bradley (1971) has described a mathematical controversy of +1760 to +1770 between Daniel Bernoulli and Jean d'Alembert over inoculation. This was one of the first full-dress attempts at a mathematical epidemiology, involving not only mortality but morbidity. Investigating the overall demographic effects of variolation in Europe is extremely difficult because of the fragmentary nature of the data, and beset with physiological complications. For example, pregnant women are exceptionally susceptible to smallpox, and men may frequently become sterile after a non-fatal attack. Razzell has made out a strong case for believing that the general result of inoculation was immensely beneficial to the population. He has successfully destroyed the illusory watershed which conventionally separated 'unscientific' variolation from 'scientific' vaccination.[126]

In +1743 inoculation became compulsory for all Foundling Hospital children, in +1748 Richard Mead gave his weighty support to the technique, and in +1755 it gained the official and authoritative approval of the Royal College of Physicians. For a dozen years before that time it had been practised very successfully by James Kirkpatrick from America, and for a dozen years afterwards by the Sutton family, Robert and his son Daniel.

Contrary to a frequently expressed opinion, there was no decline between +1728 and +1740, but it is true that France, Germany and other European countries were a good deal slower than England in taking up inoculation. Miller (1957) suggests that in France the attitude to new knowledge was more class-bound, contrary to the empirical approach of the English, who were perfectly willing to learn from craftsmen and 'old women'. A good case of this would be William Withering's recognition in +1785 of digitalis, previously a folk medicine, as an important cardiotonic.[127] In France it took the *philosophes* to break through the barriers against the use of knowledge coming from below, from hedge-doctors and barbarians.

In +1793, on the eve of the Jennerian discovery, John Haygarth published his *Sketch of a Plan to Exterminate the Smallpox from Great Britain, and to Introduce General Inoculation*. It was a bold suggestion, foreshadowing the total eradication of smallpox

---

[124] Stearns & Pasti (1950), p. 118.

[125] The average figure later attained was at least of the order of 1 in 60, then 1 in 100.

[126] Luckin (1977). Razzell (1977a) has suggested that during the +18th century the epidermal site and arm-to-arm inoculation were tacitly selecting for a less virulent 'cool' strain of virus. As the century went on inoculators gradually abandoned the deep incisions of their predecessors.

[127] Garrison (1929b), pp. 356–7.

by the World Health Organization. Whether or not this could ever have been done without vaccination is doubtful, but the words of Klebs are worth pondering: 'Plentiful suggestions are to be found everywhere which lead one to infer that variolation, without the advent of vaccination, might have furnished the world with an equally safe and perhaps more efficient method of preventive immunisation . . .'[128]

One thing is sure: smallpox inoculation provided the capital for the column of evidence, as Miller puts it, that supported the doctrines of contagion and specificity.[129] If taking a drop of lymph and scratching it into the skin of the patient brought on a mild attack of smallpox and subsequent immunity, there could be no escape from the conclusion that the active substance was a *venenum sui generis*, specific for the disease in question. In Thomas Fuller's words (+1730): 'The Particles which constitute the material and efficient Cause of the Small-Pox, Measles, or other venomous Fevers, are of specific and peculiar Kinds; and as essentially different from one another as Vegetables, Animals and Minerals of different Kinds are from one another.'[130]

Thus was the doctrine of contagion finally and firmly established, even though its universal acceptance had to await the coming of modern bacteriology. 'Peccant humours' and breakdowns of internal *krasis* could not adequately explain the phenomena. The 'perverse and violent pneumata' of the Chinese (p. 127), even though not envisaged as particles, were not so very different.

The situation in the Russian Empire must be considered separately. Five years after the Treaty of Nerchinsk settled the borders between Russia and China in +1689, a number of Russian students were sent to the Chinese capital to learn the technique of inoculation for smallpox. They also enquired into a new quarantine system for smallpox which had been introduced by the government in Peking, with a special official in charge. An English medical man in the Russian service, Thomas Harwin, participated in this mission. One does not know how far the Chinese technique penetrated in Russia after the return of the students, but Wu Yün-jui 吳雲瑞 (*1947*) believes that they transmitted it to Turkey.[131] In fact this must have happened a good deal earlier, presumably by way of Sinkiang. Paradoxically, when the Russian imperial family desired inoculation in +1768, they called in a physician from the West, Thomas Dimsdale, to do it, and made him a baron for his pains.

(8) VACCINATION

Smallpox inoculation (variolation, engrafting or transplanting – there were many names for it in English) had its drawbacks. A severe reaction could follow, leading even to death

---

[128] Klebs (1913a), p. 83. On Haygarth see Fenner (1977) and Henderson (1976). The added efficiency would arise from the lifelong immunity given by inoculation, in contrast with the limited protection conferred by vaccination.

[129] (1957), p. 259.   [130] Pp. 95–6.

[131] *Kuei-ssu tshun kao* 癸巳存稿 (1833), ch. 6, pp. 163–4, and ch. 9, p. 250, record the mission in detail. See also Cordier (1920), vol. 3, pp. 273ff., and Wu Yün-jui (*1947*). [The primary source does not say that the Russians were taught inoculation. It states that they were trained as 'smallpox doctors' (*tou I* 痘醫), a skill that may or may not have included those techniques. The discussion of Russian–Chinese relations in Yü's *Kuei-ssu lei kao* 癸巳類稿, ch. 9, pp. 332–9, mentions smallpox, but also does not allude to inoculation. – Editor]

in some cases. Of course a mortality of 1 or 2 per cent was far better than the rate for unchecked naturally occurring smallpox, which was 20 to 30 per cent, and could run higher in severe epidemics.[132] By the end of the +18th century the deaths were only 1 in 200, and indeed some estimates gave only 1 in 600.[133] Even so this was inferior to the results with vaccination, which made serious illness or death exceedingly rare.

The worst trouble with the variola inoculation was the unreliability and variability of the strains.[134] Secondly, the inoculated patient could pass the virus on to persons in close contact. Although isolation was desirable it was not often achieved. Inoculation did, in unpropitious circumstances, start epidemics, but experts agree that 'scab virus seems to lack epidemic potential'.[135] Yet the active virus is long retained, if in reduced form, and can be demonstrated in scabs held at room temperature for more than three years,[136] so the Chinese attenuation process made good sense. Since the immediate risk was more obvious than the advantage of lifelong protection, any method promising greater safety was likely to be generally adopted in due course.

The story of Edward Jenner (+1749 to 1823) and the beginnings of vaccination has been told so often that it would be nugatory to give much detail here.[137] A country doctor in Gloucestershire, he took notice of the belief, widespread among the rural people, that dairymaids who contracted the light smallpox-like disease of cowpox from the udders of the cows they milked would never suffer from smallpox. He was the first qualified medical man to take this belief seriously,[138] and then to inoculate cowpox lymph and test the inoculated subjects with smallpox lymph. Jenner's brief writings became classical.

He published his *An Inquiry into the Causes and Effects of the Variolae Vaccinae* at his own expense in +1798, after the Royal Society declined it. Apart from a sequel in the following year, this was all his authorship, but the new technique spread quickly throughout the world.[139]

Jenner was not the first person to try cowpox lymph as a protection against the ravages of smallpox. In +1774 there was a severe outbreak of smallpox in Dorset around Yetminster. Benjamin Jesty, a prosperous local farmer and cattle breeder, knew of the general belief. As he had contracted cowpox in his youth, he did not fear the smallpox at all. But his wife and two sons did not have this defence, so he inoculated them with

---

[132] This was evident until recently in India with variola major. African variola minor does not go above 15 per cent and may be as low as 0.1 per cent. We are grateful to Dr Frank Fenner for helpful discussions.

[133] See Simpson (+1789), Woodville (+1796), Downie (1951).   [134] As Baxby (1978) makes clear.

[135] Dixon (1962), p. 298. It is fairly well established that epidemics consequent upon inoculation occurred at Boston in Massachusetts, in England and in Afghanistan.

[136] C. H. Kempe, in Beeson & McDermott (1963), p. 389.

[137] There have been several biographies, notably Baron (1827) and Simon (1857), together with the more personal one of Fisk (1959), but all are hagiographical. The works of Creighton (1889, 1891–4) and Crookshank (1889) are learned and useful, but hostile to Jenner and vaccination. On these two writers see Miller (1957), pp. 285–6. A good modern account is Parish (1965).

[138] There is a distinct parallel with William Harvey as well as William Withering. Anatomists before Harvey 'had neither the structure, action or name that we attribute to "valve"'; French (1994), pp. 350–61.

[139] There was of course plenty of opposition, based largely on the circumstance that a domestic animal was used as the source; Baxby (1978). But vaccination triumphed, and direct smallpox inoculation was made illegal in England in 1840.

cowpox lymph, and it took. This brought him considerable local unpopularity, but he lived it down. A smallpox inoculation of the sons fifteen years later demonstrated that they were still immune. A similar incident occurred in +1791 in Holstein, where a young man, Peter Plett, then tutor to a family at Schönwaide, inoculated its children success-fully with cowpox lymph. Neither Jesty nor Plett followed up systematically, nor did they draw any general conclusions.[140] It would be interesting to know what observations were made earlier in Europe on the occurrence of smallpox-like diseases in animals, and whether any advantage was taken of them.

Chinese and other East Asians made such observations, and acted on them. The value of the resulting therapies remains uncertain. To begin with, Sun Ssu-mo 孫思邈, in his *Chhien chin fang* 千金方 (Essential prescriptions worth a thousand, finished by +659), speaking of a certain prescription used in the treatment of epidemic smallpox in man, says that it is equally effective for the six domestic animals (the cow, horse, sheep, dog, pig and hen).[141] Sun evidently believed that smallpox (which he called *wan tou chhuang* 豌豆瘡, lit. 'pea-pox') occurred in these creatures. The same opinion appears in a curi-ous story recorded *ca.* +1270 by Yeh Chih 葉寘 in his *Than chai pi hêng* 坦齋筆衡. It concerns a veterinary surgeon (*niu i* 牛醫) whom the famous poet Su Shih 蘇軾 (+1036 to +1101) called in when one of his water buffaloes was ill. The veterinary was non-plussed, but Su's wife née Wang, a woman of great knowledge and experience, inter-jected that she knew what was the matter. The animal was suffering from a kind of smallpox rash (*tou pan* 痘班). She went on to prescribe the right medicines for it.[142] Here too some sort of 'cowpox' was recognised.

Next Chou Mi 周密, in his *Chhi tung yeh yü* 齊東野語 (Rustic anecdotes from east-ern Chhi), writing in +1290 about the bad smallpox epidemic of +1273, tells us that a colleague was able to save his seriously ill grandson when a gentleman he encountered on the street gave him a medicine made from the louse-flies of dogs (*kou ying* 狗蠅). Chou does not mention that anyone else used it. In +1596 Li Shih-chen, in his influ-ential *Pên tshao kang mu* 本草綱目 (The great pharmacopoeia), also recommended it without critical reservations. Whether this species, *Hippobosca capensis*, could have con-tained a related dog virus which could give cross-protection in humans is a question difficult to answer.[143]

A rather more weighty affair is the employment of cattle lice for smallpox prophy-laxis from the +15th century onwards. Li Shih-Chen has a surprising entry on these insects (*niu shih* 牛虱), identifiable today as *Haematopinus tuberculatus*. He says that they are shaped like castor-oil seeds. They are born on water buffaloes but drop off them when they have fed on their blood.[144] When roasted, ground and taken orally, they 'dispel in advance the toxicity of smallpox in children'. Li was hesitant about the value of the procedure, since it had not been recorded in the earlier pharmaceutical natural histories,

---

[140] Parish (1965), pp. 14–15.   [141] Ch. 10, p. 189a.
[142] Ch. 1, quoted in *SF*, ch. 18, p. 2b; cf. Fan Hsing-chun (*1953*), p. 112. [On the little-known Yeh see Yü Chia-hsi 余嘉錫 (*1958*), pp. 885–6. – Editor]
[143] *Chhi tung yeh yü*, ch. 8, p. 7a, b; *PTKM*, ch. 40, p. 109. On the insect see R/43.
[144] *PTKM, ibid.*; R/44; Fan Hsing-chun (*1953*), p. 111.

but medical authors had referred to it from the beginning of the Ming onwards.[145] For example, Kao Chung-wu 高仲武 in the lost and undated *Tou chen kuan chien* 痘疹管見 (Narrow examination of smallpox) said that this was a folk usage, like those of Jenner's dairymaids and Withering's foxglove.

Still more interesting, Than Lun 譚倫 described it in detail, especially in *Shih yen fang* 試驗方 (Tried and tested prescriptions) and *I chia pien lan* 醫家便覽 (Convenient exposé for the physicians). Since Than took his degree in +1457, what he said about the cattle louse method is of respectable antiquity. One louse was to be taken for each year of the child's life. They were to be dried, pounded and mixed with rice as a cake, then slowly consumed. Other methods were given. 'One can avoid for a lifetime the suffering of smallpox.'[146] If the buffalo had the virus particles of cowpox in its blood-cells, and if swallowing allowed some of them to enter the respiratory mucous membrane just as the variola virus does, then a mild vaccination-like effect might be attained, and long-continuing immunity achieved.[147]

Finally, Tai Li 戴笠 (the first of two famous physicians of this name, also known as Tai Man-kung 戴曼公) recognised monkey pox. In +1653, after the Chhing conquest, Tai settled in Japan. He became a Buddhist monk and the medical teacher of a number of men later to be distinguished as smallpox specialists, notably the Ikeda family. His famous *Tou kho chien* 痘科鍵 (Key to smallpox) may have referred to inoculation.[148] Ikeda Hitomi 池田獨美 described 'smallpox' in monkeys in *Tou kho pien yao* 痘科辨要 (Assessing priorities in smallpox treatment, 1811).[149]

Thus we note observations of five smallpox-like illnesses in animals. Only one of them, Sun Ssu-mo's statement on domestic livestock, comes before the turning-point of +1000, when smallpox inoculation first arose (by tradition; see p. 154). So the practice would probably have been known in the time of Su Shih and Chou Mi, and certainly in that of Than Lun, Kao Chung-wu, Li Shih-chen and Tai Li. Today we know that the group of ortho-pox viruses includes variola, the lost 'cowpox', vaccinia, and mousepox (ectromelia). Related viruses are known in rabbits, raccoons and gerbils, while a different group again comprises fowl pox. Still further removed are sheep, goat and swine pox, and at the extreme the myxomatosis virus of the rabbit. How far cross-protection might extend is not at all clear, but surely every piece of information about animal poxes is of great interest, both historically and virologically.[150]

What exactly happened during the early years of vaccination we may perhaps never know. As was said at the beginning, vaccinia is a virus strain quite distinct serologically

---

[145] Yang Yüan-chi 楊元吉 (*1953*), p. 20, cites the sources discussed below. [The books, otherwise unknown, are not cited on p. 20 or elsewhere in Yang's book. – Editor]

[146] Ming authors also mention therapies which could not have been of any avail. E.g. Kuo Tzu-chang in *Po chi hsi tou fang lun* (+1577) recommended reducing lice to ashes and ingesting them.

[147] Wang & Wu (1936), p. 216.

[148] Tai brought inoculation to Japan, according to Tsuji Zennosuke 辻善之助(*1938*), p. 203. Another Chinese physician, Li Jen-shan 李仁山, practising in Nagasaki in +1746, wrote there, among other books, a *Chung tou i than pi yü* 種痘醫談筆語 (Medical chats and jottings on inoculation).

[149] Ch. 4, p. 58. For Tai's many publications see *KSSM*, esp. pp. 556–7, 830 and 846; for those of Ikeda, pp. 50 and 649–50.

[150] We are indebted to Dr Frank Fenner for some of the information in this paragraph.

from cowpox virus and from variola. It is almost a creation of man, bred and maintained by passage through laboratory animals, a matter that Jenner could not have anticipated.[151]

There can be no doubt that the cowpox virus got mixed with variola at an early stage. One of the most inviting possibilities is that vaccinia is a genetic hybrid of variola and cowpox.[152] All these viruses freely exchange recombinant DNA with one another if grown in mixed cell cultures. Razzell (1965b, 1977a), who has given such good evidence of the demographic value of +18th-century inoculation, goes too far, perhaps, in saying that all later vaccines were really variola after the first few years. They certainly were mixed; variola was introduced to 'strengthen' the vaccine. He has, however, abundantly proved his case that vaccination cannot be considered in isolation from the history of inoculation in East and West.

The speed with which vaccination spread round the world has rightly been called spectacular. It quickly reached North America. People were being vaccinated in the smallest New England townships by 1805. Ships conveyed relays of boys to keep the vaccine going from Spain to South America, and at the instigation of a Portuguese merchant, from the Philippines to Macao. Russian doctors were vaccinating in Khiatka on the Chinese border by 1805. In 1812 Tartar merchants in Bokhara and Samarqand were distributing pamphlets, printed at Kazan in Arabic and Chagatay Turkish, about Jenner's discovery.[153]

The Chinese indeed warmly accepted vaccination, in spite of a certain *méfiance* in those days towards modern-Western medicine in general.[154] The first tract on vaccination was prepared by Alexander Pearson and translated into Chinese by Sir George Staunton and Chêng Chhung-chhien 鄭崇謙. It appeared at Canton in 1805 under the title *Ying-chhili kuo hsin chhu chung tou chhi shu* 喫咭唎國新出種痘奇書 (Novel book on the new method of inoculation, lately out of England). It was several times reprinted, bearing alternative titles such as *Thai hsi chung tou chhi fa* 泰西種痘奇法 (The novel inoculation method from the Far West).[155] Afterwards the literature grew quite rapidly. Chhiu Hsi's 邱熺 *Yin tou lüeh* 引痘略 (Leading out the pox, 1817) went through sixty-one editions, and also circulated in Japan. Several titles of his such as *Niu tou hsin fa* 牛痘新法 (New method for using the cowpox, *ca.* 1825) were devoted to the cowpox. As other works followed, by the latter half of the century vaccination had spread very widely.[156]

[151] From about 1840 onwards the cow, horse, sheep and buffalo have all been used.

[152] Bedson & Dumbell (1964). For other possibilities see Baxby (1978).

[153] On India, de Carro (1804); on America, Waterhouse (1800–2), White (1924); on Spanish America, Cook (1941–2); on Macao, Wu Lien-Tê (1931); on Russia, McNeill (1977), p. 256.

[154] The story can be found in Phêng Tsê-i 彭澤益 (1950); Chhen Shêng-khun (1979), p. 28; and Fan Hsing-chun (1953), pp. 135, 147ff., 182, 184. Laufer (1911) and Miyajima (1923) have also described its passage into China and Japan.

[155] William Lobscheid reprinted it at Hong Kong in 1806. There was another edition at Canton in 1817, and at Shanghai and Peking in 1828. The bibliography of these tractates is complicated, and we have not seen any detailed study of it. Older acounts are in Klaproth (1810), vol. 1, p. 111, and Rémusat (1825–6), vol. 1, p. 249.

[156] For details, Wang & Wu (1936), pp. 273–301, and Ball (1892), pp. 750ff. Yang Chia-mao 楊家茂 (1990), p. 84, lists twenty-nine early writings on vaccination by Chinese. On *Niu tou hsin fa* and other books with similar titles (e.g. Wu Jung-lun 武榮綸 & Tung Yü-Shan 董玉山 (1885)), see *LHML*, items 7947ff.

(9) The background religious tradition in China

(i) *Wang Tan*

The moment has come to examine the persistent conviction that for some five centuries before the earliest extant writings on inoculation for smallpox, it had been practised under conditions of restriction and secrecy.

The figure round whom the tradition revolved was Wang Tan 王旦 (+957 to +1017), prime minister in the reigns of two Sung emperors, Thai-tsung and Chen-tsung. He got his first office in +990 and became Hanlin Academician in +998. In +1004 he helped to negotiate peace with the Liao. Throughout his life he was a leader of the anti-war party. He was so active in appeasing China's enemies that he acquired the popular name of 'Thai-phing tsai-hsiang 太平宰相', the peace-seeking premier. His reputation also suffered due to his collusion in the fabrication of portents and the arrangement of elaborate ceremonials to gain support for unpopular policies. He was particularly good at organising the bureaucracy and the imperial examinations, and on balance went down to posterity as a virtuous prime minister.[157]

His connection with smallpox inoculation came about because his firstborn sons had died of this disease. When his youngest son, Wang Su 王素, was a child, the father sought everywhere for some means of preventing a similar calamity. He consulted all kinds of physicians and shamanic technicians (*wu fang* 巫方), till finally the gods were compassionate and sent him a divine man (*shên jen* 神人), who carried out inoculation. From then on the technique was handed down from one practitioner to another with stringent confidential precautions.[158]

Such is the account as Chu I-liang 朱奕梁 gave it in his *Chung tou hsin fa* (1808). Other books on inoculation include the same account, with numerous variations. The oldest statement may go back to about +1713, as it occurs in Chu Shun-ku's *Tou chen ting lun*. Chu calls the inoculator a *shên i* 神醫 (divine physician). Others refer to a *thien lao* 天姥 (old woman from heaven), a Buddhist master (*seng* 僧) or a *chi hsien* 乩仙 (planchette immortal). The writer of the *Chung tou chih chang* (before +1796) says that the medical attendant of Wang Tan's family was a *ku hsien, san pai chen jen* 古仙三白真人, an 'ancient immortal, a three-white perfected immortal'. The great medical historian Hsü Ta-chhun 徐大椿 in *I hsüeh yüan liu lun* 醫學源流論 (+1757) just calls him one of the immortals (*hsien* 仙), in other words, a numinous Taoist.[159] It might be relevant too that Wang Tan himself asked that he be habited as a 'monk' after his death.[160] All affirm that

---

[157] Wang's official biography is in *Sung shih* 宋史, ch. 282, pp. 9542–53.

[158] Considering the chronology of Wang Tan's life, the inoculations in his family would have taken place around +995. Many sources, however, put it after the accession of Chen-tsung in +998.

[159] Ch. 1, sect. 2, p. 120, tr. Unschuld (1989), p. 339. In this connection one recalls that the inocula were called *shên miao* and *tan miao* (see p. 131, above), both very Taoist. We learned above (see p. 134) that Yü Thien-chhih in +1727 called the 'strange man' who brought the inoculation procedure to Ning-kuo prefecture an alchemical adept.

[160] Franke (1976), vol. 3, p. 1152.

the remarkable inoculator came from Mt O-mei 峨眉, in southwestern Szechwan, famous for its associations with both Taoism and Buddhism.[161]

Chang Yen repeated Chu's anecdote in his *Chung tou hsin shu* of +1741. In +1743 the imperial *I tsung chin chien* 醫宗金鑑 incorporated this book, giving it a certain cachet of orthodoxy. The tradition was accepted on all hands.[162]

If we look more closely at the tradition of Mt O-mei, we can see that the Buddhist connection may be rather misleading. There are many evidences of a Taoist presence on and around that mountain. These include Hsien fêng shih 仙峰石 (rock of the holy immortals' peak), and buildings such as the Chiu lao hsien fu 九老仙府 (palace of the nine ancient immortals). There are caves of Taoist hermits such as Li hsien tung 李仙洞 and Ko hsien tung 葛仙洞; there was a tradition that the great alchemist-physician Sun Ssu-mo 孫思邈 did some of his alchemy at O-mei Shan, as witness a 'stone from which to admire the elixir' (*wan tan-shih* 玩丹石). All in all, there is a strong, if subordinate, element of Taoism about O-mei and its neighbourhood.[163]

There are only faint indications that inoculation may have been a good deal older than the time of Wang Tan. A very late opinion made it arise in the Khai-yüan reign-period of the Thang time (+713 to +741), but this has never received much support.[164] The great pathologist Chhao Yüan-fang, in his *Chu ping yüan hou lun* of +610 or thereabouts, emphasised preventive measures to avoid transmitting 'Cold Damage' (*shang han*) febrile disorders, in which he included the 'chalice papule' variety of smallpox. But his discussion of prevention did not single out or even allude to that variety.[165]

We have not attempted to survey generally the Chinese literature on smallpox through the centuries. It would be valuable to study closely all the medical books which discuss smallpox between the +7th and the +16th centuries, to uncover further evidence of inoculation that may be lurking there.[166] It is significant, in view of the Taoist connections that we have already emphasised, that during the Chhing period a book by one Thiao Yüan-fu 調元復 bore the title *Hsien chia pi chhuan tou kho chen chüeh* 仙家秘傳

---

[161] Cf. Fan Hsing-chun (*1953*), pp. 114, 117. Wang & Wu (1936), pp. 215–16, rely upon this account, but claim without justification that the inoculator was an Indian philosopher. Western writings garble the anecdote, beginning with Cibot (+1779), p. 393. Von Zaremba (1904), p. 205, has 'a certain person of the name of Yo-mei-schan'. Lockhart (1861), p. 237, copied by Ball (1892), pp. 683, 698ff., asserts that 'a philosopher of Go-mei Shan in +1014 inoculated for Prince Tchiu-siang', mistaking Wang's family name (lit. 'king') for a title, and his title for a personal name. Wise (1867), p. 543, makes it 'Yo-meishan, who inoculated the grandson of the Emperor Tchin-tsong'. There is a very circumstantial, indeed hagiographical, account of the episode of Wang Tan's family in McGowan (1884). We have not been able to identify its source, since he gave the title only as 'Correct treatment of smallpox', without Chinese characters. In this version the inoculator is a nun of some kind, who turns out to be an incarnation of the Goddess of Mercy, Kuan-yin. So we are probably not missing much.

[162] Ch. 60, p. 1a. See further: Chhen Pang-hsien 陳邦賢 (*1953*), p. 371; *ICK*, p. 1381; and Chhêng Chih-fan 程之范 (*1978*), p. 471.

[163] Phelps (1936), pp. 226, 248, 284–6, 337. On Taoist books associated with O-mei, see Vol. 5, pt 3, pp. 75–7 and *passim*.

[164] Wu Jung-lun & Tung Yü-shan (*1877*).

[165] See p. 127, above. Fan Hsing-chun (*1953*), pp. 113–14, declined to accept a Thang origin, but one would be credible if Chhao had inoculation in mind. [There is not a hint of inoculation in Chhao's book. – Editor]

[166] Particularly the *i an* 醫案 case-history literature of Chin, Yüan and Ming times has never been adequately investigated.

痘科真訣 (True instructions on smallpox, privately handed down by the adepts of the immortals). It seems to be lost.[167]

(ii) *Religious connections of inoculation*

The inoculation discovery was surrounded by a wide aura of religious and magical conceptions and practices strongly Taoist in character down to quite late times. Standing back and looking over the whole tradition, one can distinguish several separate ingredients in this magma. First there was the pantheon of divine beings, office-holders in the Taoist celestial bureaucracy, provided as of right with temples here below. There were also votive temples for the Original Inoculator, and other successful practitioners afterwards. The Taoist connection was strengthened by the charms, incantations and cantraps used in cases of smallpox, and the altars for worship set up when someone fell ill of it. It was natural that secrecy and circumspection should govern the use of these armaments of apotropaic medicine, but other therapeutic methods had their confidential aspects too (see p. 118).

The most prestigious alchemists and the most learned scholar-physicians used 'forbidden prescriptions' (*chin fang* 禁方), but they also played a great part, as could well be imagined, in the practice of 'hedge-doctors' or medical pedlars. The probable mediaeval origin of smallpox inoculation implied religious overtones, and a considerable amount of hocus-pocus too, though as in other cases not all hocus-pocus was fraud.[168] We shall briefly take up each of these aspects in turn.

There is much to be learnt from the structures of organised religion in cultures very different from those of the Western world. The Taoist pantheon of divinities, which mirrored the pattern of the civilian bureaucracy on Earth, is very well known.[169] Its Celestial Academy of Medicine (Thien i yüan 天醫院) included three types of demiurges: first the ancient Chinese deities and deified genii such as Phan-ku, Fu-Hsi, Shên-nung and Huang-ti; secondly Yao-wang 藥王, the King of Medicines, originally a Buddhist spirit of healing, Bhaiṣājya-rāja bodhisattva; and thirdly divinities who specialised in certain illnesses. There was a female Comptroller of Exanthematic Diseases, Pan-chen niang-niang 斑疹奶奶. The much more dangerous Comptroller of Smallpox was also a woman, Tou shên niang-niang 痘神娘娘. Her sons, gods in this division, may have typified four main aspects of the disease. Pan Shên 瘢神 certainly had to do with the fatal haemorrhagic form; Chen Shên 疹神 may have represented either variola minor or measles (morbilli); Sha Shên 痧神 dealt not with cholera or diarrhoeas, as the name might indicate, but with scarlet fever, chickenpox or German measles; and Ma Shên 痳神 superintended pustules and pockmarks.[170]

So greatly feared was the scourge of smallpox that there grew up among the immortal Taoist deities a quite separate Celestial Academy of Smallpox (Thien tou yüan 天痘院),

---

[167] Wylie (1867), p. 103; not in *LHML*.

[168] See Hsu's (Hsü Lang-kuang) (1952) classical study of religion and exorcistic magic alongside modern inoculation and scientific medicine during a cholera epidemic in a Szechuanese town during the Second World War.

[169] Vol. 5, pt 2, p. 110; Doré 1914–29, vol. 10, pp. 53ff., esp. p. 59.

[170] Werner (1932), pp. 505ff., 354, 46, 408 and 300; Doré (1914–29), vol. 10, pp. 739, 748, with late illustrations of the goddess and her four sons.

presided over by a male god (Tou shên 痘神).[171] According to the legend of his deification, his name had been Yü Hua-lung 余化龍. At the end of the Shang (ca. −1030),
defending Tung-kuan against the invading Chou forces, he used infected garments to
sow smallpox among them. Later they took an overwhelming revenge and killed him.[172]
Like Tou shên niang-niang he was assisted by sons, five instead of four; they were his
assistants for the four quarters of space and the centre.

Even this was not the end of the smallpox deities, for there were special shrines to a
deified magistrate, a Marshal Chang, supposedly born in +703, whose district was spared
the smallpox epidemics which raged all around.[173] So he was elevated to the Heavens
as Superintendent of Epidemics and Protector of Children from Smallpox, presumably
a brevet appointment. Had he had something to do with inoculation?

There were votive temples for those who had. The prefectural gazetteer of Huchow
湖州 tells us that towards the end of the Ming a young man named Hu Phu 胡璞 of that
city (now Wuhsing, Chekiang) ran away from his family in +1644, irresistibly driven
to become a physician. He carried out many inoculations before disappearing in +1712.
He was supposedly seen in +1723 in Nanking, though that would hardly have been
possible. At least since the time of Khang-hsi's accession in +1662 there were votive
temples in Soochow and Huchow dedicated to the 'immortal teacher of inoculation'
(Chung tou hsien shih 種痘仙師) and to the 'mountain recluse of O-mei Shan' (Sung
O-mei shan jen 宋峨眉山人). The image often looked very like that of Shun-yang tsu-
shih 純陽祖師, i.e. the famous adept and alchemist Lü Tung-Pin 呂洞賓, whose dates
possibly belong in the +8th century.[174] Here perhaps is another thread linking inoculation with the activities of the Taoist alchemists.

Today we have the benefit of numerous researches in the Taoist literature, in some
cases by those who have themselves participated as ordained clergy in the rites and ceremonies of that religion. We know that an abundance of its texts have been preserved,
both within the Taoist patrology (Tao tsang 道藏) and outside it in the form of prayers
and rubrics handed down in priestly families. Where the specialised votive temples were
concerned, there must have been a wealth of practices and incantations which were never
written down. Those would have been the secret preserve of Taoists who conducted the
affairs of the supplicants. Between these and the confidential formulae of exorcists and
physicians just on the other side of the borderline between the priests and the medical practitioners, there would have been no sharp distinction. Writing of the secret
manuals (chin ching 禁經) in the early Thang period, Sun Ssu-mo, himself an alchemist
as well as a physician, remarked:

[171] Werner (1922), pp. 175, 246–7; Doré (1914–29), vol. 10, p. 757.
[172] The opposing general was the legendary Chiang Tzu-ya 姜子牙. The story, from the Ming novel
Fêng shên yen i 封神演義, is apocryphal, but it remains part of the prehistory of bacteriological warfare. See
chs. 81, 82, 99, tr. Grube (1912). For a sketch of its history see Veith (1954). Lord Amherst is said to have
repeated Yü's stratagem in +1763, when he tried to organise the distribution of smallpox-infected blankets
among opposing Amerindian tribes. Long (1933), pp. 186–7; McNeill (1977), pp. 251, 349.
[173] Doré (1914–29), vol. 9, p. 634, vol. 10, p. 758; Werner (1932), p. 42. [There were a number of popular
smallpox gods with the title of Marshal, identified in different localities with different historic figures; see, for
example, Katz (1995). Note also the variant names of smallpox deities on p. 161, below. – Editor]
[174] Fan Hsing-chun (1953), pp. 113–14; on Lü, Vol. 5, pt 3, pp. 147–8 and Fig. 1360.

The Heavenly Master said: 'Of those who have received my techniques, gentleman of the higher sort will rise up to the skies as immortals; those of the lower sort will be promoted in office, while commoners will increase their longevity. These secrets are not to be transmitted even to one's son or younger brother, but only to worthy men. If no one is worthy, do not transmit it. Otherwise calamity will fall upon descendants in future generations!'[175]

These secret manuals contained all kinds of techniques for maintaining good health and defeating evil influences by 'drawing talismans and pronouncing incantations' (*hua fu nien chou* 畫符念咒).[176] Among these were the amulets for the protection of the body (*hu shên fu* 護身符), charms to drive away malign influences (*chhü hsieh fu* 驅邪符), peach-wood safeguards for the New Year (*thao fu* 桃符) and registers of magical art (*fu lu* 符籙). Many medical books have chapters on talismans, for example ch. 200 of the officially commissioned *Shêng chi tsung lu* 聖濟總錄 (General record [commissioned by] sagely benefaction, +1122), and the *Shên i phu chiu fang* 神醫普救方 (The divine physicians' formulary for universal deliverance) of Chia Huang-chung 賈黃中 (+986). In +1596 Sun I-khuei 孫一奎 complained in introducing his discussion of smallpox that 'the specialised practitioners today take [their treatments for] smallpox as secret techniques and forbidden prescriptions'. Each taught only the views of his own tradition, so that diverse views were never reconciled, and no single best method could emerge.[177] Hsü Ta-chhun wrote (+1757) that there had always been arts of this kind. Some of these restricted techniques could only be handed down by eccentric hermits, Taoist adepts, Buddhist high monks, or even gods. They had to be protected from the vulgar who, unaware of how arduous it was to obtain and master them, would take them lightly and ignore the special precautions, even the reverent frame of mind, that were mandatory if they were to be effective.[178] It begins to look as if the inoculation for smallpox was exactly the kind of thing which would have been transmitted in this way from Sung or before to Ming times.[179]

Then there was the question of the 'hedge-doctors', the itinerant medical pedlars, some of whom had remarkable knowledge and skill, but none of whom could write. The poor peasant farmers depended almost entirely on them.[180] They were called *tsou fang i* 走方醫 (itinerant doctors) because they went knocking from door to door in the countryside, *ling i* 鈴醫 (bell-doctors) because they signalised their presence by jingling bells

[175] *Chhien chin i fang* 千金翼方, ch. 29, p. 345b.
[176] Paper charms turn up from time to time in tombs. For example, one dating from +551, found at Kao-chhang (Turfan) in Sinkiang, was written with red characters on yellow paper. See Anon. (*1978b*), pp. 36–7 and fig. 13. It includes the familiar Taoist adjuration to the spirits: 'urgently, urgently, by lawful command' (*chi chi ju lü ling* 疾疾如律令).
[177] *Chhih shui hsüan chu* 赤水玄珠, ch. 27, pp. 1aff.
[178] *I hsüeh yüan liu lun*, ch. 1, pp. 78ff. In the West for many centuries physicians protected themselves to a large extent by the learned language which they developed, as is still the case today. The injudicious use by laymen of powerful medicaments given us by modern science is a problem that *chin fang* might obviate. The Chinese profession evidently felt that jargon was not enough. [The mention of drug formulae and the citation of *Huang ti nei ching, Ling shu*, ch. 48, suggest that in this passage Hsu had neither ritual healing nor smallpox inoculation in mind. – Editor]
[179] Hsü Ta-chhun says exactly this on p. 120 of *I hsüeh yüan liu lun*. [What Hsü says, without reference to any period, is only that inoculation 'has been transmitted by immortals'. The essay goes on to explain its many advantages, indicating that this phrase is not a remark about its history, but a common form of praise for its value. – Editor]
[180] One thinks of the herbalist Nick Culpeper in Rudyard Kipling's story 'A Doctor of Medicine'.

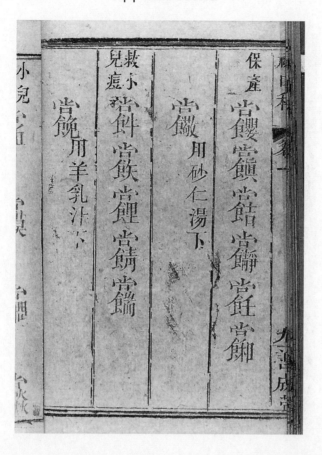

Fig. 9. Talismans used in exorcism to cure smallpox in a child. From *Chu yu shih-san kho* 祝由十三科 (The thirteen departments of apotropaic medicine, after +1456).

on the staffs which they carried, or *yung i* 庸醫, a term of contempt for run-of-the-mill practitioners. Their methods corresponded exactly with those mentioned above, *wu shu* 巫術, apotropaic arts, to dispel malign influences, and *chin fang*, confidential therapies, to restore people to health. They had their own slang for expellant drugs (*chieh* 截, 'leapers'), emetics (*ting* 頂, 'raisers') and purges (*chhuan* 串, 'penetrators'); their remedies were cheap and effective in use.

Their greatest recorder was Chao Hsüeh-min 趙學敏, the scholar who wrote a sequel to *Pên tshao kang mu*, the greatest of the pharmaceutical natural histories. In +1759 he completed *Chhuan ya* 串雅 (The penetrator improved). In it he edited the notes over thirty years of a famous itinerant healer, his friend Tsung Po-yün 宗柏雲. Tsung's work contained many useful suggestions about the prevention, as well as the cure, of diseases. Here again, the confidential therapies of oral tradition were very much to the fore, not so much because these practitioners were unwilling to record them, but because they could not write literary Chinese.

It is good that we are provided with some anti-smallpox talismans to illustrate (Figs. 9 and 10). They come from a curious book of many titles, unknown editorship,

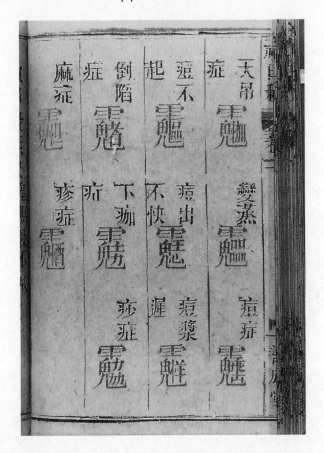

Fig. 10. Talismans for abnormal symptoms in smallpox: failure of pustules to form, slow formation, etc. From *Chu yu shih-san kho* 祝由十三科.

and uncertain date, the *Chu yu shih-san kho* 祝由十三科 (The thirteen departments of apotropaic medicine, probably after +1456).[181] The procedure is always the same: write the magic character on paper with red cinnabar ink, burn it to ashes, and have the child take them in liquid. It remains the same in Chang Hao's 張浩 *Jen shu pien lan* 仁術便覽 (Handy guide to the humane art) of +1585 and even in the *Chiang hu i shu pi chhuan* 江湖醫術秘傳 (Esoteric therapeutic methods of doctors of the *demimonde*), an anonymous manual, *ca.* 1930, of apotropaic healing.[182]

The celebrated +18th-century novel *Hung lou mêng* 紅樓夢 (Dreams of red mansions) has a passage about a case of smallpox in a child of a wealthy family:

A room had to be swept out and prepared for the worship of the Smallpox Goddess. Orders had to be given to the servants to avoid the cooking of all fried and sautéed things . . . sexual

---

[181] Ch. 1, pp. 9b, 41b. The collection supposedly originates from a stone inscription said to have been excavated in +1188, but the preface to the fourth of its books is dated +1328.

[182] Ch. 3, p. 26. There are also many examples of charms in the inoculation books themselves, e.g. *Chung tou hsin shu*, ch. 3, p. 11a, and in most other genres of medical literature.

abstinence was enjoined on the parents of the sufferer. A length of dark red cloth had to be procured and made up into a dress for the child by the combined labours of the nurses, maids and female relations most closely associated with it. Finally, a ritually purified room had to be made ready for the two doctors who would take it in turns to examine the little patient and make up her medicines, and who would not be permitted to return to their own homes until the customary period of twelve days had elapsed.[183]

This account agrees closely with those in medical texts, such as the *Tung i pao chien* 東醫寶鑑 (Precious mirror of Eastern medicine), a Sino-Korean compilation of +1610. The latter emphasises keeping the patient's room quiet and clean, avoiding impurities. 'It seems that the *chhi* of human beings moves (i.e. circulates as it should) when it smells incense, and stops when it smells something foul.' Outsiders, Buddhists and Taoists were frowned on, apparently to avoid hustle and bustle. There is nothing about red clothes or hangings, but Chang Yen's book mentions them, speaking of an altar (*than* 壇) to be used for worship on the occasion of inoculation:[184]

As soon as possible after inoculation with the germ it is necessary to select a lucky day for setting up the altar. For offerings the flesh of two animals made into five dishes, as well as five different sorts of fruit, are required. The altar must be in a clean empty room, hung and draped with red cloth. The names of the gods are to be written at the top of a piece of red paper, with the names of their boy and girl attendants at the bottom. The names of the deities are Chang tou yüan chün 掌痘元君 and Ssu tou niang-niang 司痘娘娘.

We doubt whether the emphasis on the colour red in this disease was anything more than conventional in religious ceremony, but it is curious that modern studies on the effects of coloured light have claimed that red moderates suppuration in smallpox and inhibits pock-marking because of the exclusion of rays of more chemically active wavelengths.[185] The association of the colour red with smallpox, and perhaps other epidemic diseases, is an old idea in Europe. Where it originated remains obscure, but red light was recommended in the works of many of the +13th-century European physicians such as Gilbertus Anglicus (d. +1250), Henry de Mondeville (+1260 to +1320), Bernard de Gordon (fl. +1285 to +1307) and John of Gaddesden (+1280 to +1361), whose *Rosa anglica* of +1314 was printed in +1492.[186] Most people have read the story of Edgar Allen Poe, 'The Masque of the Red Death' (1842); the work of Niels Finsen (1860 to 1904) is also generally known.[187] But chromotherapy, apart from certain special conditions such as lupus, plays little part in modern medicine.

(iii) *Sickness and sin*

There is a singular comparison to be made between China and India. It revolves round the question of whether sickness was considered a divine punishment for former misdeeds and transgressions, and how far it was accompanied by feelings of guilt or sin or

---

[183] Ch. 21, tr. Hawkes (1973–86), vol. 1, p. 424.
[184] *Tung i pao chien*, ch. 11, p. 661b; *Chung tou hsin shu*, ch. 3, pp. 9bff. One of us (G.D.L.) vividly remembers this custom from her childhood in Nanking.
[185] Vollmer (1938), Fields (1950).   [186] Garrison (1929a), pp. 155, 164–5.   [187] Castiglioni (1947), p. 1125.

shame. We have already seen how in China some believed that womb poison arose from the parents' excessive erotic pleasure in procreation (see p. 130). We have just touched upon the preparation of family altars for sacrifices and oblations to the smallpox deities. Sacrifice generally implies repentance, and repentance implies the recognition of former wrong-doing. Over against this lay the conviction of the rational physicians that the causes of smallpox were natural, and that they could cure it, or at least alleviate and moderate it, by therapeutic actions. Still more daring was the conviction of the earliest inoculators that by inducing the disease in its mildest form they could give lifelong protection against even the most dangerous forms of it.

The paradox is that this technique arose precisely in a milieu saturated with religion. But perhaps the Taoist religion was not quite so resigned to fate as other religions? Although fundamentally a Nature-mysticism, even after it provided itself with personal deities, indeed a Holy Trinity, and one which always believed that Nature's way was best, perhaps it could generate the idea of borrowing tools from Nature's workshop?

The contrast is particularly instructive if one studies the remarkable book of Chattopadhyaya (1977) on science and society in ancient India. It is about the struggle that took place in that civilisation for a rational medicine. The *Suśruta-saṃhitā* and *Caraka-saṃhitā*, greatest of the ancient Indian medical works,[188] have both of them an ambiguous character, partly scientific, partly not. Chattopadhyaya's dissection reveals an intense struggle between the theological philosophers and the rational physicians. The former wanted to attribute everything to the law of *karma*, regarding all illness as caused by ethical and moral infractions or sins in previous existences. The Indian physicians were striving for a truly scientific view of the world. They believed that diseases were not caused or sent by the gods on account of sin, but rather were due to the natural causes that they sought out. The admixture of science, magic and religion is as clear in the *Caraka* book as in China. For instance, it defines *yukti-vyapaśraya bheṣaja* as therapeutic systems based on the use of natural material substances such as drugs and diets, as against *daiva-vyapaśraya bheṣaja*, therapeutic systems based on propitiatory or penitential incantations, charms and sacrifices. This is just the distinction which we have been looking at in China between apotropaic techniques and rational scientific techniques. There, through the centuries, the former took second place to pharmacy and physical therapy. It remains paradoxical that inoculation arose among the exorcists.

A key word in the ancient Indian literature is *svabhāva*, which could be translated 'inherent nature', 'innate thus-ness', or 'the essential nature of things'. It must have had close relations with *ṛta* and even *dharma* in some senses, meaning 'the Order of Nature' or the way in which Nature works – all recalling the Chinese Tao. The physicians[189] were seeking the pattern-principles in Nature, the ultimate reasons (ultimately of course inscrutable) why things are as they are and behave as they do.

It is interesting to see how the Buddhist philosophers translated these Sanskrit words into Chinese. *Isvabhāva* became *hsing* 性, defined as embodied cause, the unchanging,

---

[188] Vast medical compilations, very difficult to date, but perhaps mainly of the +1st and +2nd centuries; Renou & Filliozat (1947–53), vol. 2, pp. 147, 150.
[189] The *svabhāva-vādins*, as their opponents called them.

independent, self-dependent, fundamental 'nature' behind the manifestation or expression of anything. Sometimes this was amplified as *tzu hsing* 自性, 'the primary germ [*verb. sap.*] out of which all material appearances are evolved, the first source of the material world of phenomena'. Other more curious locutions were *ssu-pho-pho* 私婆婆 and *tzu-thi-thi* 自體體, 'own state', essential or inherent property, innate or peculiar disposition, natural state or constitution. So much for the nature of Nature as it passed from Sanskrit into Chinese, and found a natural home in Taoist philosophy.

In China the idea of reincarnation was never so strong as in India. The piling up of sins was not so overwhelming, but it was certainly there. Conceptions very like *karma* were current in early Taoist religion.[190] In the parishes of the Celestial Masters, which flourished from the +2nd to the +6th centuries,[191] all illness was thought of as a visitation on account of previous misdeeds or transgressions. The hierarchs presided at sessions which today we should call faith-healing. Recovery (in most cases natural) was attributed to confession of faults, sometimes publicly. In the early theocratic communities of Chang Chüeh's 張角 time (*ca.* +180), the patient went into retreat to reflect on his wrong-doings, while at the same time prayers were made on his behalf, with his name and history written on papers despatched to the gods as he paid money to the public funds. Thus he gained cure and release.[192] Psychological catharsis was achieved by means of quasi-orgiastic rites such as the 'ceremony of the three primal ones' (*san yüan chai* 三元齋), the 'mud and charcoal ceremony' (*thu than chai* 塗炭齋) and the hierogrammatic celebration of 'uniting the *chhi*' (*ho chhi chai* 和氣齋).[193]

There has been much discussion, which we can do no more than mention here, concerning the question of whether Chinese culture as a whole should be thought of as a 'sin-and-guilt society' or as a 'shame society'. *Kuo* 過 can be translated 'transgressions', but how far *tsui* 罪 may be taken as 'sin' in anything like the sense of the People of the Book (Israel, Islam and Christianity) is highly doubtful, for monotheism was never characteristically Chinese. On the other hand, words involving shame are very prominent: e.g. *chhih* 恥, which was essentially connected with blushing; *ju* 辱, which derived from 'dirty', hence 'disgraced'; *hsiu* 羞, which started as 'ugly' and came to mean 'ashamed' in the sense of 'embarrassed', like its equivalent, *tshan-khuei* 慚愧. And there was always, too, *shih lien* 失臉, 'loss of face', well known to Old China Hands and combated in modern China.

Here we need only recognise that the idea of sin or *karma* as the prepotent cause of disease did make its appearance in China during a certain period. Insofar as it was thought of as shame rather than theological sin or guilt it could have come near to psychosomatic pathogenesis. An organic naturalism which refused to distinguish sharply soul and body was always typical of Chinese medicine.

Here comes the dénouement. While in India the theological conception of *karma* tended to dominate all through the centuries, in China, that more rationalistic culture, it did not gain the day in the long run. We have seen that the cure, and even more the prevention, of smallpox were closely associated through the ages with Taoist religion

[190] Maspero (1971), pp. 43, 357–8, 411–12, 415–16; Welch (1957), pp. 115ff., 120ff., 133–4; Eberhard (1967), pp. 12–13, 20; *ibid.* (1971), pp. 316, 360; Kandel (1979), p. 44.
[191] Vol. 2, pp. 155ff.   [192] Fan Hsing-chun (*1947–8*), p. 21.   [193] Vol. 2, p. 150.

and magic. In that milieu the inoculation technique appeared. Surely this must go back to the Nature-mysticism of Taoist philosophy on which the Taoist Church was founded, the idea that everything goes best when man interferes least. For a time the idea that illness was a punishment for evil-doing flourished. Indeed one wonders whether it was not essentially an Indian importation, perhaps connected with Buddhism. Before long, Taoist technical *élan* reasserted itself. The inspiring thought arose that man could borrow some of the tools in Nature's workshop and, by following her, bring about something effectual for human benefit.[194] After all, the great Taoist doctrine of *wu wei* 無為 – 'do nothing contrary to Nature' – avoided setting a hundred men to work on water-raising machinery when the same result could be achieved by taking the water off upstream and leading it by a canal along a higher contour level to its destination.[195] So also using sagely Nature's own variola virus under the right conditions could permanently protect against the worst forms of the disease. If this was 'thinking God's thoughts after him', as some pious Western Christian might have put it, who would begrudge the Taoists their candles and incense?

(10) THE ETHNOGRAPHICAL DIMENSION

Before the end of the +17th century, medical observers were recording 'folk' practices in various parts of Western Europe which later writers recognised as cognate to inoculation. Then from the middle of the +18th century onwards travellers, and later on physicians stationed in outlandish parts, reported recognisable inoculation procedures, generally involving scarification, all over the Old World and in many parts of Africa.[196] It thus became customary for historians of medicine to say with an airy sweep that inoculation for the smallpox must have originated in primitive societies everywhere, and must have been practised for countless centuries. For example, Sir George Baker wrote in +1766 that 'It cannot but be acknowledged that the Art of Medicine has, in several instances, been greatly indebted to accident, and that some of its most valuable improvements have been received from the hands of Ignorance and Barbarism.'[197]

Sinologists might raise their eyebrows at this statement. When Baker wrote, Europeans were ostentatiously admiring the polite nature of Chinese society. Beyond that, the historiographic basis of his assertion is faulty. If one finds a technique in a preliterate or semi-literate society, one cannot simply assume that it has existed there from time immemorial, especially if it exists also in a literate one in which its origin is documented. In the present case textual authority goes back earlier in China than anywhere else, and contains no hint of foreign origin. The possibility plainly presents itself that the discovery was made first there, and radiated out to all those other peoples among whom it turns up later.

In order to test this supposition, we must look at what is known about possible routes of contact. When this is done, the conclusion will present itself that a dating of +1500

---

[194] We study many examples of this phraseology at length in Vol. 5, pt 5, pp. 293–6.　　[195] Vol. 2, pp. 68ff.
[196] The best guides are the reviews of Klebs (1913a, 1914) and the book of Miller (1957). We note individual reports as we proceed.
[197] Baker (+1766), p. 1.

in China, and even more so that traditional dating of +1000, would have allowed ample time for the knowledge to spread throughout the rest of the Old World and Africa.

Through the Turkic peoples of Central Asia and Asia Minor the technique reached the Greeks and Western Europe (pp. 116ff.). The celebrated Old Silk Road, which we have mapped elsewhere, is the obvious route.[198] The age-long presence of the Uighur Turks in Sinkiang (Chinese Turkestan, as it used to be called) makes it easy to see a transmission westwards precisely through them. In +1726 the Jesuit d'Entrecolles wrote:

But if it was not so long before the conquest of China by the Tartars [i.e. the Manchus, +1644] that people decided to popularise this 'artificial smallpox', was it in this same Empire that the invention was born, or was it received from neighbouring Countries? If we are to believe some of the English Gentlemen, the Greeks of Constantinople obtained the secret from the Countries lying around the Caspian Sea, which might lead to the thought that the Chinese got it from the same source by means of the caravans of Armenian Merchants which have come for many years into this Empire. However, this same Conjecture would equally prove that it was from China that this secret passed to the Inhabitants of the lands around the Caspian.

He pointed out that the northern steppe peoples such as the Manchus had known nothing of the discovery before they came to China. So also Lockhart, a century and a half later, remarked that:

Inoculation has been known and practised in China since the time of the Sung dynasty, about eight hundred years ago. It was first introduced into Europe by Lady Montague, the wife of the British Ambassador at Constantinpole in 1721, and doubtless had found its way to Turkey across the centre of Asia from China. The Turks who lived on the Chinese frontier must have carried the knowledge of it when they moved westward.

Wise, writing about the same time, preferred a more roundabout course. He thought 'it is probable that the Chinese communicated their method of inoculation to the natives of the west, and it was followed more readily by the Mohammedans than the Hindus, from its being more in accordance with their character and belief, and they conveyed their knowledge of this mitigating remedy to the Turks, by whom it was introduced into Europe'.[199]

In any case, smallpox inoculation was being done in India during the +18th century.[200] The earliest descriptions, such as that of Chais in +1754, supposed that the practice had existed from remote antiquity. Later on in the century there were travellers such as Sonnerat (in India from +1774 to +1781), who gave an account of the smallpox goddess Śitālā, and physicians such as Holwell (+1767), who delineated the activities of the priestly inoculators, itinerant at certain seasons of the year.[201] Later travellers from the West included Ward, and later physicians, Ainslie, who summarised the knowledge available.[202] While the world was fairly well informed as to what preceded the coming of vaccination in the early 19th century, the literary evidence was very scarce. Indianists agree

---

[198] Vol. 1, pp. 181ff.; Vol. 5, pt 4, pp. 406–7.

[199] D'Entrecolles (+1726), pp. 312–13; Lockhart (1861), pp. 237ff., reproduced in Ball (1892), p. 752; Wise (1867), vol. 2, p. 546.

[200] Garrison (1929a), p. 71; Castiglioni (1947), p. 89.

[201] Moore (1815), pp. 26ff., summarises Sonnerat and Holwell.

[202] Ward (+1795), vol. 1, p. 174; ibid., vol. 4, p. 339; Ainslie (1830).

that the ancient medical books mention neither smallpox nor inoculation. A putative reference to the former, in the *Bhāva-prakāsa* of +1558 by Bhāvamiśra, a medical writer of worth, may be an interpolation.[203] The only other textual reference we have found is to a work entitled *Sacteya-grantha*, attributed to Dhanvantari. He was a real physician of the +6th century, whose name is associated with one of the oldest medical dictionaries in the world, the *Dhanvantari-nighaṇṭu*.[204] But as the reference was adduced in support of the use of cowpox in ancient India, it was probably a misunderstanding, and is hardly likely to withstand critical examination.

As for the possibilities of Sino-Indian contact, there is no need to recapitulate what is common knowledge. Numerous seaways and land routes took Indian missionaries and Chinese Buddhist monks back and forth, ways of communication which were never closed during the past two millennia. Smallpox inoculation could have travelled in either direction. Even some eminent Chinese historians of medicine, such as Wang & Wu (1936), p. 216, Fan Hsing-chun (*1953*) and Chhen Pang-hsien (*1937*) have been inclined to think that smallpox inoculation came to China from India.[205] All in all, Seiffert & Tu Chêng-Hsing (1937) were unjustified in concluding from the Mt O-mei story that smallpox inoculation was an Indian export to China by way of Tibet. Until substantial traditions and reliable texts can be found in India to establish the discovery before +1500 or indeed +1000, the Chinese evidence holds the floor.

Africa seems at first sight more difficult to align with Chinese influence. Reports have been coming in ever since Cook in +1780 found inoculation in Senegambia. Welson & Felkin have given accounts for the Baris, Stahlmann for the Somalis, Levaillant and Michaux for the Ugandans, and there are many others.[206] Bruce found the technique among the Nubians, and many others have described its use in the Maghrib, in Tunis, Tripoli and Algeria.[207] Des Barres & Tanon (1912) have studied the Lower Niger, Rosenwald (1951) Tanganyika, Pankhurst (1965) Ethiopia, and Imperato (1968, 1974) Mali (where scarification was the rule).

Klebs himself, when he wrote that 'the earliest traces of variolisation are found in . . . Africa', fell into a common fallacy.[208] A 'primitive' or preliterate people, social scientists used to assume, were fixed at an early stage of human social evolution. This notion has long ago been abandoned, as has the further assumption that all the tools, techniques and practices of such peoples survive from the most ancient historical times.

Study of contacts with more advanced civilisations has regularly discredited such claims. Knowledge of East Africa goes back in Chinese literature to the +9th century.

---

[203] Jaggi (1969–73), vol. 4, pp. 41–2, 128; Klebs (1914), p. 11.

[204] Von Schrötter (1919), and Ainslie (1830), pp. 66–7, derived from Calvi Virambam (1819). On the dictionary, Sarton (1927–48), vol. 3, p. 1730; Jaggi (1969–73), vol. 4, pp. 2, 28, 49.

[205] Vol. 1, pp. 170ff., 176ff., 206ff.

[206] We have been unable to examine these sources. Anon. (1913) enumerates them without bibliographical details, as does Klebs (1913a); they are not in his bibliography (*ibid.* (1913b)).

[207] Bruce (+1790), vol. 1, p. 516; on North Africa, Gros (1902) and other references in Klebs (1913b), p. 65; on Arabia, Russell (+1768). Turkish influence accounts for origins in North Africa, but the picture south of the Sahara is not so simple.

[208] Klebs (1913a), p. 70.

Since the +13th centuy, the Chinese have strongly pursued trade there. Sino-African relations culminated in the great voyages of the treasure-ship fleets under the command of the admiral Chêng Ho 鄭和 in the first half of the +15th century. These expeditions carried many medical officers.[209] The most obvious hypothesis to evaluate is that African peoples learnt of the Chinese methods of smallpox inoculation from these visitors. Half a century of intensive contact, and three or four subsequent centuries of less frequent interaction, offer a broad spectrum of possibilities. A statement frequent in medical writings, namely that inoculation occurs more among the peoples of eastern than of western Africa, strengthens this impression.[210]

As a pendant to this story, African slaves destined to work the plantations took knowledge of inoculation to the New World. Cadwallader Colden recognised this in +1753, as did an odd anonymous Dutch work of +1757 which tabulated the advantageous effects of inoculation on morbidity and mortality in parallel columns. This practical knowledge from Africa seems to have helped the early and successful introduction of inoculation into America.[211]

Finally we come to early inoculation-like practices in Europe. They included wearing clothes from smallpox patients, handling their bedding, rubbing pustule lymph or scabs upon various parts of the body, and occasionally pricking with pins or scarification. They frequently involved 'buying' smallpox scabs, a transaction that had counterparts in China (see p. 138). These methods were first reported by Bartholinus & Vollgnad (+1671) for Germany, then by Bartholinus (+1673) for Denmark, then H. Grube (+1674) again for Germany, and Schultz (+1677) for Poland.[212] Soon after the Turkish inoculation began to spread in England, the *Philosophical Transactions* published letters on local practices by Perrot Williams and Richard Wright, two physicians in Wales (+1723). The first of these was addressed to Dr Jurin, Secretary of the Royal Society, who later organised the first statistical study of the technique's value (see p. 147). Later on, Kennedy (+1715) and Monro (+1765) gave analogous information about Scotland. Throughout the +18th century reports continued to come in. Ribeiro Sanches (a pupil of Boerhaave) in +1764 and S. G. Gmelin in +1771 found similar practices, one of which involved inhalation, in Russia. In +1769 Bacheracht wrote similarly about Sweden.[213]

We suspect that all these 'folk' usages were pale echoes or reflections of what had been done long before in China. If their penetration so far west should seem surprising, we might recall our enquiry into the travels of the characteristic Chinese still. It spread from late mediaeval times through the Slavonic culture-areas to the furthest confines of Ireland, where one would expect the heritage to be Hellenistic.[214] Unless evidence indicates that proto-inoculation, commonly reported in Europe from about +1670 onwards, originated independently, the simplest hypothesis is that such methods were anomalous results of

[209] Vol. 4, pt 3, pp. 491, 494ff. [210] E.g. Klebs (1914), p. 8.
[211] Colden (+1757); Anon. (+1757); Kittredge (1912) on Cotton Mather. [212] Miller (1957), pp. 42ff.
[213] A century later Grot and Hubert reported inoculation in Kamchatka by scarifying with fish bones; Klebs (1914), p. 7, without bibliographical details.
[214] Vol. 5, pt 4, pp. 103ff., 110ff.

radiation from that part of the world where smallpox inoculation has the oldest indubitable history, i.e. China. Stories by informants that a given practice had been done 'time out of mind' merely indicate the presence of folklore. They do not prove historical origin in a particular place.

(11) CONCLUSION

Inoculation for smallpox is a highly important topic for the world history of medicine and science, because it was the very first immunological procedure. As Weichardt said, 'Variolation was the foundational experiment for effective immunisation.'

The general pattern that we have described differs from any other that we have encountered in our volumes. Proceeding backward, there are four key turning points: +1800, approximately the time when Edward Jenner's heterologous cowpox vaccine inaugurated the era of almost complete safety in immunisation; +1700, when the Turkish practice of inoculation was introduced to England and thereafter to all Europe and North America; +1500, when the practice emerged from the shadows of secrecy and began to be written about in Chinese medical books; finally, +1000, the start of the procedure according to persistent (and, as we believe, rather trustworthy) Chinese tradition. We have described above (p. 156) the ambiance of Taoist religion, magic and medicine in which inoculation seems to have had its origin. The documentary evidence goes back much further than in any other civilisation, and before that there is still a background of five centuries of cryptic confidential practice.

Two centuries, then, were available, if not seven, for smallpox inoculation to reach the Ottoman Turks in time for them to hand the discovery on to the Europeans. The Old Silk Road was a ready means of westward transmission. Similarly, there were no impediments to the passage of the technique to India. But the data in both these cases, and in all the others, are ethnographic. No textual evidence establishes historical datings. One cannot rule out a series of independent origins, varieties of proto-inoculation that fell short of depositing the virus on a mucous membrane or introducing it into the epidermis and skin capillaries. Without further discoveries in the accounts of mediaeval travellers, multiple discovery is impossible to prove. The Chinese medical literature has yet to be scoured for pertinent documents. The only hope for data outside China is in India, but there the difficulty of dating documents, were they to be found, is well known.

In sum, the most judicious conclusion seems to us to be that the inoculation for smallpox did indeed originate in a Taoist milieu in or before the early Sung, perhaps as early as Sui, and that from that focus it diffused outwards, sometimes as a developed practice but often in dilute and fragmentary forms, throughout the Old World and many parts of Africa. It had from two to seven or more centuries to do this.

We think that McNeill did a grave injustice to the learned and articulate physicians of China when he wrote:

Even if, as seems probable, smallpox inoculation had been demographically significant in China and other parts of Asia for centuries before +1700, it had been a matter of folk practice analogous to the innumerable other customs and rules of hygiene that human beings had everywhere worked out and justified to themselves by a variety of naive and ingenious myths.

But we like well enough his further remark: 'Wherever the practice first developed, one may easily suppose that caravan traders heard of it, tried it, and thereafter propagated it as a folk practice throughout the parts of Eurasia and Africa where caravan traffic constituted the main form of long-distance trade.'[215]

APPENDIX: EDITOR'S NOTE

This chapter on the origins of immunology presents a great deal that is novel. It centres on two independent arguments. The first (p. 134) reveals that a medical author first mentioned variolation in China in +1549 and that, in the decades that followed, inoculation became widespread. The +1549 citation is an important discovery, and the authors amply document their narrative of what followed, first in China and then round the world.

The second (p. 156) argues that Taoists secretly practised variolation, perhaps as early as the +8th century, but at least from the time of the Sung premier Wang Tan, *ca.* +1000. Although this epochal innovation was not abortive, no one mentioned or even alluded to it until the +18th century. This silence is due to esoteric Taoist traditions connected to Wang. Physicians without exception observed the taboo because they were Taoists. This second set of claims deserves more extended evaluation than a few footnotes can afford.

Wang Tan

Both Wang Tan and his son Wang Su, who was inoculated in the story, were high officials. The Sung History gave both biographies, which would have been based in part on family documents. Neither account mentions this anecdote.[216] No one mentions it, in fact, until +1713, more than a century after Wan Chhüan wrote about inoculation. The idea that the secret tradition began in the +8th century first appears in 1884.

As for Wang's Taoist connections, the most obviously relevant is that he 'asked that he be habited as a "monk" after his death'. The biography notes that Wang asked a friend 'to have his hair cut and dress him with monk's clothing before putting him in the casket'. The tonsure is unmistakably Buddhist. Sung Taoist monks kept their hair long, but Buddhists shaved their heads. The biographer explains Wang's request as due to a feeling of defeat caused by his political compromises, an obvious motive for posthumously joining the *sangha*.

---

[215] McNeill (1977), pp. 254, 255.
[216] *Sung shih*, ch. 282, 320. Franke (1976), vol. 3, pp. 1147–53, summarises the debate over Wang Tan's character.

Another component of the argument is the names that various versions use to describe the inoculator. Various late authors call him a divine man, an immortal, etc. *No account uses an unambiguous term for 'Taoist'*, but he is called a Buddhist master. Joseph Needham, like his sources of the 1950s and 1960s, believed that 'immortal', 'divine man' and so on are peculiarly Taoist terms. He asserts that when Hsü Ta-chhun called the practitioner an immortal, he meant that he was 'a numinous Taoist'.[217]

A scholar of religion today would see all of these terms as very broad in scope, and primarily allied to popular religion, as I have noted in the Introduction. In fact, Chu I-liang's assertion that Wang consulted 'shamanic technicians' (*wu* 巫) points to popular priests; *wu* is a term of contempt that literati used for them and anyone they could not be bothered to distinguish from them.

Needham also gives considerable weight to the +18th-century tale that the inoculator was 'a *ku hsien, san pai chen jen* 古仙三白真人', i.e. 'an ancient immortal, a three-white perfected immortal'. 'Three-white' is not Taoist, but a Buddhist term that refers to keeping three dietary disciplines as well as to observing certain spiritual disciplines when moving about, when thinking and when chanting sutras.[218]

The frequent association of Wang's inoculator with Mt O-mei in Szechwan, 'famous for its associations with both Taoism and Buddhism', inspires a learned argument meant to attach it firmly if 'subordinately' to the former. Of the 'Taoist' hermits, immortals and alchemists whom legend places at O-mei, some have Taoist associations and some do not. But one can hardly conclude that, because the mountain collected tales of people with Taoist associations along with those linked to popular religion and Buddhism, anyone connected with it by legend is likely to be a Taoist.

We must finally consider the weight of massed authority. Joseph Needham claims that the tradition of Wang Tan and the O-mei connection 'was accepted on all hands'. That claim is excessive. Yü Thien-chhih in +1727 obviously did not accept it. Yü, as the text makes it clear (p. 134), wrote in some detail that inoculation arose in Anhwei between +1567 and +1572, and spread from there.

The later sources offered as examples of universal acceptance do not all serve that purpose. *I chi khao* (Tamba 1819, cited as *ICK* in note 162) cannot serve as evidence, since it is merely a verbatim compilation of bibliographical data. In this case the compiler is merely quoting the preface to *Chung tou hsin shu*, not because he agrees with it, but because he reproduces all available prefaces.

The most substantial histories of medicine published in Chinese since 1949 also do not reflect a consensus. In his monograph of 1953 the historian Fan Hsing-chun (note 161) did indeed repeat the story. He changed his mind in his 1986 history. There he says flatly that 'in the mid +16th century, in the Lung-chhing era of the Ming (+1567 to +1572), variolation was finally invented (*fa-ming* 發明)'. So does Chao Phu-shan 趙璞珊. Chia Tê-tao 賈得道 also considers only the Ming adequately supported, and gives new

---

[217] Hsü does not in fact refer to the Wang Tan story, but merely says that variolation 'has been transmitted by immortals'; whether he means one or more, and at what point in its history, is impossible to say.

[218] *Kao seng chuan* 高僧傳, ch. 23, p. 7.

evidence for it. Li Ching-wei 李經緯 and Li Chih-tung 李志東 present the Thang, Sung and Ming hypotheses, and go on: 'as for which explanation is reliable, most scholars, considering that the grounds of its argument are adequate and that there is collateral evidence, approve the hypothesis that it began in the +16th century; this of course is undeniable'. They add that although the notion of a Thang alternative remaining hidden for a millennium 'truly lacks credibility', it ought not to be entirely ruled out because of earlier Thang experiments on scarification to treat, not to prevent, warts and moles – scarcely a compelling point. This discussion does not pause over the Sung story. The history by Liu Po-chi 劉伯驥 accepts the Sung anecdote without comment or hesitation, and that of Ma Po-ying 馬伯英 with only a little: 'It appears *comparatively believable* that around the +11th century, in Szechwan and Honan, variolation was already being carried out.' These diverse assessments do not add up to unanimity on the parts of 'all hands'.[219]

THE RELIGIOUS CONNECTIONS OF INOCULATION

The conclusion of the argument asserts 'we have seen that the cure, and even more the prevention, of smallpox were closely associated through the ages with Taoist religion and magic'. It traces the motivation for preventive inoculation to 'the Nature-mysticism of Taoist philosophy on which the Taoist Church was founded, the idea that everything goes best when man interferes least'. But the argument bases the association not on concrete and pertinent evidence, but on very diverse inferences at a high level of generalisation. The Supreme Purity, Celestial Masters, Divine Empyrean and other movements, 'the Taoist Church' of the Thang and Sung eras, seen in the light of historical and liturgical documents, was no Nature-mysticism living out the dream of Chuang-tzu. Their leaders, like their Buddhist counterparts, were socially and politically active. As for the few individual devotees we know about in the Sung with an interest in therapy, one can say only that some were medical minimalists and others were not. The bearing of Taoist movements on medicine is a matter most fruitfully explored, not by reasoning about disembodied isms, but by studying concrete linkages in individuals and clearly defined groups.

There is another focus on the 'wide aura of religious and magical conceptions and practices strongly Taoist in character . . . strengthened by the charms, incantations and cantraps . . . and the altars for worship' that inoculators and curers used. Joseph Needham ties this point to the larger one by emphasising the 'secrecy and circumspection' that governed these and other aspects of therapy. Particularly cogent is 'forbidden prescriptions' (*chin fang* 禁方), described in this section as a kind of secret formula used at every social level of therapy. *Chin fang* is indeed a broad and vague term, but the discussion in this section does not follow Chinese usage. Formulae so labelled, for instance,

---

[219] Fan Hsing-chun 范行準 (*1986*), p. 213; Chao Phu-shan 趙璞珊 (*1983*), pp. 226–7; Chia Tê-tao 賈得道 (*1979*), pp. 226–7; Li Ching-wei 李經緯 & Li Chih-tung 李志東 (*1990*), pp. 245–51, esp. p. 247; Liu Po-chi 劉伯驥 (*1974*), p. 284; Ma Po-ying 馬伯英 (*1994*), pp. 809–12, esp. p. 811, my italics.

do not ordinarily contain special terminology, and do not necessarily have any Taoist significance.[220]

The example with warnings about dissemination does happen to have a Taoist connection. It is preserved in a book of the physician-alchemist Sun Ssu-mo. But the quoted remark is not Sun's. Chs. 29–30 of his *Chhien chin i fang*, from which it comes, are a manual of ritual interdictions. Some are for healing, but these specialised liturgies have little in common with the *chin fang* described elsewhere in this section. This undated manual was, before Sun's time (the mid +7th century), transmitted among Taoist initiates of the Celestial Masters, but T. H. Barrett (1991) has argued that it was circulating among laymen in Sun's time.

In general, before printing became the norm, such warnings might accompany any medical book, or any other book on which a lineage of transmission from master to disciple was founded. They did not mark a special class of books. What prompted the 'forbidden' label on certain books of the Sung and earlier was that they were, or claimed to have been, published *despite* such warnings. It advertised their esoteric nature.[221]

Other examples of 'the confidential therapies of oral tradition' come from Chao Hsüeh-min's +1759 repertory of an itinerant healer, Tsung Po-yün, and an anonymous 20th-century manual of apotropaics. Needham notes of the former that its confidentiality about methods is 'not so much because these practitioners were unwilling to record them, but because they could not write literary Chinese'. This argument is also difficult to evaluate, since there are a great many written *chin fang*, and no reason to see the genre as quintessentially oral. By the Chhing, literacy was very widespread, as Rawski (1979) has shown.

Neither of these examples transcribes a previously oral tradition. Chao's book makes it clear that he began with Tsung's own written repertory. Chao simply excerpted from it, no doubt improving the style as he went, but bowdlerising freely and unabashedly the items he chose in order to make the popular remedies look respectable to his élite colleagues. The second example (Anon. (*ca. 1930*)) is, in fact, an unacknowledged re-arrangement of the 'esoteric' but widely available Ming *Chu yu shih-san kho* (see p. 160, above) adapting its content to the circumstances of an apotropaic healer in a modern Chinese city. The compiler's additions are, if far from belletristic, quite literate. Neither book claims a connection to any Taoist movement.

The detailed catalogue of medical gods in 'the Taoist pantheon of divinities' (p. 156) obviously includes gods of popular and Buddhist origin. It is simply the collectivity of gods popularly worshipped. The various epidemic gods were not created by Taoist masters, but by ordinary people. Needham frankly speculates that 'where the specialised votive temples were concerned, there must have been a wealth of practices and incantations which were never written down. Those would have been the secret preserve of Taoists who conducted the affairs of the supplicants' (p. 157). No one has identified in the voluminous Taoist scriptures secret preserves bearing on smallpox before the Sung. One cannot deduce from unwritten and unknown liturgical practices a therapeutic

---

[220] See for instance Li Chien-min's (*1997*) special study.   [221] See Sivin (1995c).

tradition that secretly passed down inoculation for centuries without leaving a record in either medical documents or esoteric Taoist writings that have survived in great bulk.

SICKNESS AND SIN

The final component of the search for early Taoist origins of variolation suggests, with the aid of comparisons from India, that Taoism was a religion less 'resigned to fate' than others. Lu & Needham contrast this affirmative attitude to the one that motivated the 'preparation of family altars for sacrifices and oblations to the smallpox deities'. Such ritual offerings, part of the inoculation process, imply 'repentance, and repentance implies the recognition of former wrong-doing'. Few students of Chinese popular religion today would agree with this generalisation, which equates worship, in the Judaeo-Christian mode, with sacrifice. Specialists find expressed in the content of liturgy a different motive: offerings establish or reassert a relation of mutual obligation. Sinologists generally agree that some combination of guilt and shame play a role in popular religion, although they do not agree about how these two are balanced. But these emotions do not explain the structure and functions of liturgy.

Joseph Needham contrasts ritual repentance with the Taoist activism of 'the rational physicians' who believed that therapy could alleviate or cure smallpox. He describes the 'hedge-doctors' sometimes as ritualists, sometimes as 'rational' activists. They are, he believes, unshakably Taoist because they appeal to 'immortal Taoist deities', but so, he believes, are physicians convinced by Taoist 'Nature-mysticism' that they may presume to borrow 'tools from Nature's workshop'. Readers are left wondering where the inoculators of the 'five hidden centuries' fit.

CONCLUSION

This is a fragmented argument, for it appeals to so many senses of the protean word 'Taoist' that it is impossible for them to converge toward a single hypothesis. Readers familiar only with the religious picture as most Sinologists saw it a generation or more ago may well find one or another claim sympathetic. But the explanation is too universal to answer concrete questions. The hypothesis that every author in China maintained silence about inoculation from the +8th to the +16th century poses just such a concrete question.

Does the argument about the five hidden centuries hold up in the light of fuller knowledge? We find in the sources cited no clear connection between Wang Tan and any Taoist movement or master of his time. None of the arguments that inoculation lurked for centuries in a secret Taoist matrix is based on primary evidence or even on early assertions. Once one sets aside the assumption that every inoculator and indeed every physician was a Taoist adept, pledged to confidentiality, there is no answer to why no one ever betrayed this semi-millennial intrigue. Finally, how do we assess the argument that Taoism, free from the disabling remorse of other religions, inspired rationalist therapists to triumph over disease? Although problematic, it is original and thought-provoking.

How it bears on a conspiracy of silence, however, never becomes clear. We are left asking 'Which Taoism?'

The tale of Wang Su's inoculation has been told often over the centuries. It has most often figured in the charter myths of later inoculators. Historians have examined it again and again, with inconclusive results. This section contains the most detailed examination of the evidence by any historian, and the broadest argument for a Sung origin of variolation. Lu & Needham were of course distracted by obsolescent views of Chinese religion. Nevertheless, if their peerless combination of learning and analytical power could not establish the point, we can finally let it go.

(e) FORENSIC MEDICINE

From the very dawn of civilisation, as soon as any ethical conceptions of justice arose, it must have been incumbent on magistrates and juristic officials everywhere to differentiate between accidental death, murder and suicide. Soon other, similar questions began to arise as well, questions concerned with poisoning, wounding, arson, miscarriage, disputes over virginity, and the like. Did a man drown by being pushed into the water, or was he already dead? The relative participation of jurists and physicians in developing it is a very interesting story. The advances of anatomy, physiology and pathology bore markedly on its evolution.

In Europe, broadly speaking, the physicians entered the field, writing learned treatises on all these topics, in the days of Vesalius (+1514 to +1564) and Galileo (+1564 to +1642), when distinctively modern science was born. We shall look at the history of this genre of writing in detail (p. 196), but our focal point will be a book published in China centuries earlier.

The greatest work on forensic medicine before the Scientific Revolution was the *Hsi yüan chi lu* 洗冤集錄 (The washing away of wrongs), written not by a physician but by a jurist learned in medicine, Sung Tzhu 宋慈, in +1247. There is no clear evidence that in ancient Egypt, Babylonia or Greece medical knowledge was used officially as proof in courts of law. The Hippocratic corpus does discuss such medico-legal questions as the relative fatality of wounds in different parts of the body, the possibility of super-foetation, the average duration of pregnancy, the viability of embryos before full term and malingering. Greek courts of law must have considered these matters from time to time, but left no record of them. Nor do physicians appear to have been summoned to give opinions on such cases in Roman times, though it seems that a medical expert, studying the twenty-three stab wounds in the body of Julius Caesar (d. −44), declared that only one, in the chest, was fatal.[1] As long as the science of medicine accommodated so much magic and mystery, it is not to be wondered at that courts of law did not emphasise what its practitioners said.

The irrational, from the beginning, was less prominent in Chinese medicine. We have noted in Section (a) that ancient Chinese medicine, as we see it in the lives and pronouncements of such men as I Ho 醫和, I Huan 醫緩, Pien Chhüeh 扁鵲 and Shun-yü I 淳于意, was rational through and through. The yin and yang, the Five Elements and the particular associations of bodily functions with viscera were not alignable with modern science, but that is natural for the time in which they were devised. Although magistrates trying cases were generally reluctant to solicit the opinions of medical practitioners, they themselves were expected to know enough about the human body to settle forensic questions.

---

[1] Smith (1951), p. 600.

When we write of Chinese judges, we mean local officials whose broad duties included investigating and trying criminal cases. Juristic officials also included specialists in legislation and review of cases in higher echelons of government. Both groups were drawn from the scholarly or literati class. They considered themselves socially and intellectually separate from physicians, who supported themselves by curing the living.[2] Humanistic education had looked askance at scientists and technicians for many centuries. In the Northern Sung this ceased to be a deterrent. Many literati actively concerned themselves with the sciences and the technologies. By the Ming such breadth was no longer common.

The classical scholars from the beginning read many medical books, and often learned therapy to ease the old age of their parents.[3] Similarly, Sung scholars often preferred to be treated by a social peer rather than by a medical pedlar. One can see therefore how the scholarly jurists felt they could get on well enough without the evidence of the physicians.[4] They tended to call them in only to resolve practical questions such as what was responsible for a particular wound, or whether death was due to medical malpractice.

By the early +12th century, as we have shown in Section (c), the government medical colleges were turning out well-trained practitioners. Chinese physicians lacked the guild organisation of their European colleagues, through which they might have pressed for more forensic assignments. But perhaps the best explanation is that their aim was the care of the living, the cure of disease, the restoration to health; they were never interested in examining corpses to help to decide legal cases.

As we shall see (p. 191), apart from the judge himself, the only 'expert' at most medi-aeval Chinese inquests was the *wu tso* 仵作, or ostensor, a man of low social status and little education. As he often earned his living as an undertaker, he may well have had considerable experience in handling corpses, but no medical training or interests.

In due course we shall show that, beginning in the −3rd century, there was a long tradition in China of medico-legal treatises, interweaving with a related genre of legal case-books for the benefit of judges. Why should the Chinese have codified the procedures of forensic medicine so early? The answer is indeed interesting.

After the end of the Warring States period, Chinese civilisation never again became an aristocratic military feudalism, as Europe was until the Renaissance. On the contrary, it has been called a bureaucratic feudalism, in which the empire was ruled by an imperial court acting through a vast array of officials in a civil service − the mandarinate if you like. To put it in a nutshell, the idea of the career open to every talented young man was not a Napoleonic innovation. The Chinese civil service (discussed in Section (c)), which chose its officers by examination from the +1st century on, embodied it.

[2] Medical practice was not a route to high official status. See Section (c). On popular ridicule of physicians as reflected in plays and novels, see Li Thao (*1948*) and Idema (1977).

[3] As the common saying has it, 'no one's child can avoid knowing [something about] medicine' (*wei jen tzu chê pu kho pu chih i* 為人子者不可不知醫). For example, Huangfu Mi 皇甫謐, who wrote the *Huang ti chia i ching* 黃帝甲乙經 in the +3rd century, originally became interested in medicine because his mother was ill.

[4] Van der Sprenkel (1962), p. 74, studying Chhing legal institutions, hazards other reasons.

The Chinese system was centred on criminal law. Although it was highly developed, the early philosophers did not estimate it highly.[5] Higher juridical authorities reviewed local magistrates' decisions, and one could appeal to them, but in the pyramid of administration there was no fundamental distinction between juridical and administrative functions. This was rather like the system which prevailed, for example, in British colonial Africa, where the District Officer was also the judge of first instance.

The mediaeval Chinese district magistrate had to be not only a judge but also the prosecutor, the defending counsel, the chief detective superintendent, and a Sherlock Holmes as well. The process depended on individual wisdom and deep insight into motives. To some extent the government codified it, but great discretion was always explicitly left to the official. Any member of the mandarinate could be drafted into the position of judge, and specialise thereafter in juridical cases. This was true of Sung Tzhu himself, always renowned as a judge.

These characteristics do not explain the special passion for justice, the washing away of wrongs or unjust imputations (*yüan* 冤), which inspired Sung Tzhu and his predecessors and successors. While Chinese society was bureaucratic, it was also essentially Confucian, dominated by an ethical system which neither had nor required supernatural sanctions. It was a this-worldly morality, springing from belief in the fundamental social nature of man, and all the more compulsive because it depended on no higher being.[6] The scholars were rather sceptical and detached in matters of religion, however much the common people might run after Buddhist monks and Taoist priests. What they responded to at their best was a moral imperative defined 2,000 years before Immanuel Kant. It made the clearing of the innocent, the comforting of widows and orphans, and the bringing of the cruel and violent to justice a consuming urge in men such as Sung Tzhu and his predecessors, including the renowned Judges Ti and Pao. Ti Jen-chieh 狄仁傑 (d. +700) was known to everyone through the ages from song and story, and Pao Chêng 包拯 (d. +1062) was equally renowned in popular tradition. These considerations may go some way to explaining why forensic medicine arose so early, perhaps earlier than in any other part of the world, in Chinese civilisation.

(1) SUNG TZHU AND HIS TIMES

Sung Tzhu was born in +1181 into an unimpeachably official and Confucian family; his father had been a middle-ranking judge. In his youth he was taught by Wu Chih 吳稚,

[5] For general disquisitions on Chinese law and its history, we must mention the books of Escarra (1936), Chhü T'ung-Tsu (1961), van der Sprenkel (1962) and Bodde & Morris (1967), and the illuminating papers by Bodde (1954, 1963). On Chhing law, see Alabaster (1899). We have touched upon the philosophy of law at several points in our consideration of human law and laws of Nature in China and the West in Vol. 2, pp. 518ff. On the conception of law in general, see Hart (1961).

[6] [Official contempt for popular religion was a fact, although it became highly ambiguous when the State recognised many local cults from the N. Sung on. Still, despite the political domination of values that Sinologists conventionally call 'Confucian', officials, their families, local governments and the imperial government through history recognised and worshipped 'higher beings'. All officials recognised divine sanctions in making and carrying out laws. One might add that a passion for justice is quite prominent in the songs and stories of un-educated commoners. The issue needs to be argued with much more nuance. – Editor]

one of the followers of Chu Hsi 朱熹, that greatest of Neo-Confucian philosophers. In +1205, at the age of twenty, he entered the Imperial University, whose Chancellor in due course praised him for his literary ability. Graduating as Presented Scholar, the highest ranking, in +1217, he was appointed Sheriff of a sub-prefecture in Chekiang. As his father then died, he could not take up the post. In +1226, however, he became Assistant Magistrate at Hsin-fêng 信豐 (Kiangsi). During the ensuing years he made a great name for himself by putting down peasant rebellions, dealing with their root causes by distributing food from government granaries, but his success offended his superiors, and he was repeatedly impeached. In +1238 he was sent into the field again, to Nan-chien 南劍 (Fukien), where he found that the wealthy were hoarding rice. He classified the people into five groups, and relieved the extremities of the poorest by redistributing the excess property of the richest.

Sung's first full-time judicial post came in +1239 when he was appointed Judicial Commissioner of Kuangtung. There he liberated many innocent prisoners, and in eight months decided some 200 cases of murder, suicide and accidental death. As a judge in Chhang-chou 常州 (Kiangsu) he was involved in revising the local history and geography. In +1245 he became the Prefect. In +1247, while Judicial Commissioner of Hunan, he completed his immortal *Hsi yüan chi lu*. In +1249, as Prefect of all Kuangtung, he died at the age of sixty-four. He appears to have been an official of the highest quality, combining his compassion with a level-headed, indeed scientific (so far as the word could be applicable in his time), approach to the medico-legal problems in which he was particularly interested.[7]

(2) THE *HSI YÜAN CHI LU*

The *Hsi yüan chi lu* was the first systematic treatise on forensic medicine in any civilisation. Sung was driven to write it because in spite of the many advances before his time, including the compilation of selected cases (see the next Subsection), and the introduction of standard forms and diagrams, there was still much inefficiency and dishonesty in inquest procedures. He wished to end the resulting injustices.

Sung's preface of +1247 survives in the oldest extant edition, that of +1308. It vividly states the situation in his time.[8]

In judicial affairs, capital cases are the most important. In capital cases, most important are the data initially collected (*chhu chhing* 初情).[9] For these, holding inquests is most important. The

---

[7] In 1957 his tomb was found in the hills near the village of Chhang-mao-tshun 昌茂村 in Chien-yang county in Fukien. There are biographies of him by Chu-ko Chi 諸葛計 (*1979*); Jen An 軔庵 (*1963*), pp. 68ff.; Yang Wên-ju 楊文儒 & Li Pao-hua 李寶華 (*1980*); Kao Ming-hsüan 高銘暄 & Sung Chih-chhi 宋之琪 (*1978*); and Miyasita Saburō in Franke (1976), vol. 3, p. 990. Much has been written about Chinese forensic medicine: Chhen Pang-hsien 陳邦賢 (*1937*), p. 189, Sung Ta-jen 宋大仁 (*1957a*), Shih Jo-lin 施若霖 (*1959*), Chhen Khang-i 陳康頤 (*1952*) and Chang Hsien-chê 張賢哲 & Tshai Kuei-hua 蔡貴花 (*1971*). Some of these authors had access only to post-Yüan versions of the *HYL*. Sung and Chang & Tshai contain detailed chronologies of Sung Tzhu's life. The gazetteer on which Sung worked was *Chhung hsiu Phi-ling chih* 重修毘陵志 (+1268). On Chu see Vol. 2, pp. 455ff.

[8] Pp. 1–3, tr. auct. adjuv. McKnight (1981), pp. 37–8. Sung's book was not a government publication. It was much smaller than the later official enlargements.

[9] In any doubtful case, repeated investigations were required; McKnight (1981), p. 5.

power to grant life or death, egress or admission; the decision [lit. 'trigger'] that countenances or rejects wrongful accusation: on the basis [of inquests] everything within the scope of the law is decided. They elicit the greatest care on the part of judicial officials.

Recently in circuit and district cities the responsibility of conducting inquests has been delegated to newly-appointed officials and military officers, who lack deep experience but are expected suddenly to cope. Moreover, the deceitfulness of ostensors and the corruption of petty clerks give rise to all sorts of fantasies and falsifications, confusing the case beyond the possibility of investigation. Even someone percipient, [limited by] his one mind and two eyes, lacks the means to apply his wisdom. How much worse it is if he looks on from afar, holding his nose in disdain!

Four times I have been a legal official. Lacking other talents, I devoted myself wholly to criminal cases, pondering them again and again, never daring to be in the least dilatory or complacent. If I saw clearly that a case was specious, I would immediately remand it. If I could not decide what was doubtful and what reliable, I continued to turn it over in my mind. I mainly feared acting precipitately, subjecting the deceased to investigation in vain. I often reflected that wrong decisions in criminal cases mostly derive from discrepancies in the initial stages. Errors in conclusions based on the investigation are all traceable to superficial inquiries and tests.

I have collected a number of recent books on the subject, from the *Nei shu lu* 內恕錄 (Record of inward empathy) onwards. I have selected the most important items, corrected them, supplemented them with my own perceptions, and made the whole into a book. I have named it 'Collected Writings on the Washing Away of Wrongs'. The blocks for the printing were carved at the office of the Judicial Commissioner for Hunan. I wish to show the book to my colleagues, so that they may test it in practice.

The matter resembles that of master physicians discussing the ancient techniques. If you first clearly understand the circulatory system and the difference between the patefact and subdite signs (*piao li* 表裏), and sooner or later use [this knowledge] when inserting the acupuncture needle, you will never miss. Thus [this book], in washing away wrongs and in doing good, will have as great utility [as those of medical writings that] bring the dying back to life.

This preface the Judicial Commissioner of Hunan signed with all his official ranks and titles.

Sung Tzhu emphasised the themes that run throughout the subsequent literature: the heavy load that rests on the conscience of the judge in seeing that true justice is done; the unreliability of the assistants and attendants, only too eager to accept bribes and to falsify the evidence; and finally the equivocality of the tests, assays and verifications which the men of old had accepted. It may be significant that Sung ended his preface by an explicit comparison with what the physicians were trying to do. He was responding to the special status of medical evidence in criminal law. It was not yet possible to build entirely systematic links between medicine and law, but they were equally high-minded in their aims.

(3) FORENSIC MEDICINE IN CHINA BEFORE SUNG TZHU

Forensic medicine had a long history before Sung's time. His book mentions, in addition to the *Nei shu lu*, of unknown date, three legal writings which he found useful. All are lost. None of the titles look like those of books, nor are their authors known. They may have been simply essays by his Northern Sung predecessors. One was entitled

Shên hsing shuo 慎刑説 (Discussion on the special care needed in sentencing); another was *Wei hsin phien* 未信篇 (Examples of what can hardly be believed); and the third was *Chieh an shih* 結案式 (Form for judicial decisions), apparently used for examining personnel who would conduct inquests.[10]

A well-known book earlier than that of Sung was the *Chê yü kuei chien* 折獄龜鑑 (Magic mirror for solving cases). Chêng Kho 鄭克 produced it in +1133, just after the Northern Sung fell to the Chin Tartars. Earlier than that, Ho Ning 和凝 and his son Ho Mêng 和㠓 published the *I yü chi* 疑獄集 (Collection of doubtful cases, between +907 and +960).[11]

The books of the Ho family were by no means the first of this genre. In the +6th century, Hsü Chih-tshai 徐之才 wrote *Ming yüan shih lun* 明冤實論 (Concrete discourse on bringing unjust imputations to light).[12] Hsü's work has not survived, though it may be possible to reconstruct it from quotations.

As we have already intimated, the medico-legal treatises were preceded by others which might be called 'case-books'. In +1211 Kuei Wan-jung 桂萬榮 compiled a remarkable work entitled *Thang yin pi shih* 棠陰比事 (Parallel cases from under the pear tree). This was a collection, arranged categorically, of cases decided by eminent judges.[13] It was printed in +1222 and again in +1234, so Sung may have known it. Kuei (probably *ca.* +1170 to +1260) was a judge himself. In +1208 he became Administrator for Public Order in the prefectural administration of Chien-khang 建康, near modern Nanking. Kuei was a successful administrator, a devoted Neo-Confucian philosopher and an amateur of medicine.

From his book we may take an outstanding example of the experimental approach to forensic medicine, recorded earlier in the *I yü chi*. Sung must have known it:

When Chang Chü 張舉 of the Wu dynasty [+220 to +280] was magistrate of Chü-chang, a woman murdered her husband and thereafter set fire to the house so that it burned down. She falsely stated that her husband had burned to death. Her husband's family suspected her, and accused her before the authorities. The woman denied it, and would not confess her crime. Chang Chü then took two pigs. One he had killed; the other he let live. Then he had both of them burned in a shed on a heap of faggots. On examining [the two burned pigs for differences, he found that] the pig previously killed had no ashes in its mouth, while the mouth of the pig burned alive was full of them. He verified that there were no ashes in the dead man's mouth. Confronted with this evidence, the woman indeed confessed.

As van Gulik commented, it is interesting that Chang Chü chose pigs for his experiment, for it is well known that the anatomy and general size of a pig are similar to those of the human body.

---

[10] Chung Hsü 仲許 (*1956*). None of these titles is found in the bibliographies of the dynastic histories.

[11] *Chê yü kuei chien* was re-issued in +1261 with a preface by Chao Shih-tho 趙時彙; see Sung Ta-jen 宋大仁 (*1957b*). The many important Ming writings on forensic medicine include Chang Ching's 張景 *Pu i yü chi* 補疑獄集 (*ca.* +1550?).

[12] Chhen Pang-hsien (*1937*), p. 189. On cases from *I yü chi*, *Chê yü kuei chien* and *Thang yin pi shih* see Wang Chi-tsu 汪繼祖 (*1958*).

[13] The author relied particularly on the *I yü chi* and the *Chê yü kuei chien*.

This is not the only time that the theme of experimentation occurs in the *Thang yin pi shih*. For example:

> When the Academician-in-waiting, Hsü Yüan 許元, at the beginning of his career [*ca.* +1043], was Intendant of Exchange, he was greatly distressed because most of the official boats [proved unworthy because the shipwrights] used fewer nails than they charged for. [Since all the badly constructed boats had sunk, there was no means of checking how many nails had been used; thus the contractors could practise this deceit indefinitely.] One day Hsü Yüan gave orders that a newly constructed boat should be burned. [After the ashes were raked through,] he had the nails weighed. It turned out that they were only a tenth as many as had been paid for. Hsü Yüan then fixed the amount of iron nails to be used.[14]

From this one can see that criminal trials and inquests were not the only problems facing the officials of the mediaeval Chinese bureaucracy.

(4) THE BAMBOO SLIPS OF CHHIN

Sung Tzhu's predecessors were collections of cases; it was he who had the originality and the organising spirit to systematise forensic medicine. All subsequent books on the subject have taken *Hsi yüan chi lu* as their blueprint.

Recent archaeological research has succeeded in tracing a concern with forensic medicine back to the Chhin time. In 1975, several Chhin tombs were opened at Shui-hu-ti 睡虎地 near Yün-mêng 雲夢 in Hupei. The archaeologists found, in tomb no. 11 of this group, 1,155 bamboo slips, including records of legal cases. The deceased, almost certainly a judge in the state of Chhin, was born in −262 and died in −217. We do not know his family name, but only his given name, which was Hsi 喜. There were twenty-five pro forma documents, entitled *Fêng chen shih* 封診式 (Models for sealing and invest-igating), written on ninety-eight bamboo strips and buried to the right of the skull. In date they must run from −266 to about −246. The dead man must have greatly prized them.

Of the twenty-five specimen cases, four are germane to our theme. Let us examine them, conflating the translations of Bodde and McLeod & Yates:

1. 5.19 (E20). Death by robbery with violence

> Report: A thief-catcher A belonging to guard-station X made an announcement, saying 'At place Y within this jurisdiction there has been a death by murder. The victim was a single male with bound-up hair [i.e. an adult], but his identity is not known. I have come to report.'
> We immediately ordered the prefectural clerk B to go and make an examination.
> Report by the prefectural clerk B:
> Together with prison-bondsman C we accompanied A to make the examination. The male corpse was lying in house Z straight out, facing upwards, and with its head to the south. On the left temple there was one knife wound, and on the back of the head was one some four inches long and one inch wide, with blood oozing from one to the other; the lacera-tion was just like those made by an axe. The chest, temples, cheeks and eye-sockets of the corpse all exuded blood, which covered the body down to the ground, so that nowhere could the length or breadth of the bloodstains be established. Other than this, the corpse was intact.

---

[14] *Thang yin pi shih*, tr. van Gulik (1956), pp. 102–3, 90–1; *HYCL*, p. 75.

The clothes were of unlined cloth, skirt and jacket, one item each. The back of the jacket showed two vertical cuts, and was blood-stained, as also was the middle of the skirt. West of the body there was a pair of lacquered Chhin-patterned shoes, one of them a little more than six paces away from the corpse, the other ten paces. When the shoes were tried on its feet they fitted.

The ground being very hard, we could not see evidence of the assailant's footprints. The victim had been able-bodied, in the prime of life, light-complexioned, and five feet one inch tall, with hair a foot and a half long. On his abdomen there were two old moxibustion scars.

From the place where the man's corpse was to A's guard-station X was some 100 paces, and to the farmhouse (*thien shê* 田舍) of D, a member of the rank and file [i.e. a commoner] of E village, was 200 paces.

We ordered A to cover the corpse with its cloth skirt, to bury it at place F and to await further orders, after bringing the jacket and shoes to the judicial office. The people at A's guard-station and those living near C are to be questioned to ascertain on what day [the man] died, and whether or not shouts of 'bandit' (*khou* 寇) were then heard.[15]

This wonderful document may mark the very birth of forensic medicine. For the −3rd century, the accuracy of description which was demanded of the examiners is very striking. One would like to know what the thief got away with, and whether he was brought to justice; but these documents are pro forma examples.

2.   5.20 (E21). Death by Hanging

Report: A, the chief of X village, stated that 'villager B, a commoner, has died by hanging in his home; the cause is unknown, and I have come to report'.
We immediately ordered the Clerk C to go and make an examination.
Report of the Clerk C:
Together with prison-bondsman D, I accompanied A, as well as B's wife and daughter, to examine the corpse of B. It was suspended from the cross-beam of the north wall within the eastern bedroom of the house facing south, by a hempen rope as thick as one's thumb. The noose was tied round the neck, and knotted at the nape. Above it the rope was twice looped around the cross-beam and then tied in a knot, leaving a tail end some two feet long. The top of the head was two feet under the cross-beam, and the feet were suspended two inches above the floor. The head and back touched the wall, and the tongue protruded slightly; below, excreted faeces and urine had soiled both his legs. Upon releasing the rope, the breath issued with a gasp from the mouth and the nose. The rope left a compression bruise[16] [around the neck] except for a two-inch space at the nape. Elsewhere the body showed no signs of knife wounds or traces of [damage by a] wooden [cudgel] or rope. The cross-beam was one span in circumference, three feet long, and projected westwards two feet beyond the edge of an earthen platform [on the ground below]. From the top of this it was possible [for the deceased] to lead and bring together the rope around the beam. The ground being hard, no human footprints could be detected. The rope was ten feet long. The clothes consisted of an unlined jacket and skirt of unbleached silk, two items, and the feet were bare.

---

[15] Anon. (*1976, 1977*); Bodde (1982), McLeod & Yates (1981), Hulsewé (1978), p. 193, and *ibid.* (1985), pp. 183ff. All these papers are based upon a wealth of learned considerations, which we perforce omit. The cases cited belong to Hulsewé's Group E; transcribed Chinese text in Anon. (*1976*), pp. 35ff. The first numbers given above are those of McLeod & Yates; the second are those of Hulsewé.
[16] Cf. Gordon & Shapiro (1975), p. 96.

We immediately ordered A and the daughter of B to convey C's corpse to the Judicial Office.
When examining [a body] it is essential to make first a thorough and careful inspection of the evidence. Assuming he or she was alone at the place of death, one should note where the rope comes together [around the victim's neck] and see whether at this meeting-point there are any marks of the rope's passage [around the neck]. Check whether the tongue sticks out or not. How far away are head and feet from where the rope joins [near the beam above] and from the floor [below]? Have excrement and urine been voided?

Then release the rope and see whether there is a gasp from the mouth and nose. Inspect the form of the bruises left by the noose, and try whether it can be released from the head. If it can, then remove the clothing, and examine the entire body from head to perineum. But if the tongue does not protrude, if mouth and nose do not spurt air, if the imprint of the rope leaves no bruise, and if the knot of the rope is so tight that it cannot be loosened – then the true cause of death is difficult to determine. If a long time has elapsed since the death, the mouth and nose will sometimes be unable to gasp.

When someone commits suicide, there must be a reason. Those who have lived together with the dead person should be questioned in order to elicit this.

Again we are amazed by the precise detail and the quantitative exactitude with which the examiners scrutinised dead bodies in the −3rd century. They recognised the possibility that the supposed suicide was really murder; and that the victim was strung up after death so as to simulate suicide, having first perhaps been poisoned. Although we have no reason to believe that Sung Tzhu saw these texts, he was the heir of a very long tradition.

3.   5.22 (E23). A Miscarriage

The denunciation of C by A, both adult women commoners of X village, reads:
A had been pregnant six months. Yesterday she had a fight with C, who lived in the same ward. They seized each other by the hair, and C knocked A down, so that she fell against the screen inside the gate. Gentleman D, who also lived in the ward, came to the rescue and separated A and C. As soon as A got home, she felt ill with abdominal pains. Last night the child miscarried and was aborted. Now A, bearing the wrapped-up foetus, has come to denounce C.

We immediately ordered the prefectural clerk B to go and arrest C, and then to examine the sex of the baby, how much hair had grown, and the appearance of the placenta. We also ordered bondswoman E, who has already borne several children, to examine the appearance of A's emissions of blood from her vagina as well as her wounds. We also interrogated members of A's household about the condition in which she reached home, and the kind of pains which then came on.

Report by prefectural clerk B:
Bondswoman E and I have been ordered to examine A's foetus. At first, when wrapped in cloth, it was like a mass of [coagulated] blood, the size of a fist, and unidentifiable as a child. When placed in a basin of water and agitated, the [coagulated mass of] blood was [seen to be] a foetus. Its head, body, arms, hands and fingers, legs down to the feet, and toes, were of human type, but eyes, ears, nose and sex were unidentifiable. When it was taken out of the water, it again had the appearance of clotted blood.

Another report form reads:
Bondswomen E and F, who have each had several children, were ordered to examine A. Both stated that there is dried blood beside her vagina, and that the bleeding is still

continuing slightly, but it is not menstruation. E had a child, but it miscarried, and the bleeding then was just like A's now.

This passage shows how the medico-legal experts of the −3rd century were trying to attain the same sort of accuracy as in the other cases, but could not do so owing to inadequate embryological information. Clerk B garbled his account, but no doubt he was doing his best. Immersing the alleged foetus in water would not determine its characteristics. If it really were in the sixth month, all its aspects would be clearly distinguishable without immersion; nor would it again assume the appearance of a mass of clotted blood upon removal from water.[17]

One possibility is that what the text calls a foetus was actually a large clot of blood, which the mother expelled incidentally, not as a result of her fight with the other woman. If this were so, Clerk B could hardly have imagined the human form of the foetus. One can only conclude that the medico-legal experts in the −3rd century were trying to do their best with this problem also, but without reliable embryological information could not get very far.

We turn to the last of the cases we shall adduce from the bamboo slips of Judge Hsi.[18] It is different from all the others, because it deals with a particularly dangerous disease, against which the only defence, in ancient and mediaeval times, was isolating the patient from the rest of the population in leprosaria (*li so* 癘所 or *li chhien so* 癘遷所). Lepers apparently were executed only for capital crimes.[19] For a leper to go about at random was criminal behaviour, as in this example.

4.   5.18 (E19). Leprosy (*li* 癘)[20]

Report: The headman A of X village has brought in villager B, a rank-and-file commoner. His denunciation reads: 'I suspect leprosy, and I have come along with him [to the judicial office].'

We interrogated B. His statement reads: 'At the age of three, I suffered from sores on my head, and my eyebrows fell out. The nature of the illness could not be determined. I am not guilty of any offence.'

We ordered the physician C to examine him. His statement says: 'B has no eyebrows. The bridge of his nose is gone, and his nostrils are rotted. When I lanced what is left of his nose, he did not sneeze. His elbows and knees down to the soles of both feet are defective, and suppurating at one place. His hands have no hair [on them]. When I asked him to shout, his voice and breath were feeble. It is leprosy.'

This passage shows how much concern the ancient Chinese had for public health. It also bears on the origins of leprosy in the Old World. Whether the disease existed in

---

[17] Chinese counted the period of gestation as ten months, so the foetus may have been five months old or less. See the description of foetal development in *HYCL*, p. 44. The discussion in Bodde (1982), p. 9, based on consultation with the gynaecologist Michael Mennuti, is illuminating.

[18] *HYCL* contains parallel entries for the first three of these cases. Sung does not discuss leprosy, presumably because judicial courts in his time no longer considered cases of such diseases. Nor is it mentioned in *HYL*. Harper (1977), p. 104, also translates the next report.

[19] See Bodde (1982), p. 11. Other bamboo slips (D101, D102) mention leper colonies, and say that lepers unwilling to live there must be executed by drowning; Hulsewé (1985), pp. 154–5.

[20] Also called *lai* 癩. Note that in this case a medical man was consulted. The discussions in McLeod & Yates (1981), pp. 152ff., and Bodde (1982), pp. 8ff., are instructive.

the Middle East or Egypt before the −2nd century remains unclear, despite much discussion.[21] The earliest extant Western description of low-resistance leprosy is that of Celsus (−25 to +37). He called the disease 'elephantiasis', as did his contemporary Pliny the Elder (+23 to +79), who averred that it had appeared in Italy only about a hundred years before. The oldest Western clinical description is by Aretaeus of Cappadocia, about +150.

We have elsewhere invited attention to a passage in the *Analects* where Confucius visits a disciple afflicted with an 'evil disease' (*o chi* 惡疾), which nearly all the commentators identify as leprosy. The sage is willing only to touch the man's hand through a window. This description is ambiguous. A page in the *Huang ti nei ching, Su wên* 黃帝內經素問 (probably −1st century) offers a much more tenable identification. It describes the anaesthesia, destruction of the 'pillar of the nose' and skin discoloration. These are just as characteristic as the symptoms described in the bamboo slips.[22]

The first non-Chinese writer to mention the destruction of the nasal septum in leprosy was the Arabic philosopher and physician Ibn Sina (Avicenna, d. +1037). A test for lepromatous anaesthesia was not recorded in the West until the time of Arnold of Villanova (d. +1312). We can conclude that the disease was present in China at least by the −3rd century, a couple of hundred years before it was recognised and recorded in the Mediterranean region.[23]

All this indicates a close connection with the doctrines of the Legalist School (*fa chia* 法家), which we have contrasted with Confucian conceptions of law and justice.[24] In this day and age, we may well have doubts about the Legalist theory of deterrence, and indeed it did not stop the acts investigated on the bamboo slips of Judge Hsi.

However there was something proto-scientific about Legalism, some soil out of which medico-legal investigations could grow. The laws, said the Legalist philosophers, should be clearly written down, exactly stated, and not subject to individual interpretation. They should have something about them like the indifference of Nature, like 'the rain that falls both on the just and on the unjust'. Hence the scientific objectivity, one could almost say, with which human actions must be probed, elucidated, judged and if necessary, punished. Since human nature may, in certain circumstances, be deceitful, the magistrate must use to the maximum the knowledge of the medical science of his day, in order to establish what exactly had happened. A fascination with the quantitative and its measurement also comes out again and again in Legalist writing.

These perhaps were the aspects of Legalism that favoured the appearance and growth of forensic medicine. If in the end history gave the victory to the Confucians, it was because they allowed for a certain humanistic element in ideal justice, recognising human beings to be not robots but the most highly organised and complex organisms known, the springs

---

[21] See, for instance, the discrepant views of various authorities in Kiple (1993), esp. pp. 251, 273, 337 and 834–9. Most ancient sources label diverse skin diseases with words that historians tend to translate as 'leprosy'.
[22] Lu Gwei-Djen & Needham (1967), pp. 12, 15; *Lun yü*, 6, 8, tr. Legge (1861), p. 52; *HTNC/SW*, ch. 12 (*phien* 2), 1.1.
[23] For a famous case of leprosy in +11th-century China, see Wilhelm (1984), p. 280. Liu Pan 劉攽, a friend of the poet Su Shih 蘇軾, died of it in +1088.
[24] Vol. 2, pp. 3, 204ff.

of whose actions are very deep and subtle.[25] Yet the application of science and medicine to law may well turn out to have been a characteristic contribution of Legalism.

(5) EARLIER EVIDENCE

The very advanced state of the legal formulations of Chhin argues a long antecedent development, but if we press our investigation of forensic medicine further into the mists of antiquity, there is not much to be found. Of law and legal actions much is said, but of medico-legal expertise, not much. For example, one may have a look at the *Yüeh ling* 月令 (Monthly ordinances), a work which must date from the −3rd century. One finds the following entry for the first month of autumn (the lunar month before the one that includes the equinox):[26]

In this month orders are given to the proper officers to revise the laws and ordinances, to put the prisons in good repair, to provide handcuffs and fetters, to repress and stop villainy, to maintain a watch against crime and wickedness, and to do their best endeavour to capture criminals.

Orders are [also] given to the judicial officials-in-charge to investigate superficial lesions (*shang* 傷), to examine open [or bleeding] wounds (*chhuang* 瘡), and to look for damaged bones and sinews (*chê* 折) as well as those which are broken (*tuan* 斷).[27]

In judging cases, it is essential for officials to be correct and fair; those who have committed the crime of killing others must receive the most severe punishment.

This is the oldest document of forensic medicine in Chinese literature.[28]

Through Chinese history criminals were supposed to be executed only in the autumn, to accord with the severity of the season. Thus for the second month of autumn the text goes on to say:[29]

Orders are given to the proper officers to revise with strict accuracy [the laws about] the various punishments. Beheading and [the other] capital executions must be according to [the crimes committed], without excess or defect. Excess or defect out of such proportions will bring on itself the judgement [of Heaven].

Finally, for the last month of autumn:[30]

In accordance with [the season] they hurry on the decision and punishment of criminal cases, not wishing to leave them any longer undealt with.

[25] [This discussion reflects the Whiteheadian organicism used to characterise Neo-Confucianism in Vol. 2, pp. 472–93. One may argue that, since we do not know what individuals and what social processes were responsible for the beginnings of forensic medicine, it is impossible to assign them to one disembodied ism or another. − Editor]

[26] Legge (1885), p. 285. In −239 the tractate was incorporated whole into *Lü shih chhun-chhiu* 呂氏春秋. A couple of centuries later it was also incorporated in the *Li chi* 禮記, whence Legge translated it. We cite the latter version.

[27] *Li chi*, ch. 6 (*Yüeh ling*), pp. 72b–73a. Legge (1885) seems to believe this passage refers mainly to the governors of prisons. Tshai Yung's 蔡邕 commentary makes it clear that the words refer also to judges in charge of inquests. Most commentators agree. See *Yü chu pao tien* 玉燭寶典, ch. 7, p. 6 (p. 424), and Chu-ko Chi (*1979*).

[28] Chang Fu 張慮, brother-in-law of Kuei Wan-jung, compiler of the case-book *Thang yin pi shih*, wrote a commentary on the *Yüeh ling* entitled *Yüeh ling chieh* 月令解; van Gulik (1956), p. 7.

[29] Pp. 75a–b, tr. Legge (1885), p. 288.   [30] P. 81, tr. Legge (1885), p. 295.

Such a text demonstrates that a legal system was in force and that there were punishments for evil-doers, but does not tell us about the medico-legal aspects of the magistrates' activities.

Records of legal cases go back to the beginning of the −1st millennium in China. For example, in 1975 an excavation at Chhi-shan 歧山 unearthed from a tomb of about −820 a bronze ladle with a handle (*i* 匜), bearing an inscription that records a court case.[31] It relates many such details as the basis on which sentence was given, how it was reduced, and how the guilt was eventually dealt with. A famous judge, Pai Yang Fu 白揚父, heard the case. One Mu Niu 牧牛 filed a suit against his official superior, named Chen 僬. Mu lost and was sentenced to be branded or tattooed (*chhing hsing* 黥刑). He was pardoned, and did not even have to pay a fine. He brought and lost another suit. As a result he was sentenced to a thousand strokes of the whip. This was first reduced to 500 and eventually commuted to a fine, paid in bronze to Chen. The latter used the metal to have the inscribed ladle cast. Sometimes these fines were paid in armour. The inscription does not say what the suits were about, so we do not know whether there was a criminal aspect in the matter.

Another case comes from the −10th century.[32] The record, a declaration of a judgement, is inscribed on a bronze cauldron with three legs. A military officer with the personal name of Chhi 旂 alleged that his inferiors, the defendants, had not followed him to fight for the High King of Chou. The judge, Pai Mao Fu 白懋父, eventually fined the chief defendant 300 *huan* of bronze (*ca.* 30 kg), but the defendants would not pay until Pai Mao Fu ordered them to be banished. It was out of this bronze that the vessel was cast.

In these beginnings of legal and judicial proceedings there is nothing about legal medicine, which thus, it seems, arose in the State of Chhin in the Warring States period − perhaps in connection with the Legalist School.

(6) THE DEVELOPMENT OF THE SUBJECT IN YÜAN AND MING

As forensic medicine vigorously evolved, Sung Tzhu's text became a form for recording innovations. The book was tacitly enlarged in many subsequent editions enumerated in the Appendix [to this Section] (p. 200), particularly the government-commissioned work of +1662 that we describe below. Scholars, not only European but Chinese, who relied on these editions praised Sung for ideas of a later date (p. 188).

Other writings on forensic medicine closely followed upon that of Sung. Another writer, probably Chao I-chai 趙逸齋, wrote a *Phing yüan lu* 平冤錄 (Redressing of wrongs). Since the work has not survived and the ascription is uncertain, we can only say that it is later than Sung's, probably of about +1255. After the Yüan conquered the whole of China, Wang Yü 王與 wrote on the basis of these two books, with modifications and additions, a treatise entitled *Wu yüan lu* 無冤錄 (The avoidance of wrongs,

[31] Shêng Chang 盛張 (*1976*), p. 42; Thang Lan 唐蘭 (*1976a*); Phang Phu 龐樸 (*1981*).
[32] Thang Lan 唐蘭 (*1976b*), p. 33.

+1308).[33] The three books were sometimes collectively printed as the *Sung Yüan chien yen san lu* 宋元檢驗三録 (Three treatises on inquest procedure from the Sung and Yüan periods).

A few details about the transmission to Korea and Japan are in order. The *Wu yüan lu* was printed in Korea in +1384. There and in Japan it became the basis of forensic medicine.[34] An annotated version, *Hsin chu wu yüan lu* 新注無冤録 (+1438), long remained the fundamental Sino-Korean text on the subject. Liu I-sun 柳義孫, who contributed a preface of +1447, may well have been its editor. It was transmitted to Japan, some say by the close of the +16th century, and became the standard work there. Kawai Jinhyō 河合甚兵衛 prefaced it in +1736, and it was published as *Wu yüan lu shu* 無冤録述 (The avoidance of wrongs, faithfully transmitted) in +1768.

(7) The development of the subject in Chhing times

In 1827 an excellent scholar named Chhü Chung-jung 瞿中溶 identified the twenty medical and other books, from *Huang ti nei ching* to *Chhi hsiao liang fang* 奇效良方 (Excellent and remarkably effective prescriptions, +1470), used in compiling the +1662 edition.[35] Together with the newly discovered Yüan xylograph of the original book, this helps us distinguish between Sung Tzhu's own text and ideas added to it, especially in the government editions. Recently Kuan Chhêng-hsüeh (*1985*) listed many errors by scholars who did not distinguish these additions. Certain antidotes, such as the use of eggs and emetics to deal with arsenic poisoning, were not in the original text, for example. Such mistakes, handed down in recent Chinese medical textbooks and encyclopaedias, are reminders of the need to consult original texts. This is why it was quite acute of Sydney Smith to say: 'It would be of great interest if some student of Chinese history could unearth and translate an undoubted mediaeval text of the *Hsi yüan lu*' – which McKnight in 1981 duly did. Smith added, despite the uncertainty of the Giles translation on which he was forced to rely, 'I am left with the conviction that in mediaeval times Chinese forensic medicine was far in advance of contemporary European practice.'[36]

After the mid 19th century, books continued to be written on the subjects discussed in the *Hsi yüan chi lu*, but they were of little importance. European anatomy and forensic medicine gradually entered China, leading to the synthesis of the present day.[37]

In the late edition (1843) that Giles used for his translation, the *Hsi yüan chi lu* consists of four chapters. The first deals with distinguishing real and counterfeit wounds,

---

[33] *Wu yüan lu* distinguished suicide by hanging and murder by strangling more carefully than before. It also corrected the age-old error about the relative positions of the trachea and oesophagus, maintaining the former is anterior to the latter; Chung Hsü (*1956*).

[34] The story has been told in detail by Chia Ching-thao 賈靜濤 (*1981b*) and Sung Ta-jen (*1957b*).

[35] *Pên* 5, ch. 6A, p. 3a; Sung Ta-Jen (*1957b*), p. 281. On the +1662 ed., see the Appendix [to Section (*e*)], p. 200.

[36] Smith (1951), p. 602, echoed by Gradwohl (1954), p. 7: 'It was very far ahead of anything which existed anywhere else in the world when it was first written.'

[37] For an account of this development see Chung Hsü (*1956*) and the book of Kuo Ching-yüan 郭景元 (*1980*). It can be followed further in Chhen Khang-i 陳康頤 (*1952*) and Chia Ching-thao 賈靜濤 & Chang Wei-fêng 張慰豐 (*1980*). [These texts do not indicate that a synthesis of traditional-Chinese and modern-Western forensic medicine has formed, but that the latter has replaced the former in the law courts. – Editor]

examining decomposed corpses, examining bones to determine whether they were injured before or after death, and observing the decomposition of the body at different seasons. The second group deals with wounds, whether inflicted by the hand or foot, or with weapons generally; suicide with weapons; suicide by strangulation; and burning, whether before or after death. The third group discusses miscellaneous suspicious appearances, and the results of falling from a height or being crushed to death. It also takes up many kinds of poisons and their effects. The fourth group considers methods of restoring life, e.g. after hanging, drowning, scalding or freezing, and continues the pharmacological notes on poisons.

One cannot do better than quote the opening chapter of the *Chhung khan pu chu Hsi yüan lu chi chêng* 重刊補注洗冤錄集證 (Washing away of wrongs, with collected and supplemental notes, reprinted, 1837):

There is nothing more important than human life; there is no greater crime than that punishable by death. A murderer must give life for life; the law shows no mercy. If such punishment is not appropriate, regret is inescapable. Thus the validity of a confession and the sentence passed depend on a satisfactory examination of the injury. If it is of a *bona fide* nature, and the confession of the accused tallies with it, then life will pay for life. Those who know the laws will fear them, crimes will become less frequent among the people, and due value will be regularly attached to human life. If an inquest is not conducted in good faith, the wrong of the murdered person is not redressed, new wrongs are raised up against the living, two or more lives will be forfeit instead of one, and both sides will be aroused to vengeance of which no one can foresee the end.

In serious cases involving human life, when death has not already ensued, everything will depend on the officials' personally examining the victim without delay, noting down injuries and wounds in dangerous locations, and judging their severity. They fix a 'death-limit' in the hope that medical skill may effect a cure. If re-examination after death is requisite to determine an appropriate penalty, the unpleasantness of a complete dissection may be avoided.

Where death has already resulted from the wounds, there is still greater need for dissection the same day, while it is still easy to record and mark on the form the severity and size of each injury. If it is delayed, the body decomposes. Guarding against the infliction of false, or the tampering with real, wounds is the great key to murder cases. The judge, accompanied by the ostensor [see p. 191] are to hasten to the spot with all speed, so that the guilty parties may have no time to concoct evasive schemes.

If death has just taken place, first examine the top of the head and the back, the ears, the nostrils, the throat, the anus and the vagina, in fact all places where anything might be inserted, on the chance that a sharp-pointed instrument has been used there. If nothing is found, proceed to examine the body systematically.

If examination of the body is necessary, first carefully interrogate the relatives of the deceased and the neighbours to establish the identity of the perpetrator. Bid them state clearly who struck whom with what weapon, and in what part of the body. Set down clearly the deposition of each.

Then with your assistants, the ostensor, the petitioner, and the accused, proceed together to where the body lies, and examine it and make out the report in the prescribed manner. Mark whether the cause of death involved a vital spot, or whether the injury is perceptible on the skin or in the flesh, or penetrates to the bone. Note its colour [blue, red, purple or black], circumference or length, with exact dimensions; whether inflicted by the hand or foot, or by some weapon; whether severe or not, whether recent or old. Verify the marks of injury so that each is clear. Fill out and annotate with your own hand the form that records particulars of the corpse. Do not leave it to your assistants. Do not let yourself be deterred by the smell (lit. '*chhi*') of the corpse, so that you sit at a distance, your view intercepted by the smoke of fumigation, letting the ostensor call out the wounds and your subordinates enter them on the form, so that they hide what is important and report what is trivial, and add or subtract details.

Considering the diversity of deaths from self-strangulation, cutting one's throat, taking poison, burning, drowning and all sorts of other causes, it is essential to minutely scrutinise [the corpse] in order to reveal every detail so that the verdict will be reliable. If this be neglected, the ostensor and the secretary will twist the words of the report; the culprit will devise some means for escaping punishment, the relatives of the deceased will appeal against your decision, middlemen will raise a fuss, and hoodlums will play their tricks. In such circumstances verdicts are often discredited. If another inquest is ordered [to clear up the uncertainties], it may become necessary to steam the corpse and scrape the bones, an outrage on the dead and a great burden on the living. These are the evils attendant on a dilatory or perfunctory discharge of your duties.[38]

Several points could do with some explanation. Nothing analogous to the widely used 'death-limit' (*pao ku* 保辜 or *shou ku* 守辜), so far as we know, occurred in Western mediaeval law. The terms might be better translated, perhaps, as 'period of responsibility' or 'liability time-limit'. The wounded person was handed over to the care of the assailant's family for a specified period. The idea was that the guilty party would be so anxious to avoid the penalty for murder that he would get the best medical advice that money could buy, and that the family would nurse the victim back to health. If the victim did not recover within the term, the assailant was held responsible; but if death occurred, even half a day after the limit, that was not considered his fault. This system protected both parties.[39]

Secondly, although the filling-in of standard forms by the judge during an inquest has a modern ring to it, Chêng Hsing-i 鄭興裔 introduced them in +1174 in *Chien yen ko mu* 檢驗格目 (Inquest forms for examination of the corpse). Chêng, a Judicial Commissioner, was struck by the muddle, dishonesty and miscarriage of justice at inquests. He devised a serially numbered form on which many aspects of the case had to be filled in, notably the name of the victim and assailant, the date and time of the offence, the distance of the place from the judge's office, the number of wounds and the cause of death, together with the names of the officials concerned.[40] In +1204, a judge introduced anterior and posterior outline diagrams of the human body, so that the coroner could mark the locations of wounds or contusions with coloured inks to enhance visibility. The Yüan edition of *Hsi yüan chi lu* included them. The forms had to be completed in triplicate, one to go to the office of the Prefect, one to be given to the victim's relatives, and one to be kept by the Judicial Commissioner. Such forms arose naturally out of the passion for justice and the bureaucratic character of Chinese society.

[38] Tr GI, p. 61, mod. auct. On vital spots see p. 192, below.

[39] For example, when the extremities were wounded by an instrument other than a knife, the period was twenty days; for a knife wound, boiling water or fire, thirty days; for other wounds, ten days. When broken bones or miscarriage, the period was fifty days. Such were the prescriptions laid down in the *Tou ou lü* 鬥毆律 (Regulations pertaining to fighting); *HYL*, ch. 1, p. 10a.

The practice is mentioned in the +1st-century dictionary *Chi chiu phien* 急就篇, p. 67a. For further details, see van Gulik (1956), p. 92, and Alabaster (1899), pp. 229ff.

[40] *Chien-yen i lai chhao yeh tsa chi* 建炎以來朝野雜記 (Miscellaneous records of officials and commoners, between +1127 and +1130), pt 2, ch. 11, pp. 474a–b. A similar account in *Sung hui yao* 宋會要 adds details of a second inquest when necessary; *tshê* 170 (*hsing fa*), pp. 5a–6b.

The filling-in of forms is recent in Western Europe,[41] but +18th-century sea-captains whose ships needed repairs on the China coast were surprised to see magistrates along the coast filling in forms: how much wood was required, how many carpenters, and so on. This is just another indication that the Chinese bureaucracy was far ahead of Europe for many centuries.

Lastly there is the question of the ostensor (*wu tso* 仵作). Eventually becoming a lowly functionary, not an appointed official, he (or she if the cadaver were female) drew attention to the wounds on the body and called out their nature to the judge. He also prepared the body for examination and did any simple procedures that were required. Much emphasis was placed on washing the body with vinegar, no doubt as an antiseptic.[42] As for the 'old woman' (*tso pho* 坐婆 or *wên pho* 穩婆) who in mediaeval China assisted in examining female corpses, she was usually a midwife. In early Chhin times she had generally been a bondswoman (*li chhen chhieh* 隸臣妾).

Late +15th-century pictures of anatomical dissections in Europe show the professor in his robes, seated aloft in a stall, reading from the book. Down below, a menial demonstrator or prosector actually did the dissection. An ostensor used a wand to point out the parts to the students, who did not dissect.[43] 'Ostensor', 'he who shows', describes the functions of the *wu tso*.

An early name for the ostensor was *hsing jen* 行人 or 'runner', because he made haste to get to the place where the dead body was found. When not on duty, he might well be an undertaker, a seller of coffins, or possibly a barber or butcher, underpaid, socially despised and uneducated. One source emphasises 'Never base your report only on what the ostensor says.'[44]

From +1268 onwards, the law gave the *wu tso* more responsibility. While the judge still had to participate in the examination of the body, the *wu tso* had to certify that the matters established were true. In cases of poisoning, medical men were now called in. The *wu tso* were at first commoners, but before the end of the Chhing dynasty, special training schools were made for them. Eventually they got the title of Inquest Clerk (*chien yen li* 檢驗吏 or *chien yen kuan* 檢驗官), and were placed in the ninth rank of the bureaucracy.

[41] See Gradwohl (1954), pp. 60, 76, 101–3. The practice may go back to the standardisation of medical reports in the books of de Blégny (+1684) and Devaux (+1703), see p. 9. On Ambroise Paré (+1575) and reports see p. 197, below.
  A key passage occurs in Walter's (+1750) account of Admiral Anson's voyage around the world between +1740 and +1744, entry for 19 December +1742. A 'mandarine' deputed by the Viceroy of Canton visited Anson's ship the *Centurion* in Macao roads, and took down all the particulars about the necessary repairs on previously prepared forms. For knowledge of this we are indebted to Mr G. F. Peaker and Dr Ulick R. Evans.
[42] van Gulik (1956), pp. 60, 171–2. The term was used in Sung Tzhu's book (*HYCL*, p. 23 and *passim*) and in earlier works such as *Chê yü kuei chien* of +1133. The history of the ostensor and his work has been written by Chia Ching-thao 賈靜濤 (*1982*) and Chung Hsü 仲許 (*1957*). Ho Mêng had a good deal to say on the subject in *Yü thang hsien hua* 玉堂閑話.
[43] See for instance the illustration from the Italian *Fasciculo di medicina* printed at Venice in +1493, the first page of an Italian translation of the *Anothomia* (+1316) of Mondino de Luzzi (*ca.* +1270 to +1326), in Singer (1957), fig. 35, p. 7.
[44] *Hsi yüan lu chieh* 洗冤錄解 (1831), cit. Chia Ching-thao (*1982*). In Sung and Yüan, this official was often called '*wu tso hsing jen*'; *HYCL*, pp. 40–1.

Finally, the 'middlemen' (*sung-shih* 訟師) might be thought of as irregular forerunners of the legal profession. Especially in the Chhing era, when many who had passed the local examinations had no hope of an official appointment, some served as intermediaries, with no legal standing, in criminal cases, raising written objections to one point or another on behalf of one party or the other. This was not exactly an occupation, but it provided a living for a certain number of people. Their knowledge of the law was perhaps less important than their ability to write literate briefs.

(8) Matters of medical interest

Now when we come to the specifically medical content of these treatises, there is much of great interest, because the government insisted that the investigators closely examine the body of anyone who died in uncertain circumstances.[45] As we have seen, the *Hsi yüan chi lu* of +1247 consolidated all previous knowledge in forensic medicine. Authors of its various revisions rearranged the book in many ways, varying the title. We shall distinguish in what follows between the older work as *Hsi yüan chi lu*, and the greatly enlarged *Lü li kuan chiao chêng Hsi yüan lu* 律例館校正洗冤錄 (The washing away of wrongs, collated and verified by the Codification Office) of +1662 as *Hsi yüan lu* (citing the 1843 edition that Giles used).

Sung and his successors carefully list the 'vital spots', sites where any trauma, contusion or shock will be exceptionally perilous, leading to internal injury or death, sometimes with no external sign. They were given names like those of acu-points. These loci of particular danger, which might become the site of a mortal lesion (*chih ming chih chhu* 致命之處),[46] were sixteen on the front of the body (or twenty-two if all the bilateral points were counted) and six on the back (or ten with bilaterals), making thirty-two in all. Modern medical science has confirmed many of these, such as the fontanelles of the skull, the occipital and cervical regions, the parts just above and below the sternum, the perineal region and scrotum.[47]

The book warns the examiner when examining a body not to be deterred by the unpleasant state of the corpse, but in every case to make with his ostensor a systematic examination from the head downwards. It advises him particularly to identify counterfeited wounds, wounds caused by such means as blows with the fist, kicks, various sorts of weapons, strangulation and drowning. Accidental drowning has to be distinguished from death caused by being held under the water, and from suicidal drowning. It explains how to detect that death by strangling has been passed off as suicide, and how to distinguish a drowned person and one thrown into the water after death. It clarifies the difference between post-mortem and ante-mortem burning and between traces of

---

[45] Alabaster (1899) considers Chhing practice in cases of manslaughter (pp. 288ff.), murder (pp. 292ff.), suicide (pp. 303ff.) and assault (pp. 346ff.).

[46] The phrase occurs in *HYCL*, but the charts of vital spots are given only in *HYL*. See Lu & Needham (1980), figs. 79, 80, where the date should be corrected to the +17th century. Pp. 302ff. discuss the vital spots in relation to the so-called 'martial arts' (*wu shu* 武術).

[47] Camps & Cameron (1971).

various poisons. It exhorts the examiner to study the place, the surroundings and the circumstances with the utmost care; nothing can be overlooked as unimportant.[48]

The *Hsi yüan lu* describes many tests and experiments, some very sensible in terms of modern medical science, others combining proto-science and folk beliefs. For example, there are some remarkable techniques for making wounds, severe contusions, bruises, or bone fractures appear clearly on the surface of the body. In particular, filtering the direct light of the sun through red-oiled umbrella paper made certain discolorations more noticeable.

Under the heading 'A discussion of bones and blood-vessels in the body, and the location of vital spots' the text says:

When examining bones it is necessary that the day be clear and bright. First wash them with water, and then tie them together with hemp twine [to form a skeleton] before laying them out on a mat. Then dig a pit in the ground measuring five feet long, three feet wide and two feet deep, burning wood and charcoal in it until the ground is very hot. Clear out the ashes and put in two pints of good wine and five pints of strong vinegar. While it is still steaming lay the bones in the pit and cover them with a straw mat, letting them remain there for two to four hours.[49] When the ground has cooled, remove the mat, and take the bones to a level well-lit place where they can be examined under a red silk or paper umbrella which has been waterproofed with oil. If there are places on the bones which have been struck, then there will be traces of a red colour. The two ends of any fracture will show a blood-red halo. If one holds a bone up to this light and it shows a vivid red colour, the injury was inflicted before death. If on the other hand the bones are broken but show no traces of red colour, the fracture occurred after death . . .[50]

This use of a light-filter was ancestral to the employment of infra-red light and similar aids in modern forensic medicine.[51]

The practice is an ancient one. Near the end of the +11th century, Shên Kua 沈括 reported an intriguing story in his *Mêng chhi pi than* 夢溪筆談 (Brush talks from Dream Brook):

When the Erudite of the Court of Imperial Sacrifices, Li Chhu-hou (李處厚), was magistrate of Shên district in Lu prefecture, the case arose of a person who had beaten someone to death. Li went to examine the injuries. He spread the dregs of wine, an infusion of ashes, and such substances, over the bones, but found no trace of injury. An old fellow asked to meet him, and said 'I am a former clerk of the local office. I know that your examination turned up no traces [of wrongdoing]. They are easy enough to reveal. Hold a new red oiled [paper or silk] parasol [over the body] in full sunlight. Pour water over the cadaver and the traces will become visible.' Li did as he said, and the injuries were clearly visible. Since that time officials between the Huai River and the Yangtze have regularly used this technique.[52]

---

[48] Most of these discussions were greatly enlarged in *HYL*.

[49] This must have cleaned the bones thoroughly.

[50] *HYCL*, section 18, pp. 53–4; tr. McKnight (1981), p. 102, mod. auct. Cf. Yang Fêng-khun 楊奉琨 (*1980*), pp. 45ff., 136, and the parallel passage from *HYL*, ch. 1, pp. 49a–56b, tr. GI, p. 66.

[51] Walls (1974), pp. 69ff. For another use of a light-filter umbrella, see section 8 (p. 41), tr. McKnight (1981), p. 81.

[52] *MCPT*, ch. 11, pp. 10a–b (item 209), tr. auct. The ash infusion must have been slightly alkaline, removing oil and whitening the bones.

Since the same story occurs in the *Yü hsia chi* 玉匣記 (Jade box records), about +880, 'examining corpses in red light' (*hung kuang yen shih* 紅光驗尸) goes back at least as far as the Thang. The account recurs in the *Chê yü kuei chien* of +1133 and in Fang I-chih's 方以智 *Wu li hsiao chih* 物理小識 (Little notes on the principles of the phenomena) of +1643.[53]

Wang Chin-kuang 王錦光 (*1984*), who unearthed these references, experimented with light filters. He found that red glass or red nylon, which let through only the longest wavelengths of visible light (around 6500 Å [650 nm]), have effects similar to those stated. Light so filtered will reveal even the deepest superficial veins. Bruises otherwise invisible appear purplish in colour.

We also find in *Hsi yüan chi lu* a great deal of osteological and anatomical description. It tests for genetic kinship by dropping blood from a living individual onto the bones of a dead one:

To identify relatives using the technique of dripping blood on bones: You have for example A, a deceased father or mother, whose bones are available. B comes forward claiming to be a son or daughter. How are you going to find out? Let the claimant cut himself or herself slightly, drawing one or two drops of blood, and let them drip on to the bones of the parents. If B is indeed son or daughter, and only then, the blood will sink into the bone. This is the point of the common saying 'drip blood to establish a relationship'.

The *Hsi yüan lu* elaborates slightly:

Children may identify the bones of their parents in the following manner: Let the inquirer cut himself or herself with a knife and drip the blood on to the bone. If there is an actual relationship, and only then, the blood will soak into the bone. Caution: should the bones have been washed with salt water, even if there is a relationship, the blood will not soak in. This is a trick to guard against.[54]

This blood-dripping test (*ti hsüeh* 滴血) is much older than the thirteenth century. It figures in the +3rd-century story of Chhen Yeh 陳業, who identified the body of his brother in this way after a shipwreck. It is also mentioned in the biography of Hsiao Tsung 蕭綜, King of Yü-chang 豫章 of the +6th century, and the folk ballad 'Mêng Chiang-nü 孟姜女', which deals with events of the −3rd century.[55]

Possibly these procedures had no rational basis, but one can never be certain, because some immunological property might be involved; after all, blood groups are used today

[53] Ch. 3, pp. 24a–b.

[54] *HYCL*, section 18, p. 53, tr. McKnight (*1981*), pp. 101–2, mod. auct.; *HYL*, ch. 1, pp. 57a–60b, tr. GI, p. 73.

[55] Chung Hsü (*1957*) and Li Jen-chung 李仁眾 (*1982*) enumerate other cases as well. The first record is in *Liang shu* 梁書, ch. 55, p. 824, the second in *Kuei-chi hsien hsien chuan* 會稽先賢傳. In the ballad the test was done as between husband and wife, a use which Sung Tzhu and the later medico-legal experts would have rejected. Later brother–sister relationships were tested by observing whether drops of blood from both 'congealed' (i.e. precipitated) in absolutely pure water (*HYL*, ch. 1, p. 55a). The ballad in any case is evidence that the test was known. [The precursor of this very popular ballad probably first appeared in the Thang, roughly a thousand years after the events it narrates. See Nienhauser (*1986*), pp. 88–9 and *passim*. – Editor]

in tests of relatedness.[56] It may be dangerous to write such things off as superstition, because the progress of science sometimes reveals explanations which one did not see at first. In any case, this ancient test is ancestral to the determination of genetic relationship by immunological means today.

There is also an interesting statement about the geobotanical effects on vegetation after a cremation has taken place. This comment, in the section on examining the site of murder, is part of a passage on verifying a murderer's confession that he has destroyed a corpse by burning it:[57]

Where a body has been burned, the grass will be dark in hue and unctuous, taller than [the grass] surrounding it. These characteristics will not change for a long time, because the fat and grease of the body deeply penetrate the roots of the grass. As the days pass it will remain luxuriant. If the spot is on a hill or a wild place next to a meadow, where tall grasses normally grow, it will grow taller still, and will ultimately resemble the shape of a human being. If [the burning] took place on stony ground, when you burn the site, you can depend on the crumbling and splitting of the stones [to outline the victim's form]; it will be easy to discern.

The *Hsi yüan lu* also describes death caused by certain diseases such as rabies or tetanus that are rather easy to recognise, as well as poisoning by all kinds of substances, including arsenic, soda, and other favourite recourses of mediaeval suicide. It also has much to say on haemostatic substances, antidotes, emetics and all kinds of first-aid practices in cases of poisoning. It incidentally recommends animal experiments with food suspected of being poisoned.[58] Experimentation was characteristic of Chinese forensic medicine right back to the Han, a matter of no mean interest.

The book also describes carbon monoxide poisoning. This is not unexpected because rooms were heated with charcoal pans.[59] On the other hand, the book contains much on resuscitation and reanimation, including what we might now call the 'kiss of life', and getting respiratory reflexes to start again.[60] All the methods belong to the late +17th century.

All in all, the Khang-hsi editors were well informed on medical matters. The writers knew the medical terminology of diseases and drugs, and were aware of their uses. They referred frequently to *chhi*, the 'pneuma' of which the physicians often wrote. Like their predecessors, they cited medical books such as the *Ching yen fang* 經驗方 (Tried and tested prescriptions, +1025) of Chang Shêng-tao 張聲道. They also referred to a *Wu tsang shên lun* 五臟神論 (Divine discourse on the five solid viscera). This was perhaps the title of a chapter in some version of the *Huang ti nei ching*. They drew as well on an older medico-legal work, the anonymous *Nei shu lu*. Neither of these titles is listed by the bibliographical encyclopaedists.

[56] Our friend the late Professor Hermann Lehmann informed us that, apart from the blood group O, all the others would precipitate on the bone. They would also sink into the bone of a closely related person if death was recent. In the *Hsi yüan lu* editions, many reported cases testify to the effectiveness and accuracy of the tests (e.g. *HYL*, ch. 1, pp. 57aff.). On the agglutinins and erythrocyte antigens used in forensic medicine, see Walls (1974), pp. 133ff., 155ff. Tests of relationship are now done by DNA 'finger-printing'; see Jeffreys *et al.* (1985) and Hill *et al.* (1986).

[57] *HYL*, ch. 1, pp. 61b–62a; not in *HYCL*.

[58] *HYL*, ch. 3, p. 19a, ch. 4, p. 17a, ch. 3, p. 36b; tr. GI, pp. 92, 103, 94.

[59] *Ibid.*, ch. 3, p. 52a, tr. GI, p. 97. The *khang* 炕, a platform for sleeping and sitting on that channelled hot air from the stove, particularly in North China, was another source of danger.

[60] *HYL*, ch. 4, pp. 1aff., tr. GI, pp. 98ff.

(9) Some comparisons with Europe

We come back in the end to the point at which we started, namely the development of forensic medicine in Europe. There was not much role for legal medicine in ancient Greece, though perhaps more in Rome. In Hellenistic Egypt 'public physicians' reported dubious cases to the local authorities.[61]

The Byzantines did more than the early Greeks and Romans. In the Justinian enactments between +529 and +564 we find many striking new things. The Code of Justinian provided for the regulation of medicine, surgery and midwifery; the classes of physicians that were to be recognised; the limitation of their numbers in any one town; and the penalties to be imposed for malpractice.[62] The status and function of the medical expert in legal procedures emerges more clearly than before in the famous phrase *Medici non sunt proprie testes, sed majus est judicium quam testimonium*. Sydney Smith has freely interpreted it to mean that the medical expert is not properly used if he is regarded simply as a witness appearing for one side or the other; his function is rather to assist the judiciary by impartial opinion based on specialised knowledge. As he says, this dictum is still true.[63]

The Justinian Code required judges to consult medical experts in a considerable variety of legal cases involving questions of sterility, pregnancy, impotence, rape, poisoning, legitimacy, mental disease and survivorship. Smith was probably right in saying that the Justinian enactments were the closest point to defining forensic medicine in the ancient world. But still there was no systematic treatise.

Roman and Byzantine law was generally more concerned with property, inheritance, wills, guardianship and the like, than with criminal actions leading to inquests. In China, on the other hand, law seems to have been primarily criminal law, while other matters were if possible attended to by elders or leaders of the community. Disputants took civil matters to the courts only if there was no other way to resolve them.[64] Of course, if an affray resulted from disagreement, the magistrate would take it in hand.

The legal systems of the European Middle Ages incorporated little of the Justinian heritage. Hunnisett's book on *The Mediaeval Coroner* (1961), the best of its kind, covers the period approximately from the Norman Conquest to the end of the +15th century. You will look in vain in the index for anything indicating medical evidence; the words 'physician', 'medical' and 'doctor' do not occur.

This period of darkness seems to have ended in +1507 when the Bishop of Bamberg and John of Schwarzenberg issued a comprehensive, systematic penal code, which was soon adopted by a number of the neighbouring German states. In +1532 and +1553, developing from the Bamberger Code, the still more influential *Constitutio criminalis*

---

[61] See Ackerknecht (1950–1), Gorsky (1960), Greene (1962) and Kerr (1956), pp. 340ff. On Egypt, see Amundsen & Ferngren (1978), who consulted a profusion of papyrus fragments.

[62] Cf. Thomas (1975, 1976); Scott (1932); Pothier & de Bréard-Neuville (1818); and Tissot (1806).

[63] Smith (1951), p. 601. Professor Peter Stein has informed us after a thorough search that the sentence is not in the *Corpus juris civilis* of Justinian. It may well come from some later commentary, but the origin has not been identified.

[64] For civil law in late imperial China see Allee (1994).

carolina or Caroline Code was promulgated by the Emperor Charles V to be observed throughout the empire. Both these early Germanic codes for the first time clearly and definitely required that judges obtain expert medical testimony for their guidance in cases of murder, wounding, poisoning, hanging, drowning, infanticide, abortion and in other circumstances involving the person.[65]

Now it can hardly be a coincidence that between these two versions, in +1543, Andreas Vesalius published his splendid monograph *De fabrica humani corporis*.[66] Vesalius paved the way for all autopsies of later times, putting his own hand to the business of dissection, doing away with the demonstrators and ostensors of former times. He was also a great creative artist, and the plates which he published were extraordinarily influential. He was one of those who, together with Galileo, brought modern science into being.[67]

It was in this atmosphere that medical scientists began to realise that their evidence would be indispensable in guiding judges to right judgements. Perhaps the first to realise this was the eminent French surgeon Ambroise Paré. His 'Traicté des rapports, et du moyen d'embaumer les corps morts' was never published separately, but appeared in his collected works, *Les oeuvres de M. Ambroise Paré, Conseiller et Premier Chirurgien du Roy*, published in Paris in +1575.[68] The collection was reprinted many times thereafter and translated into English in +1634, where the treatise is entitled 'How to make reports, and to embalme the dead'.

The first comprehensive treatise on medico-legal matters was by an Italian physician, Fortunato Fedele, whose *De relationibus medicorum* appeared at Palermo in +1602. Although Fedele was a contemporary of Wang Khên-thang,[69] neither knew that the other existed, let alone what he was doing. Before long the book of Fedele was put into the shade by a much more impressive work, that of Paolo Zacchia, whose *Quaestiones medico-legales* appeared at Rome in +1628. In this book he deals with all the questions likely to perplex judges and their assistants. The book long maintained a great reputation, though its knowledge of anatomy and physiology was still poor. The science of forensic medicine had now been born.[70]

All the European countries henceforward competed in the study of this subject. One may mention the German book of Rodericus à Castro, *Medicus politicus*, published at Hamburg in +1614; and the French one of Nicolas de Blégny, *La doctrine des rapports de chirurgie*, published at Lyon in +1684. By now the Harveian doctrine of the circulation of the blood was becoming generally accepted, and medicine was entering its modern phase. We need not follow the further history of the subject into modern and contemporary times.

---

[65] Smith (1951), p. 602.   [66] See Singer (1957), pp. 110ff.

[67] Gross anatomy had of course been well developed in China since ancient times; the weights of organs and the lengths of blood-vessels were given in *HTNC/LS*; Yamada (1991). We suspect that it was not so much the lack of anatomy that made Chinese forensic medicine fall behind that of the West in the 19th century, as the lack of microscopy, physiology and biochemistry.

[68] Doe (1937), p. 107.

[69] On Wang see p. 201. The simultaneous appearance of these two treatises seems to be another instance of a curious phenomenon. Although the Scientific Revolution happened only in Europe, during the late +16th and +17th centuries many remarkable examples of the ethos of modern science arose spontaneously in China; Needham (1983).

[70] Smith (1951), pp. 602ff.

It now only remains to ponder McKnight's interesting suggestion that Sung Tzhu's example might have influenced late mediaeval Europe, especially Norman England.[71] The law of the Anglo-Saxons and Germans in early times had indeed required rather precise descriptions of wounds, because of the monetary recompense to be paid for them. All the members of the court were supposed to view the wounds. The English system forms a closer parallel. From +1194 onwards, it became customary that the 'crowner' selected in each county held inquests on all those who had died through homicide or accident. A jury was empanelled and an oath exacted from the foreman of the jury, all *coram populo*. Time limits for deaths from wounds were established, but the victim was not handed over, as in China, to his assailant to be cured if possible. McKnight adds that the resemblance of the English practices to the Chinese becomes even more striking when we read the list of faults of which coroners were accused. It was said that, among other misdeeds, they often sent proxies to the viewing, delayed or tarried in their work, accepted bribes, or entered information into the record which either did not derive from the inquest or was patently false.

At this time Norman England was in close contact with the Norman Kingdom of Sicily, where Roger II (+1098 to +1154) and Frederick II (+1194 to +1250) were both extremely interested in intellectual matters. The Sicilian court was in close contact with the Islamic countries further east. Thus there might have been some connection. We have suggested in Section (c) that the idea of qualifying examinations for physicians came to Norman Sicily, and thence to Salerno and all Europe, from China by way of Baghdad; for it was there, in +931, that the caliph instituted examinations for all qualifying medical students.

Until more is known about the practices in Arabic lands regarding the investigation of homicide, accidental death and suicide, we can draw no conclusion about influence from China. It would have to have been a good deal later than the idea of qualifying examinations in medicine.

Last of all, we must clear up an apparent contradiction. On the one hand, we said that inattention to medical evidence in cases of death or injury was characteristic of the Middle Ages in Europe, before the rise of modern science. We also noted that a commentary on the Justinian Code gave the medical expert the function of advising the judge. This is not paradoxical when one appreciates that the former statement applies especially to northern Europe, while the Byzantine attitude was perpetuated in the south, especially in Italy. It did not bear fruit until +1252, when Bologna was the first city in Europe to establish 'a service of expert medical investigation in all cases of offences against the human being'.[72] It was a striking coincidence that this should have followed so closely the appearance of Sung Tzhu's *Hsi yüan chi lu* of +1247 at the other end of the Old World. Of course the Bolognese doctors did not prepare a systematic treatise on the subject. The physicians seem to have testified only about whether or not a wound was mortal, and inflicted before or after death. This was rather unsophisticated compared with Sung Tzhu's fine distinctions between modes of homicide. The city laws of Freiburg im Breisgau followed those of Bologna, but as one would expect, a couple of hundred

---

[71] McKnight (1981), pp. 24–5.   [72] Simili (1969, 1973, 1974), Münster (1956) and Samoggia (1965).

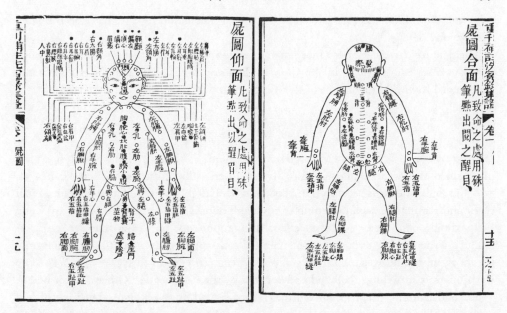

Fig. 11. Official form used in inquests. The magistrate is instructed to use red ink to fill in the circle corresponding to the vital spot responsible for death. From *Hsi yuan lu*.

years later, in +1407 and +1466.[73] At this time autopsies began to be performed in Italy, as part of the slow development of anatomical knowledge. By the +16th century modern science was beginning to evolve south and north of the Alps; yet it was no coincidence that the first book on poisons, Battista Codronchi's *De morbis veneficis*, was published in Venice in +1595. The tide of modern science was now rising, as we have noted already in the cases of Fedele and Zacchia, founders of modern forensic medicine. It may well be that China was in the circuit which led, over the centuries, to this entirely new level of accuracy and understanding.

APPENDIX: EDITIONS AND TRANSLATIONS OF *HSI YÜAN CHI LU*

(1) TRANSLATIONS

The first version in a Western language was an abbreviated text by the Jesuit Cibot (+1779). A fuller French version by Martin (1886) saw the light more than a century later. Litolf's (1909–10) translation from the Vietnamese appeared in a Hanoi journal. The first complete translation into a Western language was from the +1796 ed. into Dutch, by de Grijs (1863), in turn rendered into German by Breitenstein (1908). Shangkuan Liang-fu 上官良甫 (*1974*) mentions another German translation, by one Hoffman, but a search in Western library catalogues has failed to confirm or identify it. Perhaps the most likely translator would be the Japanologist John Joseph Hoffman, active between

[73] Volk & Warlo (1973).

1834 and 1877. There is also said to be a Russian translation, but we have been unable to learn any details.[74]

The definitive translation for more than a century was the one that Herbert Giles published as a series of articles in the *Chinese Recorder* in 1874. McKnight (1981) finally superseded it with a translation of the Yüan edition, the earliest that survives.

(2) EDITIONS

Hsi yüan chi lu 洗冤集錄 went through many editions. A Yüan version, the oldest extant, maintained the original division into five chapters. The antiquarian Sun Hsing-yen 孫星衍 republished this recension in 1808, and McKnight translated it. It contains Yüan cases and regulations as an opening excursus. There is also an annotated ed. by Yang Fêng-khun (*1980*). See also the excellent annotations in Chia Ching-thao (*1981a*).

During the Chhing dynasty, at least ten editions appeared. One of the first, certainly the most important, was the *Lü li kuan chiao chêng Hsi yüan lu* 律例館校正洗冤錄 (The washing away of wrongs, collated and verified by the Codification Office), published in +1662 and again in +1694. A committee commissioned by the Khang-Hsi emperor laid under contribution more than a dozen old medical books, rearranged the subheadings, greatly enlarged the contents, and reduced the number of chapters to four. Neither of these editions had a preface or postface, nor were any references given; see Sung Ta-Jen (*1957b*). Omitting the word *chi* 集 from the title probably began at this time. All subsequent treatises used this recension as their basis. A hundred years later, in +1770, the Chhien-Lung emperor ordered a *Chien ku thu ko* 檢骨圖格 (Standard chart for noting injuries to bones) appended to it.

The traditional coroner's manual took approximately its existing form in the *Hsi yüan lu chi chêng* 洗冤錄集證 (The washing away of wrongs, with collected critical notes), organised in +1796 by Wang Yu-huai 王又槐 and expanded in the 19th century. When Pelliot (1909) gave bibliographical details of various editions with his customary exhaustiveness, he omitted one derivative work, namely the *Hsi yüan lu chien shih* 洗冤錄箋釋 (Commentary on The washing away of wrongs, *ca.* +1602), by the eminent physician Wang Khên-thang 王肯堂. This has become very rare, but its publication was an event of some importance in view of the contemporary intervention of the medical profession in the West (see p. 197, above).

Later editions include *Hsi yüan lu chieh* 洗冤錄解 (1831), *Pu chu Hsi yüan lu chi chêng* 補註洗冤錄集證, with a commentary of 1832 by Juan Chhi-hsin 阮其新, and *Chhung khan pu chu Hsi yüan lu chi chêng* 重刊補註洗冤錄集證 with one of 1837 by Chang Hsi-fan 張錫蕃. Various provincial jurisdictions printed slightly different recensions. Giles's (1874) translation was based on an 1843 printing. Hsü Lien's 許槤 *Hsi yüan lu hsiang i* 洗冤錄詳義 (Detailed interpretation of The washing away of wrongs) of 1854 marked an important step forward in Chinese osteology. Twenty years a judge, whenever Hsü attended a suitable inquest he took along an artist attached to his office to draw the skeleton.[75] Finally we must acknowledge the fine modern annotated edition of Chia Ching-thao (*1981a*).

---

[74] Anon. (*1978a*), p. 479; Yang Wên-ju & Li Pao-hua (*1980*), p. 121; Shih Jo-lin (*1959*), p. 37.
[75] Chung Hsü (*1956*), p. 501.

BIBLIOGRAPHIES

A Chinese and Japanese books before +1800

B Chinese and Japanese books and journal articles since +1800

C Books and journal articles in Western languages

In Bibliographies A and B there are two modifications of the Roman alphabetical sequence: transliterated *Chh-* comes after all other entries under *Ch-*, and transliterated *Hs-* comes after all other entries under *H-*. Thus *Chhen* comes after *Chung*, and *Hsü* comes after *Huai*. This system applies only to the first words of titles in Bibliography A and to surnames in Bibliography B. Where *Chh-* and *Hs-* occur in Bibliography C, i.e. in a Western language, the normal sequence of the Roman alphabet is observed.

When obsolete or unusual romanisations of Chinese words occur in entries in Bibliography C, they are followed, wherever advisable, by the romanisations adopted as standard in the present work. If inserted in the title, these are enclosed in square brackets; if they follow it, they are enclosed in round brackets. When Chinese words or phrases occur romanised according to the Wade-Giles system or related systems, they are not altered.

Korean and Vietnamese books and essays are included in Bibliographies A and B. As explained in Vol. 1, pp. 21ff., dates in italics imply that the work is in one or other of the East Asian languages and is therefore to be found in Bibliography B.

LIST OF ABBREVIATIONS

The following abbreviations are used in citations. The entries that list only a title refer to journals. Those that comprise a title and date are of Chinese or Japanese books written before +1800; for full citations see Bibliography A. Those that include author and date are later publications in East Asian languages if the date is in italics, listed in Bibliography B. If the date is not italicised, they are books, essays or dissertations in Western languages, given in Bibliography C.

ACLCNCE	*Academiae Caesaro-Leopoldinse Carolinae naturae curiosorum ephemerides*
ACTAS	*Acta Asiatica* (Bulletin of Eastern Culture, Tōhō Gakkai, Tokyo)
ADVS	*Advancement of Science* (British Association, London)
AER	*Acta eruditorum* (Leipzig)
AGMN	*Sudhoffs Archiv für Geschichte der Medizin und der Naturwissenschaften*
AHAW/PH	Abhandlungen der Heidelberger Akademie der Wissenschaften (Philol.-hist. Klasse)
AHR	*American Historical Review*
AJCM	*American Journal of Chinese Medicine*
AJDC	*American Journal of Diseases in Children*
AJSURG	*American Journal of Surgery*
AM	*Asia Major*
AMH	*Annals of Medical History*
ANS	*Annals of Science*
APTH	*Archives of Physical Therapy*
AS/BIHP	*Bulletin of the Institute for History and Philology, Academia Sinica* (*Chung-yang Yen-chiu-yuan, Li-shih Yü-yen Yen-chiu-so chi-khan* 中央研究院歷史語言研究所集刊)
ASREV	*African Studies Review*
ASTH	*Archiv für Schiffs- und Tropen-Hygiene*
BBACS	*Bulletin of the British Association for Chinese Studies*
BEFEO	*Bulletin de l'École Française de l'Extrême-orient* (Hanoi)
BIHM	*Bulletin of the Institute of the History of Medicine*, Johns Hopkins University (continued as *Bulletin of the History of Medicine*)
BLSOAS	*Bulletin of the School of Oriental and African Studies*, University of London
BMJ	*British Medical Journal*
BNYAM	*Bulletin of the New York Academy of Medicine*
BSEIC	*Bulletin de la Société des Études Indochinoises*
BSHM	*Bulletin de la Société d'Histoire de la Médecine*
BSM	*Bollettino delle Scienze Mediche* (Bologna)
BSRCA	*Bulletin of the Society for Research in Chinese Architecture*
CCHG	*Charing Cross Hospital Gazette*
CCI	*Chê-chiang Chung-i tsa-chih* 浙江中醫雜志 (Chekiang provincial journal of traditional Chinese medicine)
CHIS	*Chinese Science*
CHS	*Chhien Han shu* (*ca.* +100)
CIBA/S	*Ciba Symposia*
CIMC/MR	*China Imperial Maritime Customs* (Medical Report Series)
CIMC/SS	*China Imperial Maritime Customs* (Special Series)
CINC	*Chung-i nien-chien* 中醫年鑒
CIT	Yen Shih-yün (*1990–4*)
CJST	*Chinese Journal of Stomatology*
CM	*Cambridge Magazine*

CSWT	*Chhing-shih wên-t'i* (continued as *Late Imperial China*, Pasadena, CA)
CYTT	Chiang-su Hsin I-hsüeh-yüan (*1977*)
DOT	Hucker, Charles O. (1985). References are to item numbers.
EARLC	*Early China*
ECL	*Eighteenth-century Life*
ECMM	Hu Shiu-ying (1980). References are to item numbers.
EHOR	*Eastern Horizon*
EHR	*Economic History Review*
END	*Endeavour*
EPI	*Episteme*
FEQ	*Far Eastern Quarterly* (continued as *Journal of Asian Studies*)
FLS	*Folklore Studies* (Peiping)
GI	Giles (1874)
HCTHHP	*Hang-chou ta-hsüeh hsüeh-pao* 杭州大學學報
HIT	*Hua-hsi i-yao tsa-chih* 華西醫藥雜誌
HJAS	*Harvard Journal of Asiatic Studies*
HOR	*History of Religions*
HOSC	*History of Science*
HTNC/LS	*Huang ti nei ching, Ling shu* (probably −1st century)
HTNC/SW	*Huang ti nei ching, Su wên* (probably −1st century)
HTNC/TS	*Huang ti nei ching, Thai su* (+656 or later)
HYCL	*Hsi yüan chi lu* (+1247)
HYL	*Chhung khan pu chu Hsi yüan lu chi chêng* (1837)
ICK	Tamba no Mototane (*1819*)
ISIS	*Isis*
ISTC	*Chung-hua i-shih tsa-chih* 中華醫史雜誌
JAAR	*Journal of the American Academy of Religions*
JAN	*Janus*
JAOS	*Journal of the American Oriental Society*
JAS	*Journal of Asian Studies*
JHHB	*Bulletin of the Johns Hopkins Hospital*
JHMAS	*Journal of the History of Medicine and Allied Sciences*
JHYG	*Journal of Hygiene*
JPB	*Journal of Pathology and Bacteriology*
JPISH	Japanese Journal of Medical History (*Nihon ishigaku zasshi* 日本醫史學雜誌)
JRAS/NCB	*Journal of the North China Branch of the Royal Asiatic Society*
KHS	*Kho-hsüeh* (Science)
KHSC	*Kho-hsüeh-shih chi-khan* (Chinese journal of the history of science)
KKYK	*Ku-kung Po-wu-yüan yüan-khan* 故宮博物院院刊
KSSM	Miyasita *et al.* (*1982*)
LEC	*Lettres édifiantes et curieuses des missions étrangères*
LHML	Chung-kuo Chung-i Yen-chiu-yüan Thu-shu-kuan (*1991*)
LIC	*Late Imperial China*
LSYC	*Li-shih yen-chiu* 歷史研究
MADC	*Madras Courier*
MCB	*Mélanges chinois et bouddhiques*
MCHSAMUC	*Mémoires concernant l'histoire, les sciences, les arts, les moeurs et les usages, des Chinois, par les missionaires de Pékin*
MCMP	*Miscellanea curiosa medico-physica Academiae Naturae Curiosorum*
MCPT	*Meng chhi pi than* (+1086)
MDGNVO	*Mitteilungen der deutschen Gesellschaft für Natur- und Völkerkunde Ostasiens*
MH	*Medical History*
MHSBAT	*Mémoires pour l'histoire des sciences et des beaux arts de travaux*
MOFF	*The Medical Officer*
MSOS	*Mitteilungen der Seminars für orientalische Sprachen* (Berlin)

MTR/RCP	Medical Transactions of the Royal College of Physicians (London)
N	Nature
NCC	Nan-ching Chung-i-yao Ta-hsüeh hsüeh-pao 南京中醫藥大學學報
NCR	New China Review
NIZ	Nihon ishigaku zasshi 日本醫史學雜誌
NMĴT	Chung-kuo I-yao Hsüeh-yüan hsin i chhao 中國醫藥學院新醫潮 (China Medical College (Taiwan)
NRRS	Notes and Records of the Royal Society
OC	Open Court
OSIS	Osiris
PAPS	Proceedings of the American Philosophical Society
PBM	Perspectives in Biology and Medicine
PLPLS/LH	Proceedings of the Leeds Philosophical and Literary Society, Literary and Historical Section
PMHS	Proceedings of the Massachusetts Historical Society
PMV	Progress in Medical Virology
PP	Past and Present
PPSA	PSA. Proceedings of the Biennial Meeting of the Philosophy of Science Association
PRSM	Proceedings of the Royal Society of Medicine
PTKM	Pen tshao kang mu (+1596)
PTRS	Philosophical Transactions of the Royal Society
R	Read, Bernard E. (1936). References are to item numbers.
RAI/OP	Royal Anthropological Institute of Great Britain & Ireland, Occasional Papers
RI	Revue indochinoise
RSKSA	Research in Sociology of Knowledge, Sciences and Art
S	Sinologica (Basel)
SAM	Scientific American
SF	Shuo fu 說郛 (collectanea)
SIC	Okanishi (1958)
SOC	Sociology
SOCH	Social History (London)
SPMSE	Sitzungsberichte der physikalisch-medizinischen Gesellschaft Erlangen
SPP	Sino-Platonic Papers
STHP	Shan-tung Chung-i Hsüeh-yüan hsüeh-pao 山東中醫學院學報
TCULT	Technology and Culture
TG/K	Tōhō Gakuhō 東方學報 (Kyoto)
TGM	Tropical and Geographic Medicine
TĴKHSYC	Tzu-jan kho-hsüeh-shih yen-chiu 自然科學史研究
TP	T'oung Pao (Leiden)
TPYL	Thai-phing yü lan (+983)
TRAS	Transactions of the Royal Asiatic Society (continued as Journal of the Royal Asiatic Society)
TRSTMH	Transactions of the Royal Society for Tropical Medicine and Hygiene
TS	Tōhō shūkyō 東方宗教
TSCC	Ku chin thu-shu chi-chhêng (+1726)
WWTK	Wen-wu tshan-khao tzu-liao 文物參考資料 (continued as Wen-wu)
ZAF	Zeitschrift für ärztliche Fortbildung

NEW EDITIONS

Editions follow the standard in previous volumes. Exceptions are noted in each entry, except for these series:

Standard Histories, Chung Hwa Book Co., 1974– , cited by ch. and page.
Huang ti nei ching, Su wên, Ling shu, Jen Ying-chhiu (1986), cited by ch., *phien* and subsection number.

Chê kung man lu 折肱漫錄 (Dilatory notes by one whose arm has been broken). Ming, +1635. Huang Chhêng-hao 黃承昊. The title refers to a saying to the effect that one who has suffered from a serious injury is likely to make a good physician.

Chê yü kuei chien 折獄龜鑑 (Magic mirror for solving cases). Sung, +1133. Chêng Kho 鄭克. For a translation into modern Chinese of this work, *I yü chi*, and *Thang yin pi shih*, see Chhen Hsia-tshun (*1995*).

Chen-yüan kuang li fang 貞元廣利方 (Medical formulae of the Chen-yüan reign-period for widespread benefit). Thang, +796. Emperor Tê Tsung 德宗.

Chêng chih yao chüeh 證治要訣 (Essential oral instructions on diagnosis and treatment). Ming, *ca.* +1380. Tai Yüan-li 戴元禮.

Chêng lei pên tshao 正類本草. See *Chhung hsiu Chêng-ho ching shih chêng lei pei yung pên tshao*.

Chêng tzu thung 正字通 (Comprehensive guide to correct use of characters). Ming, +1627. Chang Tzu-lieh 張自烈.

Chi chiu phien 急就篇 (Handy primer). C/Han, between −48 and −33. Shih Yu 史游.

Chiang hu i shu pi chhuan 江湖醫術秘傳. See Anon. (*ca. 1930*).

Chien ku thu ko 檢骨圖格 (Standard chart for noting injuries to bones). Chhing, +1770. Tsêng Fu 增福. See *Lü li kuan chiao chêng Hsi yüan lu*.

Chien-yen i lai chhao yeh tsa chi 建炎以來朝野雜記 (Miscellaneous records of officials and commoners between +1127 and +1130). Sung, *ca.* +1220. Li Hsin-chhuan 李心傳. In *TSCC*.

Chien yen ko mu 檢驗格目 (Inquest forms for examination of the corpse). Sung, +1174. Chêng Hsing-i 鄭興裔.

Chin kuei yao lüeh [fang lun] 金匱要略方論 (Essentials of prescriptions in the golden casket). H/Han, *ca.* +200. Chang Chi 張機. In *Chung-kuo i-yao tshung-shu*. See *Shang han tsa ping lun*.

Chin kuei yü han ching 金櫃玉函經 (Prescriptions in the gold and jade caskets). See *Shang han tsa ping lun*.

Chin-ling so shih 金陵瑣事 (Trifling affairs in Nanking). Ming, +1610. Chou Hui 周暉. Novel.

Ching shih chêng lei pei chi pên tshao 經史證類備急本草 (Pharmaceutical natural history for emergency use, classified and verified from the classics and histories). Sung, +1097. Thang Shen-wei 唐慎微.

Ching yen fang 經驗方 (Tried and tested prescriptions). Sung, +1025. Chang Shêng-tao 張聲道. Lost; see *SIC*, p. 1138.

Ching-yüeh chhüan shu 景岳全書 (Collected writings of Chang Chieh-pin). Ming, +1624. Chang Chieh-pin 張介賓.

Chou hou chiu tsu fang 肘後救卒方 (Handy therapies for emergencies). Chin, *ca.* +340. Ko Hung 葛洪. The version expanded by Thao Hung-ching 陶弘景 *ca.* +500 was entitled *Pu chhüeh chou hou pai-i fang* 補闕肘後百一方 (One hundred and one handy therapies, supplemented). Today the book is commonly called *Chou hou pei chi fang*.

Chou hou pei chi fang 肘後備急方. See *Chou hou chiu tsu fang*.

Chou li 周禮 (Rites of the Chou dynasty). Chou or Han, between early −4th and mid −2nd century. Anon. Shih-san-ching chu shu 十三經注疏 ed. See W. Boltz in Loewe (1993), p. 24.

Chu chieh Shang han lun 註解傷寒論 (The treatise on febrile diseases, with commentary). Sung, +1144. Chhêng Wu-chi 成無己.

Chu ping yüan hou lun 諸病源候論 (Aetiology and symptoms of medical disorders). Sui, +610. Chhao Yüan-fang 巢元方.

Chu yu shih-san kho 祝由十三科 (The thirteen departments of apotropaic medicine). Probably Ming, after +1456. Anon. This is the most usual title. The Chhing copy in the Needham Research Institute is called *Hsüan-yüan pei chi I-hsüeh chu yu shih-san kho* 軒轅碑記醫學祝由十三科 on the title page, *Chu yu kho thien I shih-san kho* 祝由科天醫十三科 at the head of the table of contents, and *Hsüan-yüan Huang ti chu yu kho* 軒轅皇帝祝由科 at the beginning of the prefaces.

Chuang tzu 莊子 (The book of Master Chuang). Earliest parts Chou, *ca.* −320, remainder −200 on. Attrib. Chuang Chou 莊周. Partial tr. in Graham (1981); on dating and parts see Graham (1979) and H. Roth in Loewe (1993), pp. 56−7.

Chung tou chih chang 種痘指掌 (Handy method for smallpox inoculation). Chhing, before +1796, pr. 1808. Anon. In this and other translations of titles, 'smallpox' includes eruptive diseases that traditional physicians could not distinguish from it.

Chung tou fang 種痘方 (Prescriptions for smallpox inoculation). Chhing, *ca.* +1725. Chêng Wang-i 鄭望頤.

Chung tou ho chieh 種痘和解 (Smallpox inoculation, with Japanese explication). Japan, +1746. Li Jen-shan 李仁山.

Chung tou hsin fa 種痘心法 (Heart method for smallpox inoculation). Chhing, 1808. Chu I-liang 朱奕梁. *TSCC*. *Hsin fa* from the Sung on referred to the 'central Way' of some body of knowledge or practice. The entry in *LHML*, item 7937, treats a separate, anonymous work published with this book as part of it.

Chung tou hsin shu 種痘新書 (New book on smallpox inoculation). Chhing, +1741. Chang Yen 張琰. Woodblock ed. of 1871.

Chung tou i than pi yü 種痘醫談筆語 (Medical chats and jottings on inoculation). Japan, Edo, +1746. Li Jen-shan 李仁山.

Chhao shih Chu ping yüan hou lun 巢氏諸病源候論. See *Chu ping yüan hou lun.*

Chhi hsiao liang fang 奇效良方 (Excellent and remarkably effective prescriptions). Ming, +1470. Tung Su 董宿 & Fang Hsien 方賢. Originally entitled *Thai i yüan ching yen chhi hsiao liang fang* 太醫院經驗奇效良方 (Excellent and remarkably effective prescriptions verified by experience in the Imperial Academy of Medicine).

Chhi hsiu lei kao 七修類稿 (Seven-part classified notes). Ming, after +1566. Lang Ying 郎瑛.

Chhi tung yeh yü 齊東野語 (Rustic anecdotes from eastern Chhi). Sung, +1290. Chou Mi 周密.

Chhien chin fang 千金方. See *Pei chi chhien chin yao fang.*

Chhien chin i fang 千金翼方 (Revised prescriptions worth a thousand). Thang, late +7th century. Sun Ssu-mo 孫思邈. Sequel to *Pei chi chhien chin yao fang.*

Chhien Han shu 前漢書 (History of the former Han dynasty, −206 to +24). H/Han, *ca.* +100. Pan Ku 班固 & Pan Chao 班昭.

Chhih shui hsüan chu 赤水玄珠 (Black pearl from the red stream). Ming, +1596. Sun I-khuei 孫一奎.

Chhu shih i shu 褚氏遺書 See *I shu.*

Chhuan ya 串雅 (The penetrator improved). Chhing, +1759. Tsung Po-yün 宗柏雲, ed. Chao Hsüeh-min 趙學敏.

Chhun chhiu 春秋 (Spring and Autumn Annals). Chou, anon. Tr. Couvreur (1914), Legge (1872). A chronicle of the State of Lu kept between −721 and −479.

Chhung hsiu Chêng-ho ching shih chêng lei pei yung pên tshao 重修政和經史正類備用本草 (Revised materia medica of the Chêng-ho era, classified and verified from the classics and histories). Sung, +1249. Chang Tshun-hui 張存惠. Repr., Jen-min Wei-shêng, Peking, 1957.

Chhung hsiu Phi-ling chih 重修毘陵志 (Local history and geography of Changchow). Sung, +1268. Shih Nêng-chih 史能之, assisted by Sung Tzhu 宋慈.

Chhung khan pu chu Hsi yüan lu chi chêng 重刊補注洗冤錄集證 (Washing away of wrongs, with collected and supplemental notes, reprinted). Chhing, 1837. Chang Hsi-fan 張錫蕃.

Êrh-shih-wu shih pu pien 二十五史補編 (Supplements to the 25 Histories). Kai-ming Shu-tien, Shanghai, 1936.

Fan Wên-chêng kung chi 范文正公集 (Collected writings of Fan Chung-yen). Sung, before +1186. Fan Chung-yen 范仲淹.

Fêng shên yen i 封神演義 (The enfeoffment of the gods). Ming, late +16th century? Attrib. Hsü Chung-lin 許仲琳, probably by Lu Hsi-hsing 陸西星. Novel. Partly tr. Grube (1912).

Fu yu pien 福幼遍 (For the welfare of children). Chhing, +1777. Chuang I-khuei 莊一夔.

Han shih wai chuan 韓詩外傳 (Outer tradition of explication of the Han recension of the Book of Songs). C/Han, *ca.* −150. Han Ying 韓嬰. In *Han Wei tshung-shu* 漢魏叢書. See J. R. Hightower in Loewe (1993), p. 125.

Han shu 漢書. See *Chhien Han shu* and *Hou Han shu.*

Ho kuan tzu 鶡冠子 (Book of the Pheasant-Cap Master). Parts C/Han, between −242 and −202. Author unknown. See Graham (1989b), Defoort (1997), and D. Knechtges in Loewe (1993), pp. 137–40.

Hou Han shu 後漢書 (History of the Later Han dynasty, +25 to +220). L/Sung, +450. Fan Yeh 范曄.

Huai nan tzu 淮南子 (The book of the King of Huai nan). C/Han, −139, parts later. Liu An 劉安. On history see C. Le Blanc in Loewe (1993), p. 189.

Huai Nan wan pi shu 淮南萬畢術 (The inexhaustible arts of the King of Huai Nan). C/Han, −2nd century. Attrib. Liu An 劉安. Lost. Fragments in *Thai-phing yü lan* and elsewhere. Reconstructed in *Kuan ku thang so chu shu* 觀古堂所著書 and elsewhere.

Huang ti chia i ching 黃帝甲乙經 (A–B manual [of acupuncture]). S/Kuo & Chin, finished +256/282. Huangfu Mi 皇甫謐. Named after its method of designating chapters. Also called *Chen-chiu chia i ching* 針灸甲乙經 and *Chia i ching.*

Huang ti nei ching, Ling shu 黃帝內經靈樞 (The Yellow Emperor's manual of corporeal [medicine]: the vital axis). C/Han, probably −1st century. Anon. The current versions of this and the next item are due to Wang Ping 王冰 (+672).

Huang ti nei ching, Su wên 黃帝內經素問 (The Yellow Emperor's manual of corporeal [medicine]: plain questions [and answers]). C/Han, probably −1st century. Anon.

Huang ti nei ching, Thai su 黃帝內經太素 (The Yellow Emperor's manual of corporeal [medicine]: the great innocence). Thang, +656 or later. Yang Shang-shan 楊上善. In Kosoto *et al.* (*1981*), vols. 1–3. On the date see Sivin (1998).

Huang ti pa-shih-i nan ching 黃帝八十一難經 (Manual of 81 problems [in the Inner Canon] of the Yellow Emperor). H/Han, probably +2nd century.

Hung lou mêng 紅樓夢 (Dreams of red mansions). Qing, incomplete at author's death in +1763, pr. +1792. Tshao Hsüeh-chhin 曹雪芹. Tr. Hawkes (1973–86).

Huo yu hsin fa 活幼心法 (Heart method for saving childrens' lives). Ming, +1616. Nieh Shang-hêng 聶尚恆.

Hsi yüan chi lu 洗冤集錄 (The washing away of wrongs [i.e., of unjust imputations]). S. Sung, +1247. Sung Tzhu 宋慈. Trl. in McKnight (1981). Taipei 1946 ed.

Hsi yüan lu 洗冤錄. See *Lü li kuan chiao chêng Hsi yüan lu.*

Hsi yüan lu chi chêng 洗冤錄集證 (The washing away of wrongs, with collected critical notes), Chhing, +1796, expanded in 19th century. Wang Yu-huai 王又槐.

Hsi yüan lu chieh 洗冤錄解 (Analysis of The washing away of wrongs). Chhing, 1831. Yao Te-yü 姚德豫.

Hsi yüan lu chien shih 洗冤錄箋釋 (Commentary on The washing away of wrongs). Ming, ca. +1602. Wang Khên-thang 王肯堂.

Hsi yüan lu hsiang i 洗冤錄詳義 (Detailed interpretation of The washing away of wrongs). Chhing, 1852. Hsü Lien 許槤.

Hsiao êrh pan-chen pei chi fang lun 小兒斑疹備急方論 (Prescriptions, with discussions, for emergency use in children's eruptive disorders). N. Sung, +1093. Tung Chi 董汲.

Hsiao êrh ping yüan fang lun 小兒病源方論 (Discourse on the aetiology of children's disorders, with prescriptions). J/Chin, ca. +1180. Chhen Wên-chung 陳文中.

Hsiao êrh yao chêng chih chüeh 小兒藥證直訣 (Straightforward explanations of paediatric medicines and syndromes). Sung, +1119. Chhien I 錢乙. Chung-i ku chi hsiao tshung-shu 中醫古籍小叢書 ed.

Hsien chia pi chhuan tou kho chen chüeh 仙家秘傳痘科真訣 (True instructions on smallpox, privately handed down by the adepts of the immortals). Chhing. Thiao Yüan-fu 調元復.

Hsin chu wu yüan lu (Shinju muwŏnnok) 新注無冤錄 (The avoidance of wrongs, newly annotated). Korea, +1438. Ch'oe Ch'i-un 崔致雲 et al.

Hsin hsiu pên-tshao 新修本草 (Revised materia medica). Thang, +659. Su Ching 蘇敬. See Shang Chih-chün (1981).

Hsin Thang shu 新唐書 (New history of the Thang dynasty). Sung, +1061. Ouyang Hsiu 歐陽修 & Sung Chhi 宋祁.

Hsü tzu chih thung chien chhang pien chi shih pên-mo 續資治通鑑長編紀事本末 (Unabridged comprehensive mirror for aid in government, chronologically arranged by topic). Chhing, 1903. Li Ming-mo 李明模.

I chia pien lan 醫家便覽 (Convenient exposé for the physicians). Ming, late +15th century. Than Lun 譚倫. Lost.

I ching 易經 (The classic of changes [Book of changes]). Late −9th century, with appended texts of −3rd and −2nd centuries. Anon. Tr. R. Wilhelm (1924/ 1950). See E. Shaughnessy in Loewe (1993), p. 216.

I hsien 醫先 (Antecedents to therapy). Ming, +1550. Wang Wên-lu 王文禄.

I hsüeh yüan liu lun 醫學源流論 (Topical discussions of the history of medicine). Chhing, +1757. Hsü Ta-chhun 徐大椿.

I shu 遺書 (Posthumous writings). Putatively late +5th century, but probably ca. +1226. Attrib. to Chhu Chhêng 褚澄.

I thung 醫通 (Compendium of medicine). Chhing, +1695. Chang Lu 張璐.

I tsung chin chien 醫宗金鑑 (Golden mirror of the medical tradition). Chhing, +1743. Wu Chhien 吳謙 et al.

I tsung pi tu 醫宗必讀 (Essential writings for orthodox physicians). Ming, +1637. Li Chung-tzu 李仲梓.

I yü chi 疑獄集 (Collection of doubtful cases). Thang, between +907 and +960. Ho Ning 和凝 and his son Ho Mêng 和㠓. Sometimes separated into an 'Earlier collection' (*I yü chhien chi* 疑獄前集, between +907 and +940) by the father and a 'Later collection' (*I yü hou chi* 疑獄後集, between +940 and +960) by the son. Cf. *Chê yü kuei chien*.

Jen shu pien lan 仁術便覽 (Handy guide to the humane art [of medicine]). Ming, +1585. Chang Hao 張浩.

Kao seng chuan 高僧傳 (Biographies of eminent monks). A collection that accumulated between ca. +550 and +1617. Hui-chiao 慧皎 et al.

Khai-pao pên tshao 開寶本草 (Pharmacopoeia of the Khai-pao era). Refers to two lost N. Sung compilations which can no longer be distinguished, the 'newly revised' *Khai-pao hsin hsiang ting pên tshao* 開寶新詳定本草 of +973 and the 're-revised' *Khai-pao chhung ting pên tshao* 開寶重定本草 of +974. Both were by Liu Han 劉翰 et al. See Okanishi (1958), pp. 1267–70, summarised in Unschuld (1986a), pp. 55–60.

Ko wu tshu than 格物麤談 (Simple discourses on the investigation of things). Sung, +980. Tsan-ning 贊寧.

Ku chin thu-shu chi-chhêng, Chhin ting 古今圖書集成, 欽定 (Imperially commissioned compendium of books and illustrations, old and new). Chhing, +1726. Ed. Chhen Meng-lei 陳夢蕾.

Kuan ku thang so chu shu 觀古堂所著書 (Books by the master of the Studio for Contemplating Antiquity). Chhing, between 1875 and 1908. Yeh Te-hui 葉德輝.

Kuan tzu 管子 (The book of Master Kuan). Diverse texts from −5th to mid −1st century. Anon., attrib. Kuan Chung 管仲.

Kuang chi fang 廣濟方 (Formulae for widespread benefaction). Thang, +723. Lost. Whether Emperor Hsüan Tsung 玄宗 compiled this work or commissioned it remains unknown.

Kuei-chi hsien hsien chuan 會稽先賢傳 (Biographies of former worthies of Kuei-chi). San-kuo, +3rd century. Hsieh Chhêng 謝承.

Kuei-hsin tsa chih 癸辛雜識 (Miscellaneous notes from Kuei-hsin Street, Hangchow). Sung, by +1308. Chou Mi 周密.

Kuei-ssu lei kao 癸巳類稿 (Classified notes of 1833). Chhing, 1833. Yü Chêng-hsieh 俞正燮. This and the next item were one MS, separated for printing.

Kuei-ssu tshun kao 癸巳存稿 (Leftover notes of 1833). Chhing, 1833, pr. 1847. Yü Chêng-hsieh.

Lao tzu 老子 (The book of the Old Master [?]). Oldest parts Chou, possibly ca. −350; probably compiled in late −3rd century. Anon. Tr. D. C. Lau (1982); on dating and versions see W. Boltz in Loewe (1993), p. 269.

Li chi 禮記 (Record of rites). Chou, various texts dating between −5th and −3rd century, possibly compiled late in +1st century. Anon. See J. Riegel in Loewe (1993), p. 293.

Li wei han wên chia 禮緯含文嘉 (Apocryphal treatise on the [Record of] rites; excellences of cherished literature). Mid-Han. Anon.

Liang pan chhiu yü an sui-pi 兩般秋雨盦隨筆 (Jottings from the Twofold Autumn Rain Studio). Chhing, *ca.* 1830. Liang Shao-jen 梁紹壬.

Liang shu 梁書 (History of the Liang dynasty). Thang, +629. Yao Chha 姚察 & his son Yao Ssu-lien 姚思廉.

Ling piao lu i 嶺表錄異 (Record of strange things from the deep south). Thang & Wu Tai, between *ca.* +895 and +915. Liu Hsün 劉恂.

Ling shu 靈樞. See *Huang ti nei ching, Ling shu*.

Lo-yang chhieh lan chi 洛陽伽藍記 (Record of the Buddhist temples and monasteries at Loyang). N/Wei, between +547 and +550. Yang Hsüan-chih 楊衒之. See Wang Yi-t'ung (1984).

Lun hêng 論衡 (Discourses weighed in the balance). H/Han, between +70 and +80. Wang Chhung 王充. Tr. Forke (1907–11). See T. Pokora & M. Loewe in Loewe (1993), p. 309.

Lun yü 論語 (Analects; conversations and discourses of Confucius and certain disciples). Chou (Lu), based on writings from *ca.* −465 on, compiled near end of Chou. Anon. Tr. Legge (1861), Waley (1938). Cited by section and subsection numbers from Yin-tê Index, Suppl. 16. See Loewe (1993), pp. 313–23.

Lü li kuan chiao chêng Hsi yüan lu 律例館校正洗冤錄 (The washing away of wrongs, collated and verified by the Codification Office). Chhing, +1662, +1694. Anon.

Lü shih chhun-chhiu 呂氏春秋 (Springs and autumns of Lü Pu-wei). Chou, *ca.* −239. Compiled under patronage of Lü Pu-wei 呂不韋. Ed. in Chhen Chhi-yu (*1984*).

Mêng Chiang-nü 孟姜女 (The ballad of Mêng Chiang-nü weeping at the Great Wall). Thang or pre-Thang. Anon. Tr. Needham & Liao Hung-ying (1948).

Mêng chhi pi than 夢溪筆談 (Brush talks from Dream Brook). Sung, by +1095. Shên Kua 沈括. Ed. Hu Tao-ching (*1956*). See Sivin (1995a), ch. III.

Mêng chhi wang huai lu 夢溪忘懷錄. See *Wang huai lu*.

Mêng Liang lu 夢梁錄 (Dreaming of the idyllic past), Sung, +1274. Wu Tzu-mu 吳自牧. In *Chih pu-tsu chai tshung-shu* 知不足齋叢書.

Ming yüan shih lun 明冤實論 (Concrete discourse on bringing unjust imputations to light). +6th century. Hsü Chih-tshai 徐之才.

Mo ching 脈經 (The pulse manual). H/Han, *ca.* +280. Wang Shu-ho 王叔和.

Nan ching 難經. See *Huang ti pa-shih-i nan ching*.

Nei ching 內經. See *Huang ti nei ching, Ling shu, Su wên*.

Nei shu lu 內恕錄 (Record of inward empathy). Sung, before +1247. Anon. The title is an allusion to *Li chi*, 29.3.

Nei wai êrh ching thu 內外二景圖 (Illustrations of internal and external views). Sung, +1118. Chu Kung 朱肱. Lost. *SIC*, 217.

Nêng kai chai man lu 能改齋漫錄 (Informal records from the Studio in which Reform is Possible). Sung, +1157. Wu Tsêng 吳曾, taken down and edited by his son Wu Fu 吳復.

Niu tou hsin fa 牛痘新法 (New method for using the cowpox). Qing, *ca.* 1825. Chhiu Hsi 邱熺.

Pao Phu Tzu nei phien 抱樸子內篇 (Inner chapters of the Preservation-of-Solidarity Master). Chin, *ca.* +320. Ko Hung 葛洪.

Pao Phu Tzu wai phien 抱樸子外篇 (Outer chapters of the Preservation-of-Solidarity Master). Chin, after +335? Ko Hung.

Pei chi chhien chin yao fang 備急千金要方 (Essential prescriptions worth a thousand, for emergency use). Thang, +650/659. Sun Ssu-mo 孫思邈. Frequently referred to as *Chhien chin fang*. As the author explains the title, a human life is worth a thousand ounces of gold, and so is any formula that can save it.

Pei Chhi shu 北齊書 (History of the Northern Chhi dynasty). Thang, +636. Li Tê-lin 李德林 and his son Li Pai-yao 李百藥.

Pei shih 北史 (History of the northern dynasties). Thang, +629. Li Yen-shou 李延壽.

Pên tshao kang mu 本草綱目 (The systematic pharmacopoeia; or, the pandects of natural history [mineralogy, metallurgy, botany, zoology, etc.]). Ming, +1596. Li Shih-chen 李時珍. Tabular abstract of parts on plants in R; paraphrased and abridged translations in other publications of Read & collaborators.

Pên tshao kang mu shih i 本草綱目拾遺 (Supplement to the Systematic pharmacopoeia). Chhing, comp. +1760 to 1803 or later, pr. 1871. Chao Hsüeh-min 趙學敏.

Pên tshao shih i 本草拾遺 (Gleanings from the materia medica). Thang, +739. Chhen Tshang-chhi 陳藏器.

Pên tshao thu ching 本草圖經 (Illustrated pharmaceutical natural history). Sung, +1062. Su Sung 蘇頌 *et al.*

Phing yüan lu 平冤錄 (Redressing of wrongs). Sung, *ca.* +1255. Chao I-chai 趙逸齋?

Phu chi fang 普濟方 (Prescriptions for universal benefaction). Ming, +1406. Chu Su 朱橚, a prince of the Ming imperial house.

Po chi hsi tou fang lun 博集稀痘方論 (Comprehensive collection of prescriptions for thinning out smallpox, with discussions). Ming, +1577. Kuo Tzu-chang 郭子章.

Pu chu Hsi yüan lu chi chêng 補註洗冤錄集證 (The washing away of wrongs, with collected and supplemental notes). Chhing, 1832. Juan Chhi-hsin 阮其新.

Pu i yü chi 補疑獄集 (Additions to the Collection of doubtful criminal cases). Ming, *ca.* +1550? Chang Ching 張景.

San kang chih lüeh 三岡識略 (Notes from the Three Ridges). Chhing, +1653. Tung Han 董含.

San kuo chih 三國志 (History of the Three Kingdoms, +220 to +280). Chin, *ca.* +290. Chhen Shou 陳壽.

San kuo i wên chih 三國藝文志 (Bibliographical treatise for the Three Kingdoms). Chhing, late 19th century. Yao Chen-tsung 姚振宗. In *Êrh-shih-wu shih pu pien.*

San yin chi i ping yüan lun tshui 三因極一病源論粹 (The three causes epitomised and unified: the quintessence of doctrine on the origins of illness). Sung, +1174 or slightly later. Chhen Yen 陳言.

Sha tou chi chieh 痧痘集解 (Collected commentaries on smallpox). Chhing, +1727. Yü Thien-chhih 俞天池. Also known as *Sha tou chin ching fu chi chieh* 痧痘金鏡賦集解, *Tou kho chin ching fu chi chieh* 痘科金鏡賦集解, and by a variety of other titles. *Sha tou* refers to a group of eruptive diseases of which smallpox is most prominent.

Shan fu ching 膳夫經 (Chef's manual). Thang, +856. Yang Yeh 楊曄. In Shinoda & Tanaka (*1972*).

Shan hai ching 山海經 (Classic of the mountains and rivers). Dates of parts disputed, probably between Warring States and H/Han. Anon. See R. Fracasso in Loewe (1993), pp. 357–67.

Shang han lun 傷寒論 (Treatise on febrile diseases). See *Shang han tsa ping lun.*

Shang han tsa ping lun 傷寒雜病論 (Treatise on febrile and miscellaneous diseases). H/Han, *ca.* +200. Chang Chi 張機. Divided in +1064 or +1065 into *Shang han lun, Chin kuei yü han ching,* and *Chin kuei yao lüeh. Shang han* means lit. 'damaged by cold'.

Shên chien 申鑒 (Extended reflections). H/Han, +205. Hsün Yüeh 荀悦.

Shên i phu chiu fang 神醫普救方 (The divine physicians' formulary for universal deliverance). Sung, +986. Chia Huang-chung 賈黃中. Compiled under imperial sponsorship; lost.

Shên nung pên tshao ching 神農本草經 (Pharmacopoeia of the Divine Husbandman). H/Han, late +1st or +2nd century. Anon.

Shêng chi tsung lu 聖濟總錄 (General record [commissioned by] sagely benefaction). Sung, compiled +1111 to +1117, issued +1122. Ed. Shen Fu 申甫 *et al.*

Shih chi 史記 (Memoirs of the Astronomer-Royal). C/Han, *ca.* −100. Ssu-ma Than 司馬談 & his son Ssu-ma Chhien 司馬遷. Partial tr. Chavannes (1895–1969).

Shih chin ching 食禁經 (Manual of dietary prohibitions). Sung. Kao Shên 高伸.

Shih ching 食經 (Catering guide). Sui. Hsieh Fêng 謝諷. Lost.

Shih hsing pên tshao 食性本草 (Pharmaceutical natural history: foods). Sung, late +9th century. Chhen Shih-liang 陳士良.

Shih i hsin chien 食醫心鑑 (Essential mirror of nutritional medicine). Thang, *ca.* +850. Tsan Yin 咎殷. Tung-fang Hsüeh-hui ed. of 1924.

Shih liao pên tshao 食療本草 (Pharmaceutical natural history: foodstuffs). Thang, +670. Mêng Shên 孟詵.

Shih ming 釋名 (Explanation of names). H/Han, *ca.* +200. Liu Hsi 劉熙. Lexicographical compilation. See R. A. Miller in Loewe (1993), p. 424.

Shih phu 食譜 (Menus). Thang, *ca.* +709. Wei Chü-yüan 韋巨源. In Shinoda & Tanaka (*1972*).

Shih shuo hsin yü 世説新語 (New account of tales of the world). Sung, *ca.* +430. Liu I-chhing 劉義慶. Tr. Mather (1976).

Shih wu chi yüan 事物紀原 (Records of the origins of affairs and things). Sung, *ca.* +1085. Kao Chhêng 高承.

Shih yen fang 試驗方 (Tried and tested prescriptions). Ming, late +15th century. Than Lun 譚倫. Lost.

Shou chhin yang lao hsin shu 壽親養老新書 (New book on longevity for one's parents and nurture for the aged). Sung, *ca.* +1085; original title *Yang lao fêng chhin shu* 養老奉親書; revised and renamed *ca.* +1300. Chhen Chih 陳直, rev. Tsou Hsüan 鄒鉉.

Shou shih thung khao 授時通考 (Compendium of works and days). Chhing, +1742. Compiled by imperial order under the direction of O-êrh-thai 鄂爾泰.

Shuo fu 説郛 (Hearsay). Yüan, before +1368. Thao Tsung-i 陶宗儀. Collection of unofficial literature; only 70 of the 120 ch. are original.

Shuo yüan 説苑 (Garden of discourses). C/Han, −17. Liu Hsiang 劉向.

Su wên 素問. See *Huang ti nei ching, Su wên.*

Sui shu 隋書 (History of the Sui dynasty, +518 to +617). Thang, +636. Wei Chêng 魏徵 *et al.*

Sung hui yao 宋會要 (Administrative statutes of the Sung dynasty). Chhing, *ca.* 1809. Hsü Sung 徐松.

Sung shih 宋史 (History of the Sung dynasty, +960 to +1279). Yüan, *ca.* +1345. Toktaga 脱脱 & Ou-yang Hsüan 歐陽玄.

Sung shih i 宋史翼 (Supplements [lit. 'wings'] to the History of the Sung dynasty). Chhing, *ca.* 1880. Lu Hsin-yüan 陸心源.

Sung shu 宋書 (History of the [Liu] Sung dynasty, +420 to +479). S/Chhi, +488. Shen Yüeh 沈約.

Sung Yüan chien yen san lu 宋元檢驗三錄 (Three treatises on inquest procedure from the Sung and Yüan periods). See *Hsi yüan chi lu, Phing yüan lu* and *Wu yüan lu.*

Ta Thang liu tien 大唐六典 (Six-part institutions of the great Thang dynasty). Thang, +738. Compiled under the direction of Li Lin-fu 李林甫. Chung-hua shu-chü ed. under title *Sung-pên Ta Thang liu tien* 宋本大唐六典.

Thai i chü chu kho chhêng wên [ko] 太醫局諸科程文格 (Model examination papers for diverse courses given by the Imperial Medical Service). Sung, +1212. Ho Ta-jen 何大任.

Thai-phing shêng hui fang 太平聖惠方 (Imperial grace formulary of the Thai-phing-hsing-kuo era). Sung, +982. Wang Huai-yin 王懷隱 *et al.*

Thai-phing yü lan 太平御覽 (Encyclopaedia of the Thai-phing-hsing-kuo reign-period for the emperor's perusal). Sung, +983. Li Fang 李昉.

Than chai pi hêng 坦齋筆衡 (Judicious jottings from the Candor Studio). Sung, *ca.* +1270. Yeh Chih 葉寘.

Thang hui yao 唐會要 (Administrative statutes of the Thang dynasty). Sung, +961. Wang Phu 王溥. *TSCC* ed.

Thang yin pi shih 棠陰比事 (Parallel cases from under the pear tree). Sung, +1211. Kuei Wan-jung 桂萬榮. Cf. *Chê yü kuei chien.*

Tou chen chin ching lu 痘疹金鏡錄 (Golden mirror of smallpox). Ming, +1579. Wêng Chung-jen 翁仲仁.

Tou chen chhuan hsin lu 痘疹傳心錄 (Smallpox: record of direct transmission). Ming, +1594. Chu Hui-ming 朱惠明.

Tou chen hsin fa 痘疹心法 (Heart method for smallpox). Ming, +1549. Wan Chhüan 萬全. This book is more unambiguously concerned with smallpox than many of its genre.

Tou chen kuan chien 痘疹管見 (Narrow examination of smallpox). Ming? Kao Chung-wu 高仲武. Lost.

Tou chen lun 痘疹論 (Treatise on smallpox). Sung, +1223. Wên-jen Kuei 聞人規.

Tou chen shih i hsin fa 痘疹時醫心法. See *Tou chen hsin fa.*

Tou chen ting lun 痘疹定論 (Definitive discussion of smallpox). Chhing, +1713. Chu Shun-ku (*or* Chhun-ku) 朱純嘏.

Tou kho chien 痘科鍵 (Key to smallpox). Chhing, *ca.* +1660?; pr. Japan +1730. Tai Li 戴笠.

Tou kho pien yao 痘科辨要 (Assessing priorities in smallpox treatment). Japan, Edo, 1811. Ikeda Hitomi 池田獨美 (*or* Zuisen 瑞仙). In *Huang Han i-hsüeh tshung-shu* 皇漢醫學叢書, vol. 9.

Tou kho ta chhüan 痘科大全 (Comprehensive treatise on smallpox). Also known as *Tou kho ta chhüan chin ching lu* 痘科大全金鏡錄. Chhing, +1707. Shih Hsi-chieh 史錫節. For eds. see *LHML,* item 7834.

Tsa tsuan 雜纂 (Miscellanea). Thang, *ca.* +858. Li Shang-yin 李商隱. *SF.*

Tsêng hsiu wu yüan lu ta chhüan (Chŭngsu muwŏnnok taejŏn) 增修無冤錄大全 (The complete Avoidance of wrongs, enlarged and revised). Korea, +1796. Ku Yun-myŏng 具允明.

Tso chuan 左傳 (Master Tso's tradition of the spring and autumn annals). Chou, dated by various scholars as between −450 and −200, probably late −4th century. Attrib. Tso Chhiu-ming 左丘明.

Tu chhêng chi shêng 都城紀勝 (Famous places of the old capital). Sung, +1235. Anon., signed Nai-te-wêng 耐得翁 (A patient old gentleman). Description of Hangchow at beginning of the +13th century.

Tung i pao chien (Tongŭi pogam) 東醫寶鑑 (Precious mirror of Eastern [i.e. Korean] medicine). Korea, commissioned +1596, presented +1610, pr. +1613. Hŏ Chun 許浚. In classical Chinese; includes Chinese as well as Korean remedies.

Wai thai pi yao 外臺秘要 (Arcane essentials from the Imperial Library). Thang, +752. Wang Thao 王燾.

Wang huai lu 忘懷錄 (Record of longing forgotten). Sung, between +1088 and +1095. Shên Kua 沈括. Lost, reconstructed in Hu Tao-ching & Wu Tso-hsin (*1981*).

Wei shêng chia pao 衛生家寶 (Family treasure for preserving life). Sung, early +12th century. Chang Yung 張永. Lost.

Wei shêng pao chien 衛生寶鑑 (Precious mirror for preserving life). Yüan, +1281. Lo Thien-i 羅天益.

Wei shu 魏書 (History of the Northern Wei dynasty, +386 to +550, including the Eastern Wei successor state). N/Chhi, +554, revised +572. Wei Shou 魏收.

Wên i lun 溫疫論 (Treatise on Warm-factor Epidemic Disorders). Ming, *ca.* +1642. Wu Yu-hsing 吳有性. See Hanson (1997).

Wên je ching-wei 溫熱經緯 (System of Heat Factor Disorders). Chhing, 1852. Wang Shih-hsiung 王士雄.

Wên je lun 溫熱論 (Treatise on Heat Factor Disorders). Chhing, +1746? Yeh Kuei 葉桂.

Wên ping thiao pien 瘟病條辨 (Systematic manifestation type determination in Heat Factor Disorders). Chhing, +1798. Wu Thang 吳塘.

Wu i hui chiang 吳醫會講 (Collected discourses of Soochow physicians). Chhing, +1792. Thang Ta-lieh 唐大烈.

Wu lei hsiang kan chih 物類相感志 (On the mutual stimulation of things according to their categories). Sung, late +10th century. Tsan-ning 贊寧.

Wu li hsiao chih 物理小識 (Little notes on the principles of the phenomena). Ming, mostly finished by +1643. Fang I-chih 方以智. Ning Ching Thang ed. of 1884.

Wu shih êrh ping fang 五十二病方 (Treatments for fifty-two ailments). Probably Han, −2nd century, before −168. Anon. In Anon. (*1985*).

Wu yüan lu 無冤錄 (The avoidance of wrongs). Yüan, +1308. Wang Yü 王與. In Yang Fêng-khun (*1987*).

Wu yüan lu shu 無冤錄述 (The avoidance of wrongs, faithfully transmitted). Japan, +1736. Kawai Jimbei 河合甚兵衛, ed. and tr.

Yang shêng ching 養生經 (Manual for nurturing vitality). *Ca.* +5th century. Shang-kuan I 上官翼.

Yang shêng fang 養生方 (Prescriptions for nurturing vitality). Before +610. Anon.

Yang shêng shu 養生書 (Book for nurturing vitality). +3rd century or earlier. Anon.

Yen shih chia hsün 顏氏家訓 (Mr Yen's advice to his clan). Sui, *ca.* +590. Yen Chih-thui 顏之推. Tr. Teng Ssu-yü (1968).

Yin shan chêng yao 飲膳正要 (Standard essentials of diet). Yüan, +1330. Hu-ssu-hui 忽思慧.

Yin shih hsü chih 飲食須知 (Essential knowledge about drinking and eating). Yüan, before +1368. In Shinoda & Tanaka (*1972*).

Ying-chhi-li kuo hsin chhu chung tou chhi shu 暎咭唎國新出種痘奇書 (Novel book on the new method of inoculation, lately out of England). Qing, 1805. Alexander Pearson, tr. Sir George Staunton and Chêng Chhung-chhien 鄭崇謙. No copies are known in China; three copies in England carry an English title-page as *Treatise on the European Style of Vaccination.*

Yü chu pao tien 玉燭寶典 (Jade candle treasury). Sui, +595. Tu Thai-chhing 杜臺卿. In *Sung shih i,* ch. 22.

Yü hsia chi 玉匣記 (Jade box records). Thang, *ca.* +880. Huang-fu Mei 皇甫枚.

Yü i tshao 寓意草 (Indirect and preliminary ideas [of medicine]). Ming, +1643. Yü Chhang 喻昌.

Yü lin 語林 (Grove of discourse). E/Chin, +362. Phei Chhi 裴啟.

Yü thang hsien hua 玉堂閑話 (Leisurely talks in the Jade Hall Academy). Wu Tai, *ca.* +950. Ho Mêng 和嶧.

Yu yang tsa tsu 酉陽雜俎 (Diverse offerings from the Yu-yang cavern). Thang, before +863? Tuan Chhêng-shih 段成式. See des Rotours (1948), p. civ.

Yu yu chi chhêng 幼幼集成 (Compendium for the proper care of children). Chhing, +1750. Chhen Fu-chêng 陳復正.

Yüeh ling 月令 (Monthly ordinances). Late Chou or Han, −3rd century? Anon. Incorporated into the *Lü shih chhun-chhiu* and later in the *Li chi*.

B. CHINESE AND JAPANESE BOOKS AND
JOURNAL ARTICLES SINCE +1800

Akahori, Akira 赤堀昭 (*1978*). 'Tō Kōkei to shūchū honzō 陶弘景と集注本草' (Thao Hung-ching and the Pharmacopoeia with collected annotations). Art. in Yamada (*1978*), p. 309.

Akahori, Akira (*1987*). 'Shōkanron teki rekishi 傷寒論的歷史' (The history of the Treatise on cold damage disorders). Unpubl. paper for International Conference on the History of Science in China, July 1987.

Anon. (*ca. 1930*). *Chiang hu i shu pi chhuan* 江湖醫術秘傳 (Esoteric therapeutic methods of doctors of the demimonde). Li Li, Hong Kong. Repr. as *Chiang hu i shu chhüan shu* 江湖醫術全書. Ta Fang, Taipei, 1973.

Anon. (*1976*). 'Yün-mêng Chhin chien shih wên 雲夢秦簡釋文' (Transcription of the Chhin bamboo slips discovered in the Chhin tomb near Yün-mêng). *WWTK*, 6, 11; 7, 1; 8, 27.

Anon. (*1977*). *Shui-hu-ti Chhin mu chu chien* 睡虎地秦墓竹簡(Bamboo slips from the Chhin tomb at Shui-hu-ti). Wên-wu, Peking.

Anon. (*1978a*). *Chung-kuo ku-tai kho-chi chhêng-chiu* 中國古代科技成就 (Achievements of ancient science and technology in China). Chung-kuo Chhing-nien, Peking.

Anon. (*1978b*). *Hsin-chiang li-shih wên-wu* 新疆歷史文物 (Historical antiquities of Sinkiang). Wên-wu, Peking.

Anon. (*1980*). *Chhang-sha Ma-wang-tui i hao Han mu. Ku shih yen-chiu* 長沙馬堆一號漢墓。古尸研究 (Han Tomb 1, Ma-wang-tui. Studies of the ancient cadaver). Wên-wu, Peking.

Anon. (*1982a*). *Sung shih yen-chiu lun-wên-chi* 宋史研究論文集 (Research papers on Sung history). *Chung-hua wên shih lun-tshung tsêng khan* 中華文史論叢增刊 (Supplement to Collections of essays in Chinese literature and history). Shang-hai Ku-chi, Shanghai.

Anon. (*1982b*). *Chung-kuo i-shih wên-hsien yen-chiu-so chien so lun-wên-chi* 中國醫史文獻研究所建所論文集 (Collected essays for the inauguration of the Institute for Medical History and Documents), pp. 80–1. Academy of Traditional Chinese Medicine, Peking.

Anon. (*1982c*). *Tōyō no kagaku to gijutsu. Yabuuchi Kiyoshi sensei shōju kinen rombunshū* 東洋の科學と技術。數內清先生頌壽記念論文集 (Science and skills in Asia. A Festschrift for the 77th birthday of Professor Yabuuchi Kiyoshi). Dōhōsha, Kyoto.

Anon. (*1985*). *Ma-wang-tui Han mu po-shu* 馬王堆漢墓帛書 (Silk MSS from the Han tombs at Ma-wang-tui). Vol. 4. Medical MSS. Wên-wu, Peking.

Chang Hsien-chê 張賢哲 & Tshai Kuei-hua 蔡貴花 (*1971*). '*Hsi yüan lu* yen-ko shu 洗冤錄沿革疏' (Notes on the vicissitudes of the *Hsi yüan lu*). *NMJT*, 1, 1, 29.

Chang Ju-chhing 張如青, Thang Yao 唐耀, & Shen Pheng-nung 沈澎農 (*1996*). *Chung-i wen-hsien-hsueh kang yao* 中醫文獻學綱要 (Outline of the study of Chinese medical literature). Shang-hai Chung-i-yao Ta-hsueh, Shanghai. Covers every aspect of bibliography, annotation, etc. Less on critical techniques than Ma Chi-hsing (*1990*).

Chang Phei-yü 張培玉 *et al.* (*1993*). *Pei-ching Thung-jen-thang shih* 北京同仁堂史 (History of the Shared Benevolence Drug Shop of Peking). Jen-min Jih-pao, Peking. From 1669 on.

Chang Ping-lun 張秉倫 & Sun I-lin 孫毅霖 (*1988*). 'Chiu-shih fang mo-ni shih-yen chi chhi yen-chiu 秋石方模擬實驗及其研究' (Experimental preparation and study of the ancient 'autumn stone' formula). *TJKHSYC*, 7, 2, 170.

Chang Tsung-tung 張宗棟 (*1990*). 'I-shêng chhêng-wei khao 醫生稱謂考' (Designations for physicians). *ISTC*, 20, 3, 138.

Chang Tzhu-kung 章次公 (*1948*). 'Ming-tai kua ming i chi chih chin-shih thi ming lu 明代掛名醫籍之進士題名錄' (Ming physicians who were Presented Scholars). *ISTC*, 2, 1–2, 5.

Chao Phu-shan 趙璞珊 (*1983*). *Chung-kuo ku-tai i-hsüeh* 中國古代醫學 (Ancient Chinese medicine). *Chung-kuo shih-hsüeh tshung-shu* 中國史學叢書. Chung-hua Shu-chü, Peking.

Chêng Wên 鄭文 (*1992*). 'Pei Sung Jen-tsung Ying-tsung i-liao an-chien shih-mo 北宋仁宗英宗醫療案件始末' (The whole story of the incidents involving medical treatment of the Emperors Jen-tsung and Ying-tsung in the Northern Sung period). *ISTC*, 22, 4, 244.

Chia Ching-thao 賈靜濤 (*1981a*). *Hsi yüan chi lu* 洗冤集錄(The washing away of wrongs). Shang-hai Kho-hsüeh Chi-shu, Shanghai. Annotated ed. of Yüan recension, with historical appendix.

Chia Ching-thao (*1981b*). 'Chung-kuo ku-tai fa-i-hsüeh yü hsing-chên shu-chi tsai Chhao-hsien yü Jih-pên 中國古代法醫學與刑偵書籍在朝鮮與日本' (Chinese books on forensic medicine and investigation transmitted to Korea and Japan). *ISTC*, 3, 148.

Chia Ching-thao (*1982*). 'Chung-kuo ku-tai fa-i chien-yen ti fên-kung 中國古代法醫檢驗的分工' (On the division of forensic medical duties in ancient Chinese inquests). *ISTC*, 12, 1, 13.

Chia Ching-thao & Chang Wei-fêng 張慰豐 (*1980*). 'Yün-mêng Chhin chien yü fa-i-hsüeh 雲夢秦簡與法醫學' (The Chhin period bamboo slips found at Yün-mêng and forensic medicine). *ISTC*, 10, 1, 15.

Chia Tê-tao 賈得道 (*1979*). *Chung-kuo i-hsüeh shih lüeh* 中國醫學史略 (Outline history of Chinese medicine). Shan-hsi Jen-min, Taiyüan.

Chiang-su Hsin I-hsüeh-yüan 江蘇新醫學院 (*1977*). *Chung-yao ta tzhu-tien* 中藥大辭典 (unabridged dictionary of Chinese drugs). 3 vols. Shang-hai Kho-hsüeh Chi-shu, Shanghai.

Chin Shu-wên 金樹文 (*1983*). 'Wo-kuo jen-tou chieh-chung fei shih yü Sung shih pu chêng 我國人痘接種非始於宋時補證' (Supplementary proofs that Chinese variolation did not begin in the Sung period). *ISTC*, 13, 4, 209.

Chou Tsung-chhi 周宗歧 (*1965a*). 'Chih mao ya-shua shih Chung-kuo fa-ming-ti 植毛牙刷是中國發明的' (Vegetable-bristle toothbrushes were invented in China). *CJST*, 4, 1.

Chou Tsung-chhi (*1965b*). 'Liao-tai chih mao ya-shua khao 遼代植毛牙刷考' (Study of vegetable-bristle toothbrushes of the Liao period). *CJST*, 4, 3.

Chu Chien-phing 朱建平 (*1992*). 'Chiang-ling Chang-chia-shan i-chien shu yao 江陵張家山醫簡述要' (Medical MSS on wooden strips from Chang-chia-shan, Hubei). *CCI*, 11, 511.

Chu Chhi-chhien 朱啟鈐 *et al.* (*1932–6*). *Chê chiang lu* 哲匠錄 (Biographies of Chinese engineers, architects, technologists and master-craftsmen). *BSRCA*, 1932, 3, 1, 123; 3, 2, 125; 3, 3, 91; 1933, 4, 1, 82; 4, 2, 60; 4, 3–4, 219; 1934, 5, 2, 74; 1935, 6, 2, 114; 6, 3, 148.

Chu Kho-wên 朱克文, Kao Ssu-hsien 高思顯 & Kung Shun 龔純 (*1996*). *Chung-kuo chün-shih i-hsüeh shih* (History of Chinese military medicine). Jen-min Chün-i, Peking.

Chu-ko Chi 諸葛計 (*1979*). 'Sung Tzhu chi chhi Hsi yüan chi lu 宋慈及其洗冤集錄' (Sung Tzhu and his Washing away of wrongs). *LSYC*, 4, 87. Also in Yang Fêng-khun (*1980*), p. 170.

Chung Hsü 仲許 (*1956*). 'Chung-kuo fa-i-hsüeh shih 中國法醫學史' (History of Chinese forensic medicine). *ISTC*, 8, 445; 9, 501.

Chung Hsü (*1957*). 'Yu kuan wo-kuo fa-i-hsüeh shih fang-mien êrh shih 有關我國法醫學史方面二事' (On two topics in the history of Chinese legal medicine). *ISTC*, 4, 286. On blood tests and the ostensor.

Chung-kuo Chung-i Yen-chiu-yüan. Chung-kuo I-shih Wên-hsien Yen-chiu-so 中國中醫研究院中國醫史文獻研究所 (*1989*). *I-hsüeh-shih wên-hsien lun-wên tzu-liao so-yin. 1979–1986. Ti êrh chi* 醫學史文獻論文資料索引 1979–1986. 第二集 (Index to essays and materials on the history of medicine and medical literature). Chung-kuo Shu-tien, Peking. Sequel to compendium of 1980, which covered from 1907 to 1978.

Chung-kuo Chung-i Yen-chiu-yüan 中國中醫研究院 & Kuang-chou Chung-i Hsueh-yuan 廣州中醫學院 (*1995*). *Chung-i ta-tzhu-tien* 中醫大辭典 (Unabridged dictionary of Chinese medicine). Jen-min Wei-shêng, Peking. Over 36,000 terms.

Chung-kuo Chung-i Yen-chiu-yüan Thu-shu-kuan 中國中醫研究院圖書館 (*1991*). *Chhüan kuo*

Chung-i thu-shu lien-ho mu-lu 全國中醫圖書聯合目錄 (National union catalogue of books on traditional medicine). Chung-i Ku-chi, Peking.

Chung Shao-hua 鍾少華 (*1996*). *Jen-lei chih-shih ti hsin kung-chü. Chung-Jih chin-tai pai-kho chhüan-shu yen-chiu* 人類知識的新工具。中日近代百科全書研究 (New tools for human knowledge. Studies of encyclopaedias in modern China and Japan). Pei-ching Thu-shu-kuan, Peking.

Chhen Chih 陳直 (*1958*). 'Hsi yin mu chien chung fa-hsien ti ku-tai i-hsüeh shih-liao 璽印木簡中發現的古代醫學史料' (Ancient medical sources discovered on seals and wooden strips). *KHSC*, 1, 68.

Chhen Chhi-yu 陳奇猷 (*1984*). *Lü shih chhun-chhiu chiao shih* 呂氏春秋校釋 (Critical edition of Springs and autumns of Master Lü). 4 vols. Hsüeh-lin, Shanghai.

Chhen Hsia-tshun 陳霞村 (*1995*). *Wên pai tui-chao tuan an chih mou chhüan shu* 文白對照斷案智謀全書 (Complete books of resourceful legal decisions, with classical and vernacular texts). Shan-hsi Ku-chi, Taiyüan.

Chhen Hsien-fu 陳先賦 & Lin Sen-jung 林森榮 (*1981*). *Ssu-chhuan i-lin jen-wu* 四川醫林人物 (Notable physicians of Szechwan). Ssu-chhuan Ren-min, Chengtu.

Chhen Khang-i 陳康頤 (*1952*). 'Chung-kuo fa-i-hsüeh shih 中國法醫學史' (History of Chinese forensic medicine). *ISTC*, 4, 1, 1.

Chhen Kho-chi 陳可冀 (ed.) (*1990*). *Chhing kung i-an yen-chiu* 清宮醫案研究 (Studies of the medical case records in the Chhing imperial palace). Chung-i Ku-chi, Peking. In addition to excerpts, contains 25 studies.

Chhen Kho-chi, Chou Wên-chhüan 周文泉 & Chiang Yu-li 江幼李 (*1982a*). 'Thai i nan tang. Tshung Chhing-tai huang-ti yu kuan i-yao te chu-phi (yü) khan yü-i 太醫難當。從清代皇帝有關醫藥的硃批(諭)看御醫' (The difficult lot of the palace physician. The Imperial Physician as seen from Chhing emperors' rescripts pertaining to medicine). *KKYK*, 3, 14.

Chhen Kho-chi, Chou Wên-chhüan & Chiang Yu-li (*1982b*). 'Chhing-tai kung-thing i-liao ching-yen ti thê-tien 清代宮廷醫療經驗的特點' (The special character of clinical experience in the Chhing court). *KKYK*, 3, 19.

Chhen Pang-hsien 陳邦賢 (*1937*). *Chung-kuo i-hsüeh shih* 中國醫學史 (History of Chinese medicine). Commercial Press, Shanghai.

Chhen Pang-hsien (*1953*). 'Chi-chung chi-hsing chhuan-jan-ping ti shih-liao thê chi 幾種急性傳染病的史料特輯' (Special selection of historical sources on several acute infectious diseases). *ISTC*, 5, 4, 227.

Chhen Shêng-khun 陳勝崑 (*1979*). *Chung-kuo chhuan-thung i-hsüeh shih* 中國傳統醫學史 (History of traditional Chinese medicine). Shih-pao Shu Ssu, Taipei.

Chhen Tao-chin 陳道瑾 & Hsüeh Wei-thao 薛渭濤 (*1985*). *Chiang-su li-tai i jen chih* 江蘇歷代醫人志

(Physicians of Kiangsu through history). Chiang-su Kho-hsüeh Chi-shu, Nanking.

Chhêng Chih-fan 程之范 (*1978*). 'Mien i fa ti hsien-chhü 免疫法的先驅' (Precursors of immunisation). Art. in Anon. (*1978a*).

Chhiu Hsi 邱熺 (*1817*). *Yin tou lüeh* 引痘略 (Leading out the pox). Repr., Chin-chang, Shanghai. *LHML* 7947 lists 60 eds.

Chhiu Hsi (*ca. 1825*). *Niu tou hsin fa* 牛痘新法 (New method for the cowpox). Tsung-tao-thang ed. of 1865. This appears to be a variant of the last item.

Fan Hsing-chun 范行準 (*1947–8*). 'Chung-hua i-hsüeh shih 中華醫學史' (Chinese medical history). *ISTC*, 1947, 1, 37; 2, 21; 1948, 1, 3–4, 17.

Fan Hsing-chun (*1953*). *Chung-kuo yü-fang i-hsüeh ssu-hsiang shih* 中國預防醫學思想史 (History of the conceptions of preventive medicine in China). Jen-min Wei-shêng, Peking. 2nd ed., 1954.

Fan Hsing-chun (*1986*). *Chung-kuo i-hsüeh shih lüeh* 中國醫學史略 (Outline history of Chinese medicine). Chung-i Ku-chi.

Fêng Han-yung 馮漢鏞 (*1994*). *Thang Sung wên-hsien san chien i fang chêng chih chi* 唐宋文獻散見醫方證治集 (Medical formulae, diagnoses, and therapies scattered in Thang and Sung literature). Jen-min Wei-shêng, Peking.

Fu Fang 傅芳 (*1982*). 'Chung-i ti mien i ssu-hsiang ho chhêng-chiu 中醫的免疫思想和成就' (Thought and achievements related to resistance to epidemics in traditional medicine). Art. in Anon. (*1982b*), p. 103.

Ho Pin 何斌 (*1988*). 'Wo-kuo nueh-chi liu-hsing chien shih. 1949 nien chhien 我國瘧疾流行簡史. 1949 年前' (A sketch of the history of malarial epidemiology in China to 1949). *ISTC*, 18, 1, 1.

Ho Shih-hsi 何時希 (*1991*). *Chung-kuo li-tai i-chia chuan lu* 中國歷代醫家傳錄 (Biographies of physicians in China through the ages). 3 vols. Jen-min Wei-shêng, Peking. Copies biographical notices on 22,000+ people from a wide range of sources, including gazetteers; appendix on medical teaching lineages.

Hou Shao-wên 侯紹文 (*1973*). *Thang Sung khao-shih chih-tu shih* 唐宋考試制度史 (History of the Thang and Sung examination systems). Shang-wu, Taipei.

Hu Tao-ching 胡道靜 & Wu Tso-hsin 吳佐忻 (*1981*). 'Mêng chhi wang huai lu kou-chhen – Shen Tshun-chung i-chu kou-chhen chih i 夢溪忘懷錄鉤沉 – 沈存中佚著鉤沉之一' (*Wang huai lu* recovered. Shên Kua's lost works reconstituted, 1). *HCTHHP*, 11, 1, 1.

Huang I-nung 黃一農 (*1990*). 'Thang Jo-wang yü Chhing chhu hsi li chih chêng-thung-hua 湯若望與清初西曆之正統化' (Schall and the legitimation of the Western calendar at the beginning of the Chhing dynasty). Art. in *Hsin pien Chung-kuo kho-chi shih* 新編中國科技史 (New history of Chinese science and technology), p. 465. Yin-ho, Taipei.

Hsieh Hsueh-an 謝學安 (*1983*). *Chung-kuo ku-tai tui chi-ping chhuan-jan-hsing ti jen-shih* 中國古代對

疾病傳染性的認識 (Knowledge of disease contagion in ancient China). *ISTC*, 13, 4, 192.

Hsieh Kuan 謝觀 (or Li-heng 利恆, *1934*). *Chung-kuo I-hsüeh yüan-liu lun* 中國醫學源流論 (Discussions of the history of Chinese medicine). Chhêng Chai I-she, Shanghai.

Hsiung Ping-chen 熊秉真 (*1995*). *Yu Yu: chhuan-thung Chung-kuo ti chhiang-pao chih tao* 幼幼。傳統中國的襁褓之道 (Proper care of children: the Way of infancy in traditional China). Lien Ching, Taipei. Paediatrics in the +16th and +17th centuries.

Hsu Tao-ching 胡道靜 (*1956*). *Mêng chhi pi than chiao chêng* 夢溪筆談校證 (Brush talks from Dream Brook, a variorum edition). Chung-hua, Shanghai, 2nd ed., 2 vols., 1960.

Hsüeh I-ming 薛益明 (*1997*). 'I-hsüeh khao-ho shih lüeh 醫學考核史略' (Outline history of medical personnel evaluation). *NCC*, 13, 1, 59.

I Kuang-jui 伊光瑞 (*1993*). *Nei-mêng-ku i-hsüeh shih lüeh* 內蒙古醫學史略 (Outline history of medicine in Inner Mongolia). Chung-i Ku-chi, Peking.

Ikeda Hitomi 池田獨美 (*1811*). *Tou kho pien yao* 痘科辨要 (Assessing priorities in smallpox treatment). N.p. [Tokyo?].

Ishida Hidemi 石田秀實 (*1987*). *Ki nagareru shintai* 氣流れる身體 (The body as a flow of *chhi*). Hirano, Tokyo.

Ishida Hidemi (*1992*). *Chūgoku igaku shisōshi* 中國醫學思想史 (History of Chinese medical thought). Tōkyō Daigaku Shuppan Kyōkai, Tokyo.

Jen An 靭庵 (*1963*). *Chung-kuo ku-tai i-hsüeh-chia* 中國古代醫學家 (Ancient physicians of China). Shang-hai Shu-chü, Hong Kong.

Jen Ying-chhiu 任應秋 (*1946*). 'Shu i yüan sou 蜀醫淵藪' (The aggregation of physicians in Szechwan). *HIT*, 1, 1. Repr. in Jen Ying-chhiu (*1984*), p. 262.

Jen Ying-chhiu (*1984*). *Jen Ying-chhiu lun i chi* 任應秋論醫集 (Collected medical essays of Jen Ying-chhiu). Jen-min Wei-shêng, Peking.

Jen Ying-chhiu (*1986*) (ed.). *Huang-ti nei ching chang-chü so-yin* 黃帝內經章句索引 (Phrase index to the Inner Canon of the Yellow Lord). Jen-min Wei-shêng, Peking.

Jen Ying-chhiu & Liu Chhang-lin 劉長林 (*1982*) (ed.). *Nei ching yen-chiu lun-tshung* 內經研究論叢 (Collected studies on the *Huang ti nei ching*). Hupei Jen-min, Wuchang.

Kao Ming-hsüan 高銘暄 & Sung Chih-chhi 宋之琪 (*1978*). 'Shih-chieh ti-i-pu fa-i-hsüeh chuan chu 世界第一部法醫學專著' (The first book in the world on forensic medicine). Art. in Anon. (*1978a*), p. 474.

Kao Tan-fêng 高丹楓 & Liu Shou-yung 劉壽永 (*1993*). *Ku chin hsing-ping lun chih* 古今性病論治 (Ancient and modern determination of treatment methods for venereal disease). Hsüeh-yüan, Peking.

Kao Yeh-thao 高也陶 (*1991*). 'Chung-kuo ku-tai i-hsüeh khao-shih kuan khuei 中國古代醫學考試管窺' (A personal view of medical examinations in ancient China). *ISTC*, 21, 1, 17.

Katō Shigeshi 加藤敏 (*1953*). *Shina keizaishi koshō* 支那經濟史考證 (Studies in Chinese economic history). 2 vols. Tōyō Bunkō, Tokyo.

Kosoto Hiroshi 小曽戸洋, Shinohara Kōichi 篠原孝市 & Maruyama Toshiaki 丸山敏秋 (*1981*). *Tōyō igaku zempon sōsho* 東洋醫學善本叢書 (Collected rare books on Oriental medicine). 8 vols. Tōyō Igaku Kenkyūkai, Osaka. Vol. 8 includes a variety of indexes, cross-reference lists, and concise studies.

Kuan Chhêng-hsüeh 管成學 (*1985*). *Lun Sung Tzhu yü Hsi yüan chi lu yen-chiu-chung shih-wu chi yüan-yin* 論宋慈與洗冤錄集錄研究中失誤及原因 (On errors in research on Sung Tzhu and the Washing away of wrongs, and the reasons for them). Chi-lin Ta-hsüeh, Chi-lin.

Kuan Lü-chhüan 關履權 (*1982b*). 'Sung-tai Kuang-chou hsiang-yao mao-i shih shu 宋代廣州香藥貿易史述' (A historical account of the incense and medicine trade in Canton in the Sung dynasty). Art. in Anon. (*1982a*), p. 280.

Kung Shun 龔純 (*1955*). 'Yüan-tai ti wei-shêng tsu-chih ho i-hsüeh chiao-yü 元代的衛生組織和醫學教育' (Public health organisation and medical education in the Yüan period). *ISTC*, 4, 270.

Kung Shun (*1981*). 'Nan Sung ti i-hsüeh chiao-yü 南宋的醫學教育' (Medical education in the Southern Sung period). *ISTC*, 11, 137.

Kuo Ching-yüan 郭景元 (*1980*). *Shih-yung fa-i-hsüeh* 實用法醫學 (Practical forensic medicine). Kho-hsüeh, Shanghai.

Kuo Tzu-kuang 郭子光 (*1988*). *Chung-i ko chia hsüeh-shuo* 中醫各家學説 (Theories of the schools of medicine). Kuei-chou Jen-min, Kueiyang.

Li Chi-jen 李濟仁 (*1990*). *Hsin-an ming i khao* 新安名醫考 (Famous physicians of Hui prefecture). Anhui Kho-hsüeh Chi-shu, Hui-chou. Hsin-an was the Han commandery that later became Hui prefecture.

Li Chien-min 李建民 (*1997*). 'Chung-kuo ku-tai chin fang khao-lun 中國古代禁方考論' (The transmission of secret techniques in ancient China). *AS/BIHP*, 68, 1, 117.

[Li Ching-wei 李經緯] (*1989*). *Chung-kuo jen-wu tzhu-tien* 中國人物詞典 (Dictionary of medical personages). Shanghai: Shang-hai tzhu-shu CPS. Best of its type. Over 6,000 biographies, including 800 after 1911.

Li Ching-wei & Li Chih-tung 李志東 (*1990*). *Chung-kuo ku-tai i-hsüeh shih lüeh* 中國古代醫學史略 (Outline history of ancient Chinese medicine). Ho-pei Kho-hsüeh Chi-shu, Shihchiachwang.

Li Chhung-chih 李崇智 (*1981*). *Chung-kuo li-tai nien-hao khao* 中國歷代年號考 (Study of Chinese reign periods through history). Chung Hua, Peking. The most comprehensive handbook of its kind.

Li Hsiao-ting 李孝定 (*1974*). *Chia-ku wên-tzu chi shih* 甲骨文字集釋 (Collected explanations of words in oracle script). Monographs of the Institute of History and Philology, Academia Sinica, 50. 2nd ed., 16 vols. in 7. Academia Sinica, Taipei.

Li Jen-chung 李仁眾 (*1982*). 'Ti hsüeh jen chhin ping fei shih yü *Hsi yüan lu* 滴血認親并非始于洗冤錄' (Dripping blood to determine kinship did not begin with The washing away of wrongs). *ISTC*, 12, 4, 208.

Li Liang-sung 李良松 & Kuo Hung-thao 郭洪濤 (*1990*). *Chung-kuo chhuan-thung wên-hua yü i-hsüeh* 中國傳統文化與醫學 (Chinese traditional culture and medicine). Hsia-men Ta-hsüeh, Xiamen. Medicine in every aspect of culture over the centuries.

Li Kêng-tung 李耕冬 (*1988*). 'Liang shan i tsu chi-ping kuan 涼山彝族疾病觀' (Views of disease among the I people in the Liang-shan I Nationality Autonomous Region). *ISTC*, 18, 2, 113.

Li Thao 李濤 (*1948*). 'Chung-kuo hsi-chü chung ti i-shêng 中國戲劇中的醫生' (Physicians in Chinese drama). *ISTC*, 1, 3-4, 1.

Li Thao (*1954*). 'Nan-Sung ti i-hsüeh 南宋的醫學' (Medicine in the Southern Sung period). *ISTC*, 6, 1, 40.

Li Ting 李鼎 (*1952*). 'Pên-tshao ching yao-wu chhan-ti piao shih 本草經藥物產地表釋' (Geographical origins of drugs in the Shên-nung materia medica, tabulated and explained). *ISTC*, 4, 167.

Li Yün 李雲 (*1988*). *Chung i jen-ming tzhu-tien* 中醫人名辭典 (Biographical dictionary of Chinese medicine). Kuo-chi Wên-hua, Peking.

Liang Chün 梁峻 (*1995*). *Chung-kuo ku-tai i chêng shih lüeh* 中國古代醫政史略 (Outline history of ancient Chinese medical administration). Nei-mêng-ku Jen-min, Hohhot.

Lin Chhien-liang 林乾良 (*1984*). 'I-hsüeh wên-tzu yüan-liu lun. (1) Lun chi ping 醫學文字源流論 (一)論疾病' (The history of medical terms. [1] *Chi* and *ping*). *ISTC*, 14, 4, 197.

Lin Thien-wei 林天蔚 (*1960*). *Sung-tai hsiang yao mao-i shih kao* 宋代香藥貿易史稿 (A history of the perfume trade of the Sung dynasty). Chung-kuo Hsüeh-she, Hong Kong.

Liu Chêng-tshai 劉正才 & Yu Huan-wên 尤煥文 (*1983*). *Chung-i mien i* 中醫免疫 (Avoidance of epidemics in traditional medicine). Chhung-chhing, Chungking.

Liu Hai-po 劉海波 et al. (*1982*). 'Wo-kuo ku-tai nü-i-shih 我國古代女醫師' (Female physicians in ancient China). *ISTC*, 12, 4, 221.

Liu Po-chi 劉伯驥 (*1974*). *Chung-kuo i-hsüeh shih* 中國醫學史 (History of Chinese medicine). 2 vols. Hua Kang, Taipei.

Liu Shih-chüeh 劉時覺 (*1987*). 'Ming-Chhing shih-chhi Hui-chou shang-yeh ti fan-jung ho Hsin-an i-hsüeh ti chüeh-chhi 明清時期徽州商業的繁榮和新安醫學的崛起' (On the commercial prosperity of Ming and Chhing Hui-chou and the eminence of its medicine). *ISTC*, 17, 1, 11.

Liu Shou-shan 劉壽山 (*1963-92*). *Chung-yao yen-chiu wên-hsien chai yao* 中藥研究文獻摘要 (Abstracts of research publications on Chinese drugs). 4 vols. Kho-hsüeh, Peking. Covers 1820 to 1984.

Liu Tien-chüeh 劉殿爵 (D. C. Lau) & Chhen Fang-chêng 陳方正 (1992–). *Hsien Chhin liang Han ku-chi chu tzu so-yin tshung-khan* 先秦兩漢古籍逐字索引叢刊 (The ICS ancient Chinese text concordance series). Commercial Press, Hong Kong.

Lo Chu-fêng 羅竹風 (1987–94). *Han-yü ta tzhu-tien* 漢語大辭典 (Unabridged dictionary of Chinese). 12 vols., supplementary index vol., 1995. San-lien, Hong Kong.

Lü Shang-chih 呂尚志 (1973). *Chung-kuo ku-tai i-hsüeh-chia ti fa-ming ho chhuang-tsao* 中國古代醫學家的發明和創造 (Discoveries and inventions of ancient Chinese physicians). Shanghai Books, Hong Kong.

Lung Yü-shêng 龍榆生 (1936). *Tung-pho yüeh-fu chien* 東坡樂府箋 (Lyrics of Su Shih, critically edited). Repr., Hsin Hua, Shanghai, 1950.

Ma Chi-hsing 馬繼興 (1990). *Chung-i wên-hsien-hsüeh* 中醫文獻學 (The study of Chinese medical literature). Shang-hai Kho-hsüeh Chi-shu, Shanghai.

Ma Chi-hsing (1995) (ed.). *Shen Nung pên tshao ching chi chu* 神農本草經輯注 (Pharmacopoeia of the Divine Husbandman reconstituted). Jen-min Wei-shêng, Peking. Includes over 700 pp. of studies.

Ma Chi-hsing (1996). 'Shuang-pao-shan Han mu chhu-thu ti chen-chiu jing-mai chhi mu jen-hsing 雙包山漢墓出土的針灸經脈漆木人形' (A lacquered wooden figurine with acumoxa tracts excavated from a Han tomb at Shuang-pao-shan). *WWTK*, 4, 55.

Ma Po-ying 馬伯英 (1991). 'Chung-kuo ku-tai chu-yao chhuan-jan-ping pien i 中國古代主要傳染病辨異' (Discriminations concerning the most important epidemic diseases in ancient China). *TJKHSYC*, 10, 280.

Ma Po-ying, Kao Hsi 高晞 & Hung Chung-li 洪中立 (1993). *Chung wai i hsüeh wên hua chiao liu shih* 中外醫學文化交流史 (A history of cultural exchanges between Chinese and foreign medicine). Shanghai Renmin, Shanghai.

Ma Po-ying (1994). *Chung-kuo i-hsüeh wên-hua shih* 中國醫學文化史 (A history of medicine in Chinese culture). Shanghai Renmin, Shanghai.

Maruyama Masao 丸山昌朗 (1977). *Shinkyū igaku to koten no kenkyū. Maruyama Masao tōyō igaku ronshū* 針灸醫學と古典の研究. 丸山昌朗東洋醫學論集 (Studies in medical acupuncture and moxibustion and their classics. Collected essays of Maruyama Masao on Oriental medicine). Sōgensha, Osaka.

Mei Khai-fêng 梅開豐 & Yü Po-lang 余波浪 (1985). ' "Chien-chhang pang" Chung yao yeh chien shih 建昌幫中药业简史' (Concise history of trade in Chinese drugs in Chien-chhang). *ISTC*, 15, 1, 29. On drug trade centred in southeast Kiangsi.

Miyashita Saburō 宮下三郎 (1967). 'Sō-Gen no iryō 宋元の醫療' (Medical therapy in the Sung and Yüan periods). Art. in Yabuuchi Kiyoshi (1967), p. 123.

Miyasita Saburō 宮下三郎 *et al.* (1982). *Kyō-u shooku sōsho mokuroku* 杏雨書屋藏書目錄 (Catalogue of the Apricot Rain Reading Room, Takeda Science Foundation library of medical history). Rinsen Shoten, Osaka.

Mugitani Kunio 麥谷邦夫 (1976). 'Tō Kōkei nempo kōryaku 陶弘景年譜考略' (Toward a chronological biography of Thao Hung-ching). *TS*, 47, 30; 48, 56.

Ni Yün-chou 倪雲洲 (1984). 'Shang-hai Chung-yao hang-yeh fa-chan chien shih 上海中藥行業發展簡史' (Concise history of the Shanghai medicine trade). *CINC*, 406.

Okanishi Tameto 岡西為人 (1958). *Sung i-chhien i-chi khao* 宋以前醫籍考 (Studies of medical books through the Sung period). Jen-min wei-shêng, Peking.

Okanishi Tameto (1974). 'Chūgoku honzō no shiteki tenbō 中國本草の史的展望' (Historical overview of the Chinese literature of materia medica). Art. in Okanishi, *Chūgoku isho honzō kō* 中國醫書本草考 (Studies of Chinese books on medicine and materia medica), p. 265. Minami Osaka Insatsu Sentā, Osaka.

Phan Chi-hsing 潘吉星 (1979). *Chung-kuo tsao-chih shih kao* 中國造紙史稿 (A draft history of Chinese papermaking). Wên-wu, Peking.

Phang Phu 龐樸 (1981). 'Chhi-shih-nien-tai chhu-thu wên-wu ti ssu-hsiang-shih ho kho-hsüeh-shih i-i 七十年代出土文物的思想史和科學史意義' (The meaning of the archaeological discoveries of the seventies for the history of thought and of science). *WWTK*, 5, 59.

Phêng Tsê-i 彭澤益 (1950). 'Hsi-yang chung tou fa chhu chhuan Chung-kuo khao 西洋種痘法初傳中國考' (A study of the early transmission of smallpox inoculation to China from the West). *KHS*, 32, 7, 203.

Shang-kuan Liang-fu 上官良甫 (1974). *Chung-kuo i-yao fa-chan shih* 中國醫藥發展史 (The development of medicine and pharmacy in China). Hsin-li, Hong Kong. Collection of 104 biographies.

Shang Chih-chün 尚志鈞 (1981). *Thang. Hsin hsiu pên-tshao* 唐。新修本草 (Revised materia medica of the Thang era). An-hui Kho-hsüeh Chi-shu. Critical ed. with introduction.

Shêng Chang 盛張 (1976). 'Chhi-shan hsin chhu Chen i yü-kan wên-thi than-so 歧山新出佚匜若干問題探索' (Investigation of some problems connected with the newly excavated *i* vessel from Mt. Chhi bearing the name of the official Chen). *WWTK*, 6, 40.

Shih Jo-lin 施若霖 (1959). *Chung-kuo ku-tai ti i-hsüeh-chia* 中國古代的醫學家 (Ancient Chinese physicians). Kho-hsüeh Chi-shu, Shanghai.

Shinoda Osamu 篠田統 (1974). *Chūgoku shokubutsu shi* 中國食物史 (A history of food in China). Shibata Shoten, Tokyo.

Shinoda Osamu & Tanaka Seiichi 田中靜一 (1972). *Chūgoku shokkei sōsho. Chūgoku kokin shokubutsu ryōri shiryō shūsei* 中國食經叢書。中國古今食物料理資料集成(Collected Chinese dietary manuals.

Complete materials for the foods and cooking of China, ancient and modern). Vol. I (Chin to Ming periods). Shoseki Bunbutsu Ryūtsūkai, Tokyo.

Sōda Hajime 宗田一 (1989). *Nihon iryō bunkashi* 日本醫療文化史 (Cultural history of clinical medicine in Japan). Shibunkaku, Kyoto.

Sung Hsiang-yüan 宋向元 (1958). 'Chhen Fu-chêng tui hsiao-êrh kho-hsüeh ti kung-hsien 陳復正對小兒科學的貢獻 (Chhen Fu-chêng's contribution to the science of paediatrics). *ISTC*, 3, 216.

Sung Ta-jen 宋大仁 (1957a). 'Wei-ta fa-i-hsüeh-chia Sung Tzhu chuan lüeh 偉大法醫學家宋慈傳略 (A biography of the great medico-legal expert, Sung Tzhu). *ISTC*, 8, 116.

Sung Ta-jen (1957b). 'Chung-kuo fa-i tien-chi pan-pên khao 中國法醫典籍版本考 (A study of editions of Chinese classics of forensic medicine). *ISTC*, 8, 278.

Takigawa Kametarō 瀧川龜太郎 (1932–4). *Shiki kaichū kōshō* 史記會注考證 (Records of the Astronomer-Royal, annotated critical edition). 10 vols. Tōhō Bunka Gakuin Tōkyō Kenkyūjo, Tokyo.

Tamba no Motohiro 丹波元簡 (1809). *I shêng* 醫賸 (Medical supererogations). 1st ed. Miscellaneous essays.

Tamba no Mototane 丹波元胤 (1819). *I chi khao* 醫籍考 (Studies of medical books). Pr. 1830. Repr. with added indexes as *Chung-kuo i chi khao* 中國醫籍考, Peking, 1956.

Thang Lan 唐蘭 (1976a). 'Shan-hsi Chhi-shan-hsien Tung-chia-tshun hsin chhu Hsi-Chou chung-yao thung-chhi ming-tzhu ti i-wên ho chu-shih 陝西歧山縣董家村新出西周重要銅器銘辭的譯文和注釋 (Transcription of and notes on inscriptions on Western Chou bronzes recently unearthed at Tung-chia Village, Chhi-shan County, Shênsi). *WWTK*, 5, 55.

Thang Lan (1976b). 'Yung chhing-thung-chhi ming-wên lai yen-chiu Hsi-Chou shih 用青銅器銘文來研究西周史 (Use of bronze inscriptions in studying Western [early] Chou history). *WWTK*, 6, 31.

Thao Yü-fêng 陶御風, Chu Pang-hsien 朱邦賢 & Hung Phi-mo 洪丕謨 (1988). *Li-tai pi-chi i shih pieh lu* 歷代筆記醫事別錄 (Record of medical matters in collections of jottings through the ages). Thien-chin Kho-hsüeh Chi-shu, Tientsin.

Thien Kuo-hua 田國華 (1986). 'Chieh-chhuang chien shih 疥瘡簡史 (Brief history of scabies). *ISTC*, 16, 2, 114.

Tshai Ching-fêng 蔡景峰 (1982). 'Tsang i-hsüeh ho Tsang i-hsüeh shih 藏醫學和藏醫學史。評甲白衮桑的西藏醫學 (Tibetan medicine and its history). Art. in Anon. (1982b), p. 80.

Tsuji Zennosuke 辻善之助 (1938). *Ni-Shi bunka no kōryū* 日支文化の交流 (Cultural exchanges between Japan and China). Sōgensha, Tokyo.

Wan Fang 萬方 & Lü Hsi-chhen 呂錫琛 (1987). 'Sung-tai kuan-yao-chü ti khao-chha 宋代官藥局的考察 (A study of the Sung official pharmacy system). *STHP*, 11, 3, 32.

Wang Chi-tsu 汪繼祖 (1958). 'I yü chi, Chê yü kuei chien, Thang yin pi shih ti shih li 疑獄集, 折獄龜鑒, 棠陰比事的釋例 (Examples from the Collection of doubtful cases, Magic mirror for solving cases, and Parallel cases from under the pear-tree). *ISTC*, 1, 45.

Wang Chih-phu 王致譜 (1980). 'Hsiao-kho [thang-niao-ping]) shih shu yao 消渴(糖尿病)史述要 (Brief history of consumption-thirst syndrome [diabetes]). *ISTC*, 10, 2, 79.

Wang Chin-kuang 王錦光 (1984). 'Kuan-yü hung kuang yen shih 關於紅光驗尸 (On the use of red light for examining corpses). *HCTHHP*, 11, 3, 328.

Wang Yün-mo 王筠默 (1958). 'Tshung *Chêng lei pên-tshao* khan Sung-tai yao-wu chhan-ti ti fên-pu 從證類本草看宋代藥物產地的分布 (The distribution of drug production in the Sung period as seen in the materia medica of the Chêng-ho era). *ISTC*, 2, 114.

Wu Jung-lun 武榮綸 & Tung Yü-shan 董玉山 (1885). *Niu tou hsin shu* 牛痘新書 (New book on the cowpox). Prefectural Yamen, Canton.

Wu Yün-jui 吳雲瑞 (1947). 'Chung-O i-hsüeh chiao-liu shih lüeh 中俄醫學交流史略 (Outline history of medical intercourse between China and Russia). *ISTC*, 1, 1, 23.

Yabuuchi Kiyoshi (ed.) (1967). *Sō-Gen jidai no kagaku gijutsushi* 宋元時代の科學技術史 (History of science and technology in the Sung and Yüan periods). Jimbun Kagaku Kenkyūjo, Kyoto.

Yamada Keiji 山田慶兒 (ed.) (1978). *Chūgoku no kagaku to kagakusha* 中國の科學と科學者 (Chinese science and scientists). Jimbun Kagaku Kenkyūsho, Kyoto.

Yamada Keiji (1988). 'Hen Shaku densetsu 扁鵲傳説 (The legend of Pien Chhüeh). *TG/K*, 60, 73.

Yamada Keiji (1990). *Yonaku tori. Igaku. Jujutsu. Densetsu* 夜鳴鳥。醫學。咒術。傳説 (The bird that cries at night. Medicine, magic, legend). Iwanami, Tokyo.

Yamamoto Noriko 山本德子 (1982). 'Tōdai ni okeru Taiisho no Taijōji e no shozoku o megutte – Taiisho no shokumu no shiteki hensen 唐代における太醫署の太常寺への所屬をめぐつて－太醫署の職務の史的變遷 (The subordination of the Imperial Medical Service to the Court of Imperial Sacrifices in the Thang period – a historic change in the duties of the Imperial Medical Service). Art. in Anon. (1982c), p. 209.

Yamamoto Noriko (1983). 'Chūgoku no rekishi ni okeru i 中國の歷史における醫 (The word 'physician' in Chinese history). *NIZ*, 29, 2, 99.

Yamamoto Noriko (1985). 'Kin-Gen jidai ni okeru shakai to ika no chi-i 金元時代における社會と醫家の地位 (Chin and Yüan society and the status of medical doctors). *JPISH*, 31, 2, 65.

Yang Chia-mao 楊家茂 (1990). 'Niu-tou chhu chhuan wo-kuo shih lüeh chi chhi i-i 牛痘初傳我國史略及其意義 (An outline of the early transmission of vaccination to our country and its significance). *ISTC*, 20, 2, 83.

Yang Fêng-khun 楊奉琨 (*1980*). *Hsi yüan chi lu chiao i* 洗冤集錄校譯 (Collated version of Washing away of wrongs, with vernacular translation). Chhün-chung, Peking. Based on ed. of +1308.

Yang Fêng-khun (*1987*). *Wu yüan lu chiao chu* 無冤錄校注 (The avoidance of wrongs, collated and annotated). Shang-hai kho-hsüeh chi-shu, Shanghai.

Yang Wên-ju 楊文儒 & Li Pao-hua 李寶華 (*1980*). *Chung-kuo li-tai ming i phing chieh* 中國歷代名醫評介 (Critical introduction to famous physicians of China through the ages). Shen-hsi Kho-hsüeh Chi-shu, Changan.

Yang Yüan-chi 楊元吉 (*1953*). *Chung-kuo i-yao wên-hsien chhu chi* 中國醫藥文獻初輯 (Collected documents on Chinese medicine and pharmacy, first series). Ta-te, Shanghai.

Yen Shih-yün 嚴世芸 (*1990–4*). *Chung-kuo i chi thung khao* 中國醫籍通考 (General compendium on traditional Chinese medical books). 4 vols., 1 vol. index. Shang-hai Chung-i-yao Ta-hsüeh, Shanghai. Actually a compilation of prefaces with occasional notes on authors. Lists over 9,000 books, extant and lost.

Yü Chia-hsi 余嘉錫 (*1958*). *Ssu khu thi yao pien chêng* 四庫提要辨證 (Critical studies of the Four Repositories abstracts). Orig. publ. 1937. Rev. ed., Chung-hua, Peking, 1958; repr., 1974.

Yü Ming-kuang 余明光 (*1989*). *Huang-ti ssu ching yü Huang-Lao ssu-hsiang* 黃帝四經與黃老思想 (The 'Four canons of the Yellow Emperor' and Huang-Lao thought). Hei-lung-chiang Jen-min Chhu-pan-she, Harbin.

Yü Wên-chung 于文中 (*1981*). 'Chiao-fa hsiao i 角法小議' (Notes on the therapeutic method of cupping). *ISTC*, 11, 2, 95.

Yü Yen 余巖 (Yün-hsiu 雲岫) (*1953*). *Ku-tai chi-ping ming hou shu i* 古代疾病名候疏義 (Glosses on the names and symptoms of ancient diseases). Jen-min Wei-shêng, Peking.

Yü Yung-min 于永敏 (*1987*). 'Liao-tai Chhi-tan-tsu i-hsüeh shih chien-shu 遼代契丹族醫學史簡述' (Brief account of the medical history of the Khitan people in the Liao period). *ISTC*, 17, 1, 60.

Yü Yung-min (*1990*). *Liao-ning i-hsüeh jen-wu chih* 遼寧醫學人物志 (Biographical notes on physicians of Liaoning). Liao Shên, Shênyang.

C. BOOKS AND ARTICLES IN WESTERN LANGUAGES

ACKERKNECHT, ERWIN (1948). 'Anti-contagionism between 1821 and 1867'. *BIHM*, 22, 562.

ACKERKNECHT, ERWIN (1950–1). 'The Early History of Legal Medicine'. *CIBA/S*, 11, 128.

ADAM, N. K. & STEVENSON, D. G. (1953). 'Detergent Action'. *END*, 12, 25.

ADNAN ADIVAR, ABDULHAK (1939). *La science chez les Turcs ottomans.* Maisonneuve, Paris.

ADNAN ADIVAR, ABDULHAK (1943). *Osmanli turkerinde ilim* [Science among the Ottoman Turks]. Maarif Matbaasi, Istanbul. Considerably enlarged from *ibid.* (1939).

AINSLIE, W. (1830). 'Observations Respecting the Smallpox and Inoculation in Eastern Countries. With Some Account of the Introduction of Vaccination into India'. *TRAS*, 2, 52.

ALABASTER, E. (1899). *Notes and Commentaries on Chinese Criminal Law and Cognate Topics, with Special Attention to Ruling Cases; Together with a Brief Excursus on the Law of Property, Chiefly Founded on the Writings of the Late Sir Chaloner Alabaster, KCMG, Sometime HBM Consul-General in China.* Luzac, London.

ALLEE, MARK A. (1994). *Law and Local Society in Late Imperial China. Northern Taiwan in the Nineteenth Century.* Stanford University Press, Stanford, CA.

AMUNDSEN, D. W. & FERNGREN, G. B. (1978). 'The Forensic Role of Physicians in Ptolemaic and Roman Egypt'. *BIHM*, 52, 3, 336.

ANDREWS, BRIDIE J. (1995). 'Traditional Chinese Medicine as Invented Tradition'. *BBACS*, 6.

ANON. (+1757). *De Inenting der Kinderpokjes in hare groote Voordeelen aangewezen. Uit eene Vergelykinge derzelven met die door den Natuurlyken Weg komen. Uit 't Gezag der Schryveren die voor en tegen de Inenting zich opentlyk hebben uitgelaten. Uit de Wederleggingen van alle de voornaamste Tegenwerpingen en Zwarigheden, die oort tegen dezelve Zyn ingebragt. Eindelyk getaasd door Eigen Ondervindigen.* Arrenberg, Rotterdam.

ANON. (1913). *The History of Inoculation and Vaccination for the Prevention and Treatment of Disease.* Lecture memoranda, 17th International Congress of Medicine, 1913. Burroughs Wellcome, London.

[ARBUTHNOT, JOHN] (+1722). *Mr Maitland's Account of Inoculating the Small-pox vindicated from Dr. Wagstaffe's Misrepresentations of that Practice, with some Remarks on Mr Massey's Sermon.* J. Peele, London.

BACON, ROGER (+1236 to +1245). *De retardatione accidentium senectutis.* .... Little & Withington (1928). See R. Browne (+1683).

BAKER, SIR GEORGE (+1766). *An Inquiry into the Merits of a Method of Inoculating the Smallpox, which is now Practised in Several Counties in England.* J. Dodsley, London. Also in *MTR/RCP*, 1772, 1, 275.

BALL, J. DYER (1892). *Things Chinese. Being Notes on Various Subjects Connected with China.* Sampson Low, Marston, London.

BARON, J. (1827). *Life of Dr. Jenner.* 2 vols. Henry Colburn, London.

DES BARRES, DUPONT LEROY & TANON, M. (1912). 'À propos de la protection contre la variole. Coutumes des Peuplades Noires de la bouche du Niger, et coutumes chinoises'. *BSHM*, 11, 49.

BARRETT, TIMOTHY H. (1991). 'Some Aspects of the *Chin-ching* of Sun Ssu-mo'. Unpublished MS on *Chhien chin i fang*, ch. 29–30.

BARTHOLINUS, THOMAS (+1673). 'De transplantatione morborum dissertatio epistolica'. Art. in Grube (+1673).

BARTHOLINUS, THOMAS & VOLLGNAD, H. (+1671). 'Febris ex imaginatione'. *MCMP* (Jena, 1688), 2, 264.

BATES, DON G. (ed.) (1995). *Knowledge and the Scholarly Medical Traditions.* Cambridge University Press, Cambridge.

BAXBY, D. (1978). 'Edward Jenner, William Woodville, and the Origins of Vaccinia Virus'. *JHMAS*, 34, 134.

BEDSON, H. S. & DUMBELL, K. R. (1964). 'Hybrids derived from the Viruses of Variola Major and Cowpox'. *JHYG*, 62, 147.

BEESON, PAUL B. & McDERMOTT, WALSH (ed.) (1963). *Cecil-Loeb Textbook of Medicine.* 11th ed., 2 vols. W. B. Saunders, Philadelphia.

BENEDICT, CAROL (1996). *Bubonic Plague in Nineteenth-century China.* Stanford University Press, Stanford, CA.

BENNETT, J. A. (1986). 'The Mechanics' Philosophy and the Mechanical Philosophy'. *HOSC*, 24, 1, 1.

BESENBRUCH, D. (1912). 'Zur Epidemiologie der Pocken in Nordchina'. *ASTH*, 16, 48.

BIAGIOLI, MARIO (1989). 'The Social Status of Italian Mathematicians, 1450–1600'. *HOSC*, 27, 1, 41.

BIAGIOLI, MARIO (1993). *Galileo, Courtier. The Practice of Science in the Culture of Absolutism. Science and its Conceptual Foundations.* University of Chicago Press, Chicago. A classic study of patronage.

BIELENSTEIN, HANS (1986). 'The Institutions of Later Han'. Art. in Twitchett & Loewe (1986), p. 491.

BIOT, E. (tr.) (1851). *Le Tcheou-li ou Rites des Tcheou.* 3 vols. Imp. Nat., Paris. Repr. Wên-tien-ko, Peking, 1930. Tr. of *Chou li.*

BLAKE, J. B. (1953). 'Smallpox Inoculation in Colonial Boston'. *JHMAS*, 8, 284.

DE BLÉGNY, NICOLAS (+1684). *La doctrine des rapports de chirurgie. Fondée sur les maximes d'usage et sur la disposition des nouvelles ordonnances.* T. Amaubry, Lyon. English tr., Amesbury, London, +1684.

BODDE, DERK (1942). 'Early References to Tea Drinking in China'. *JAOS*, 62, 74.

BODDE, DERK (1946). 'Henry A. Wallace and the Ever-normal Granary', *FEQ*, 5, 4, 411.

BODDE, DERK (1954). 'Authority and Law in Ancient China'. Art. in *Authority and Law in the Ancient Orient*, p. 41. *JAOS*, Supplements, 17.

BODDE, DERK (1963). 'Basic Concepts of Chinese Law. The Genesis and Evolution of Legal Thought in Traditional China'. *PAPS*, 107, 375.

BODDE, DERK (1982). 'Forensic Medicine in Pre-imperial China'. *JAOS*, 102, 1. Cf. Hulsewé (1978, 1985); McLeod & Yates (1981).

BODDE, DERK & MORRIS, CLARENCE (1967). *Law in Imperial China.* Harvard University Press, Cambridge, MA.

BODMAN, NICHOLAS CLEAVELAND (1954). *A Linguistic Study of the Shih ming. Initial and Consonant Clusters.* Harvard-Yenching Institute Studies, 11. Harvard University Press, Cambridge, MA.

BOERHAAVE, HERMANN (+1716). *Praxis medicinae Boerhaaveana, Being a Complete Body of Prescriptions Adapted to Each Section of the Practical Aphorisms of H. Boerhaave.* B. Cowse & W. Innys, London.

BOKENKAMP, STEPHEN R. (1994). *Time after Time. Taoist Apocalyptic History and the Founding of the T'ang Dynasty. AM*, 3rd ser., 7, 1, 59. On apocalyptic themes in dynastic succession from Six Dynasties on.

BRADLEY, L. (1971). *Smallpox Inoculation (Variolation). An Eighteenth-century Mathematical Controversy.* Adult Education Department, Nottingham.

BREITENSTEIN, H. (1908). *Gerichtliche Medizin der Chinesen von Wang-in-Hoai nach der holländischen Übersetzung F. M. de Grijs.* Grieben (Furnau), Leipzig. Wang Yin-huai wrote the +1796 preface to *HYL*.

BRETSCHNEIDER, E. (1881–98). 'Botanicon sinicum. Notes on Chinese Botany from Native and Western Sources'. *JRAS/NCB* (n.s.), 1881, 16, 18; 1893, 25, 1; 1898, 29, 1; repr., 3 vols., Kraus, Nendeln, 1967.

BRIDGMAN, R. F. (1955). 'La médecine dans la Chine antique, d'après les biographies de Pien-ts'io [Pien Chhüeh] et de Chouen-yu Yi [Shunyü I] (Chapitre 105 des *Mémoires historiques* de Sseu-ma Ts'ien [Ssu-ma Chhien]'. *MCB*, 10, 1.

BROOKS, E. BRUCE (1994). 'Review Article: The Present State and Future Prospects of Pre-Han Text Studies'. *SPP*, 46, 1. Review of Loewe (1993), mostly devoted to an outline of Brooks's own studies of the coming together of early texts.

BROTHWELL, DON R. & SANDISON, A. T. (eds.) (1967). *Diseases in Antiquity. A Survey of the Diseases, Injuries, and Surgery of Early Populations.* Thomas, Springfield, IL.

BROWN, THEODORE M. (1982). 'The Changing Self-concept of the 18th-century London physician'. *ECL*, 7, 31.

BROWNE, E. G. (1921). *Arabian Medicine.* Cambridge University Press, Cambridge. Repr., 1962.

BROWNE, RICHARD (+1683). *The Cure of Old Age and Preservation of Youth.* Tho. Flesher & Edward Evets, London. Tr. of Roger Bacon, *De retardandis senectutis accidentibus.*

BRUCE, JAMES (+1790). *Travels to Discover the Sources of the Nile in the Years 1768, 1769, 1770, 1771, 1772 and 1773.* G. G. J. & J. Robinson, London.

BULLOCH, W. (1930/1938). 'The History of Bacteriology'. Art. in *A System of Bacteriology in Relation to Medicine*, vol. 1, p. 15. Medical Research Council, London. Revised and enlarged as the Heath Clark Lectures at the University of London, and published with the same title, Oxford University Press, Oxford, 1938.

BULLOCK, MARY BROWN (1980). *An American Transplant. The Rockefeller Foundation and Peking Union Medical College.* University of California Press, Berkeley, CA.

BURKE, D. C. (1977). 'The Status of Interferon'. *SAM*, 236, 4, 42.

BYNUM, W. F., BROWNE, E. J. & PORTER, R. (1981). *Dictionary of the History of Science.* Macmillan, London.

CALVI VIRAMBAM (1819). 'Inoculation in Ancient India'. *MADC*, 12 January.

CAMPS, F. E. & CAMERON, J. F. (1971). *Practical Forensic Medicine.* Hutchinson Medical, London.

CAPRA, FRITJOF (1977). *The Tao of Physics: An Exploration of the Parallels Between Modern Physics and Eastern Mysticism.* Bantam, New York.

DE CARRO, J. (1804). *Histoire de la vaccination en Turquie, en Grèce, et aux Indes Orientales.* Geistinger, Vienna.

CASTIGLIONI, A. (1947). *A History of Medicine*, tr. and ed. E. B. Krumbhaar. 2nd ed., Knopf, New York.

À CASTRO, J., HARRIS, G. & LE DUC, A. (+1722). *Dissertationes in novam, tutam ac utilem methodum inoculationis, seu transplantationis.* Langerak, Leiden.

À CASTRO, RODERICUS (+1614). *Medicus politicus, sive de officiis medico-politicis tractatus.* Frobenius, Hamburg.

CHAFFEE, JOHN W. (1993). *The Thorny Gates of Learning in Sung China. A Social History of Examinations.* Cambridge University Press, Cambridge, 1985. Rev. ed., State University of New York Press, Albany, NY.

CHAN, WING-TSIT (1963). *The Way of Lao Tzu (Tao-te ching).* The Library of Liberal Arts, 139. Bobbs-Merrill, Indianapolis, IN.

CHANG CHE-CHIA (1997). 'The Therapeutic Tug of War. The Imperial Patient-Practitioner Relationship in the Era of Empress Dowager Cixi (1874–1908)'. Ph.D. diss., Asian and Middle Eastern Studies, University of Pennsylvania.

CHANG, CHIA-FENG (1996). 'Aspects of Smallpox and its Significance in Chinese History'. Ph.D. diss., History, School of Oriental and African Studies, University of London.

CHANG, CHUNG-LI (1962). *The Income of the Chinese Gentry.* Publications on Asia. University of Washington Press, Seattle, WA.

CHANG, K. C. (ed.) (1977). *Food in Chinese Culture. Anthropological and Historical Perspectives.* Yale University Press, New Haven, CT.

CHAO, YÜAN-LING (1995). 'Medicine and Society in Late Imperial China: A Study of Physicians in Suzhou'. Ph.D. diss., History, University of California, Los Angeles. Study of the period 1600–1850.

CHATTOPADHYAYA, DEBIPRASAD (1977). *Science and Society in Ancient India.* Research India Publications, Calcutta.

CHAVANNES, E. (1895–1969). *Les mémoires historiques de Se-Ma Ts'ien.* 6 vols. Leroux, Paris. Vol. 6 ed. Timoteus Pokora, publ. Maisonneuve, Paris. Tr. of *Shih chi,* ch. 1–52.

CHIA, LUCILLE (1996). 'Printing for Profit: The Commercial Printers of Jianyang (Fujian), Song-Ming (960–1044)'. Ph.D. diss., East Asian Languages and Cultures, Columbia University.

CHINA SCIENCE AND TECHNOLOGY MUSEUM & CHINA RECONSTRUCTS (1983). *China: 7000 Years of Discovery. China's Ancient Technology. Authorised Beijing Edition. China Reconstructs* Magazine, Peking. This replaced an 'unauthorised' catalogue less fixated on priorities and more historically accurate.

CH'Ü T'UNG-TSU (1961). *Law and Society in Traditional China.* Le monde d'Outre-mer, ser. 1, 4. Mouton, The Hague.

[CIBOT, P. M.] (+1779). 'De la petite vérole'. *MCHSAMUC,* 4, 392. Repr. in Dabry de Thiersant (1863), pp. 118ff., and more correctly in Lepage (1813).

CLINCH, WILLIAM (+1724). *An Historical Essay on the Rise and Progress of the Smallpox. To which is added a Short Appendix, to Prove, that Inoculation is no Security from the Natural Smallpox.* T. Warner, London.

CODRONCHI, GIOVANNI BATTISTA (+1595). *De morbis veneficis ac veneficijs.* F. de Franciscis, Venice.

COLDEN, CADWALLADER (+1757). 'Extract of a Letter . . . to Dr Fothergill, concerning the Throat Distemper'. In *Medical Observations and Inquiries* by a Society of Physicians in London, +1757, vol. 1, p. 227. Letter dated +1753.

COLOMBERO, CARLO (1979). 'Il problema del contagio nel pensiero medico-filosofico del Rinascimento italiano e la soluzione di Fracastoro'. *Atti della Accademia delle Scienze di Torino. Classe di Scienze Morali, Storiche e Filologiche,* 113, 245.

CONRAD, LAWRENCE I. *et al.* (1995). *The Western Medical Tradition. 800 BC to AD 1800.* Cambridge University Press, Cambridge.

COOK, S. F. (1941–2). 'F. X. Balmis and the Introduction of Vaccination to Spanish America'. *BIHM,* 11, 543; 12, 70.

COOPER, W. C. & SIVIN, N. (1973). 'Man as a Medicine. Pharmacological and Ritual Aspects of Drugs Derived from the Human Body', Art. in Nakayama & Sivin (1973), p. 203.

CORDIER, HENRI (1920). *Histoire générale de la Chine.* 4 vols. Geuthner, Paris.

DE LA COSTE, M. (+1723). *Lettre sur l'Inoculation de la Petite Vérole, comme elle se pratique en Turquie et en Angleterre. Adressée à Mons. Dodart . . . premier médecin du roi. Avec un appendix qui contient les preuves et répond à plusieurs questions curieuses.* Claude Labottière, Paris. Rev. *MHSBAT,* +1724, p. 1073.

COUVREUR, F. S. (tr.) (1913). *Li Ki, ou Mémoires sur les bienséances et les cérémonies. (Li chi.)* 2 vols. La Mission Catholique, Hokien fu. Repr. Belles Lettres, Paris, 1950.

COUVREUR, F. S. (tr.) (1914). *Tch'ouen Ts'iou et Tso Tchouan. Texte chinois avec traduction française.* 3 vols. Mission Press, Hochienfu. Tr. of *Chhun chhiu* and *Tso chuan.*

COWDRY, E. V. (1921). *The Office of Imperial Physicians, Peking.* American Medical Association, Chicago.

CREIGHTON, C. (1889). *Jenner and Vaccination.* Sonnenschein, London.

CREIGHTON, C. (1891–4). *History of Epidemics in Britain.* 2 vols. Cambridge University Press, Cambridge. Vol. 2 covers +1666 to 1893.

CROOKSHANK, E. M. (1889). *History and Pathology of Vaccination.* 2 vols. Lewis, London.

CROSBY, ALFRED W. (1993). 'Smallpox'. Art. in Kiple (1993), p. 1008.

CSIKSZENTMIHALYI, MARK A. (1994). 'Emulating the Yellow Emperor. The Theory and Practice of Huanglao, 180–141 B.C.E.' Ph.D. diss., Philosophy, Stanford University.

CULLEN, CHRISTOPHER (1993). 'Patients and Healers in Late Imperial China: Evidence from the *Jinpingmei'. HOSC,* 31, 2, 99. On the Chhing novel *Chin phing mei* 金瓶梅.

DABRY DE THIERSANT, C. P. (1863). *La médecine chez les Chinois*. Henri Plon, Paris.

DALTON, O. M. (1927). *The History of the Franks, by Gregory of Tours*. Oxford University Press, Oxford.

DEFOORT, CARINE (1997). *The Pheasant Cap Master (He guan zi). A Rhetorical Reading*. State University of New York Series in Chinese Philosophy and Culture. State University of New York Press, Albany, NY.

DESPEUX, CATHERINE (1981). *Taiji quan. Art martial, technique de longue vie*. G. Tredaniel, Paris.

DEVAUX, JEAN (+1703). *L'art de faire les raports [sic] en chirurgie avec un extrait des arrests, status, & reglemens faits en conséquence.* . . . Laurent d'Houry, Paris. Written +1693.

DIKÖTTER, FRANK (1992). *The Discourse of Race in Modern China*. Hurst, London.

DIMSDALE, THOMAS (+1769). *The Present Method of Inoculating for the Smallpox*. 5th ed. W. Owen, London.

DIXON, C. W. (1962). *Smallpox*. Churchill, London.

DOBELL, CLIFFORD (1932). *Antony van Leeuwenhoek and his 'Little Animals'*. Bale, London.

DOE, JANET (1937). *A Bibliography of the Works of Ambroise Paré, Conseiller et Premier Chirurgien du Roy.* . . . History of Medicine Series, 4. Van Hensden, Amsterdam.

DORÉ, H. (1914–29). *Recherches sur les superstitions en Chine*. 15 vols. T'u-se-we Press, Shanghai.

DOUGLAS, MARY (1966). *Purity and Danger. An Analysis of Concepts of Pollution and Taboo*. Routledge & Kegan Paul, London.

DOUGLAS, MARY (1975). *Implicit Meanings. Essays in Anthropology*. Routledge & Kegan Paul, London.

DOWNIE, A. W. (1951). 'Jenner's Cowpox Inoculation'. *BMJ*, pt 2, 251.

DUDGEON, JOHN (1870). 'The Great Medical College at Peking'. *Chinese Recorder*, 2, 9.

EBERHARD, W. (1967). *Guilt and Sin in Traditional China*. University of California Press, Berkeley, CA.

EBERHARD, W. (1971). *Moral and Social Values of the Chinese. Collected Essays*. Chinese Materials and Research Aids Service Centre, Occasional Series, 6. Chhêng-wên, Taipei.

ELMAN, BENJAMIN A. (1984). *From Philosophy to Philology. Intellectual and Social Aspects of Change in Late Imperial China*. Harvard East Asian Monographs, 110. Council on East Asian Studies, Harvard University, Cambridge, MA.

ENGELHARDT, UTE (1987). *Die klassische Tradition der Qi-übungen (Qigong). Eine Darstellung anhand des Tang-zeitlichen Textes Fuqi Jingyi Lun von Sima Chengzhen*. Münchener Ostasiatische Studien, 44. Franz Steiner Verlag Wiesbaden, Stuttgart.

D'ENTRECOLLES, F. X. (+1731). 'Lettre du 11 Mai, 1726, au Père Duhalde. De l'inoculation chez les Chinois'. *LEC*, 20, 304.

EPLER, D. C., JR (1980). 'Blood-letting in Early Chinese Medicine and its Relation to the Origin of Acupuncture'. *BIHM*, 54, 337.

ESCARRA, JEAN (1936). *Le droit chinois*. Vetch, Peking.

FAIRBANK, J. K. & REISCHAUER, E. O. (1958). *East Asia. The Great Tradition*. Houghton Mifflin, Boston, MA.

FARQUHAR, JUDITH (1994). *Knowing Practice. The Clinical Encounter of Chinese Medicine*. Studies in the Ethnographic Imagination, 4. Westview Press, Boulder, CO.

FARQUHAR, JUDITH (1995). 'Re-writing Traditional Medicine in post-Maoist China'. Art. in Bates (1995), p. 251.

FEDELE, FORTUNATO (+1602). *De relationibus medicorum. Libri quatuor*. N.p., Palermo. Repr., Leizig, +1674.

FEHER, MICHEL (ed.) (1989). *Fragments for a History of the Human Body*. 3 vols. Zone, New York. Collected papers on several cultures.

FEIERMAN, STEVEN (1985). 'Struggles for Control: The Social Roots of Health and Healing in Modern Africa'. *ASREV*, 28, 2–3, 103.

FÊNG YU-LAN. See Fung Yu-Lan.

FENNER, FRANK (1977). 'The Eradication of Smallpox'. *PMV*, 23, 1.

FIELDS, ALBERT (1950). 'Physiotherapy in Ancient Chinese Medicine'. *AJSURG*, 79, 613.

FILLIOZAT, JEAN (1949). *La doctrine classique de la médecine indienne*. Imprimerie Nationale, Paris.

FISK, DOROTHY (1959). *Dr Jenner of Berkeley*. Heinemann, London.

FORBES, R. J. (1954). 'Chemical, Culinary, and Cosmetic Arts'. Art. in Singer *et al.* (1954–8), vol. 1, p. 238.

FORKE, ALFRED (tr.) (1907–11). *'Lun Hêng', Philosophical Essays of Wang Chhung*. Kelly & Walsh, Shanghai. Orig. publ. as *MSOS, Beibände*, 1906, 9, 181; 1907, 10, 1; 1908, 11, 1; 1911, 14, 1.

FORTE, ANTONINO (1988). *Mingtang and Buddhist Utopias in the History of the Astronomical Clock*. Istituto Italiano per il Medio ed Estremo Oriente, Rome.

FRANK, ROBERT G., JR (1980). *Harvey and the Oxford Physiologists. Scientific Ideas and Social Interaction*. University of California Press, Berkeley, CA.

FRANKE, HERBERT (1976). *Sung Biographies*. Münchener ostasiatische Studien, 16. Steiner, Wiesbaden.

FREIDSON, ELIOT (1970). *Profession of Medicine. A Study of the Sociology of Applied Knowledge*. Dodd, Mead, New York.

FRENCH, ROGER (1994). *William Harvey's Natural Philosophy*. Cambridge University Press, Cambridge.

Fukui Fumimasa (1995). 'The History of Taoist Studies in Japan and Some Related Issues'. *Acta Asiatica*, 68, 1.

Fuller, Thomas (+1730). *Exanthematologia. Or, an Attempt to give a Rational Account of Eruptive Fevers, Especially of the Measles and Small Pox . . . to which is added, an Appendix concerning Inoculation*. C. Rivington & S. Austen, London.

Fung Yu-Lan (tr.) (1933). *Chuang Tzu. A New Selected Translation with an Exposition of the Philosophy of Kuo Hsiang*. Commercial Press, Shanghai.

Furth, Charlotte (1986). 'Blood, Body and Gender. Medical Images of the Female Condition in China 1600–1850'. *CHIS*, 7, 43.

Furth, Charlotte (1987). 'Concepts of Pregnancy, Childbirth, and Infancy in Ch'ing Dynasty China'. *JAS*, 46, 1, 7.

Furth, Charlotte (1988). 'Androgynous Males and Deficient Females. Biology and Gender Boundaries in Sixteenth- and Seventeenth-century China'. *LIC*, 9, 2, 1.

Furth, Charlotte (1995). 'From Birth to Birth. The Growing Body in Chinese Medicine'. Art. in Kinney (1995), p. 157. On schemata of conception, growth, stages of sexual activity, etc.

Furth, Charlotte (1996). 'Women as Healers in Ming Dynasty China'. Paper for Eighth International Conference on the History of Science in East Asia, Seoul, Korea, 26–31 August 1996.

Furth, Charlotte (1998). *A Flourishing Yin: Gender in China's Medical History 960–1665*. University of California Press, Berkeley, CA, in press.

Gallagher, Catherine & Laqueur, Thomas W. (eds.) (1987). *The Making of the Modern Body: Sexuality and Society in the Nineteenth Century*. University of California Press, Berkeley, CA.

Galt, Howard S. (1951). *A History of Chinese Educational Institutions*. Vol. I. *To the End of the Five Dynasties (A.D. 960)*. Probsthain's Oriental Series, 28. Probsthain, London. No further vols. published.

Garcia Ballester, Luis, McVaugh, M. R. & Rubio-Vela, A. (ed.) (1994). *Practical Medicine from Salerno to the Black Death*. Cambridge University Press, Cambridge.

Garrison, Fielding H. (1929a). *An Introduction to the History of Medicine. With Medical Chronology, Suggestions for Study, and Bibliographic Data*. 4th ed., Saunders, Philadelphia.

Garrison, Fielding H. (1929b). 'History of Drainage, Irrigation, Sewage-disposal, and Water-supply'. *BNYAM*, 5, 887.

Gatti, Angelo (+1763). *Lettre de M. Gatti à M. Roux [sur l'inoculation]*. N.p., Paris.

Gatti, Angelo (+1764). *Réflexions sur les préjugés qui s'opposent au progrès et à la perfection de l'inoculation*. 8 vols. Musier fils, Bruxelles.

Gatti, Angelo (+1767). *Nouvelles réflexions sur la pratique de l'inoculation*. Musier fils, Bruxelles. English tr. by M. Maty, Vaillant, London, +1768.

Gernet, Jacques (1959). *La vie quotidienne en Chine a la veille de l'invasion mongole 1250–1276*. Hachette, Paris. English tr. by H. M. Wright, *Daily Life in China on the Eve of the Mongol Invasion 1250–1276*. George Allen & Unwin, London, 1962.

Gibbs, F. W. (1957). 'Invention in Chemical Industries'. Art. in Singer *et al.* (1954–8), vol. 3, p. 676.

Giles, H. A. (tr.) (1874). 'The *Hsi yüan lu* or "Instructions to coroners"'. *CR*, 3, 30, 92, etc. Repr. *PRSM*, 1924, 17, 59.

Gmelin, Samuel Gottlieb (+1770–84). *Reise durch Russland zur Untersuchung der drey Natur-Reiche*. Kaiserlichen Academie der Wissenschaften, St Petersburg.

Goodall, E. W. (1937). 'Fracastorius as an Epidemiologist'. *PRSM*, 30, 341.

Goodrich, L. Carrington & Wilbur, C. M. (1942). 'Additional Notes on Tea'. *JAOS*, 62, 195.

Gordon, I. & Shapiro, H. A. (1975). *Forensic Medicine. A Guide to Principles*. Churchill Livingston, Edinburgh.

Gorsky, J. A. (1960). 'The History of Forensic Medicine'. *CCHG*, 58, 31.

Gradwohl, R. B. H. (ed.) (1954). *Legal Medicine*. Mosby, St Louis, MO. 3rd ed., ed. Francis E. Camps, Wright, Bristol, 1976.

Graham, A. C. (1979). 'How Much of Chuang tzu did Chuang tzu Write'? *JAAR*, 47, 3. Repr. in Graham (1990), pp. 283–321.

Graham, A. C. (1981). *Chuang-tzŭ. The Seven Inner Chapters and Other Writings from the Book Chuang-tzŭ*. George Allen & Unwin, London.

Graham, A. C. (1989a). *Disputers of the Tao. Philosophical Argument in Ancient China*. Open Court, La Salle, IL.

Graham, A. C. (1989b). 'A Neglected pre-Han Philosophical Text: Ho-kuan-tzu'. *BLSOAS*, 52, 3, 497.

Graham, A. C. (1990). *Studies in Chinese Philosophy and Philosophical Literature*. State University of New York Press, Albany, NY.

Grant, Joanna (1997). 'Wang Ji's *Shishan yi'an*: Aspects of Gender and Culture in Ming Dynasty Medical Case Histories'. Ph.D. diss., History, School of Oriental and African Studies, University of London.

Greene, L. B. (1962). 'Frontiers of Medical Jurisprudence. A Historical Background'. *CM*, 26, 1.

GREENHILL, W. A. (1848). *'A Treatise on the Small-pox and Measles' by Abú Becr Mohammed ibn Zacaríyá ar-Rází (commonly called Rhazes)*. . . . Sydenham Society, London.

DE GRIJS, C. F. M. (tr.) (1863). *Geregtelijke Geneeskunde, uit het Chinees vertaald.* Lange, Batavia.

GROS, H. (1902). 'La variolisation'. *JAN*, 7, 169.

GRUBE, H. (+1673). *De arcanis medicorum non arcanis commentatio.* 2 vols. Danielis Paulli, Hafnia (Copenhagen).

GRUBE, H. (+1674). *De transplantatione morborum. Analysis nova.* Gothofredum Schulze, Hamburg. See Klebs (1934), pp. 6, 65.

GRUBE, W. (tr.) (1912). *Die Metamorphosen der Götter.* 2 vols. Brill, Leiden. *Fêng shên yen i*, ch. 1–46, with summary of ch. 47–100.

VAN GULIK, R. H. (1956). *T'ang Yin Pi Shih. Parallel Cases from under the Pear-Tree. A Thirteenth-century Manual of Jurisprudence and Detection.* Sinica Leidensia, 10. Brill, Leiden.

HAAS, P. & HILL, T. G. (1928). *Introduction to the Chemistry of Plant Products.* 2 vols. Longmans Green, London.

HALL, A. R. (1989). 'Antoni van Leeuwenhoek, 1632–1723'. *NRRS*, 43, 249.

HALSBAND, R. (1953). 'New Light on Lady Mary Wortley Montagu's Contribution to Inoculation'. *JHMAS*, 7, 395.

HALSBAND, R. (1956). *The Life of Lady Mary Wortley Montagu.* Oxford University Press, Oxford.

HANSON, MARTA (1997). 'Inventing a Tradition in Chinese Medicine. From Universal to Local Medical Knowledge in South China, the Seventeenth to the Nineteenth Century'. Ph.D. diss., History and Sociology of Science, University of Pennsylvania.

HARINGTON, SIR JOHN (+1607). *The English mans docter. Or, the schoole of Salerne. Or, Physicall obseruations for the perfect preseruing of the body of man in continuall health.* I. Helme & I. Busby, London. Translation of *Regimen sanitatis salernitanum.*

HARPER, DONALD J. (1977). 'The Twelve Qin Tombs at Shuihudi, Hubei: New Texts and Archaeological Data'. *EARLC*, 3, 100. Review of Chinese publications with translations of examples.

HARPER, DONALD J. (1982). 'The "Wu shih erh ping fang": Translation and Prolegomena'. Ph.D. diss., Oriental Languages, University of California, Berkeley, CA.

HARPER, DONALD J. (1990). 'The Conception of Illness in Early Chinese Medicine, as Documented in Newly Discovered 3rd and 2nd Century B.C. Manuscripts (Part I)'. *AGMN*, 74, 210.

HARPER, DONALD J. (1998a). *Early Chinese Medical Literature: The Mawangdui Medical Manuscripts.* Sir Henry Wellcome Asian Series. Kegan Paul, London.

HARPER, DONALD J. (1998b). 'Warring States Natural Philosophy and Occult Thought'. Art. in *The Cambridge History of Ancient China*, ed. M. Loewe & E. Shaughnessy. Cambridge University Press, Cambridge.

HARRIS, WALTER (+1721). *De peste dissertatio, habita Apr. 17, 1721 in amphitheatro Collegii Regalis Medicorum Londiniensium diu accessit descriptio inoculationis variolarum.* Gul. & Joh. Innys, London. With appendix on Chinese and other methods of smallpox inoculation.

HART, H. L. A. (1961). *The Concept of Law.* Oxford University Press, Oxford.

HARTWELL, ROBERT M. (1982). 'Demographic, Political, and Social Transformations of China, 750–1550'. *HJAS*, 42, 2, 365.

HAWKES, DAVID (tr.) (1973–86). *Cao Xueqin. The Story of the Stone.* 5 vols. Penguin, Harmondsworth. Vols. 4–5 tr. John Minford.

HAYGARTH, JOHN (+1793). *A Sketch of a Plan to Exterminate the Casual Small-Pox from Great Britain, and to Introduce General Inoculation; To which is added, a Correspondence on the Nature of Variolous Contagion.* . . . 2 vols. J. Johnson, London.

HE ZHIGUO & LO, VIVIENNE (1996). 'The Channels: A Preliminary Examination of a Lacquered Figurine from the Western Han Period.' *EARLC*, 21, 81. Publ. 1997.

HENDERSON, D. A. (1976). 'The Eradication of Smallpox'. *SAM*, 235, 4, 25.

HENDERSON, JOHN (1991). *Scripture, Canon and Commentary. A Comparison of Confucian and Western Exegesis.* Princeton University Press, Princeton, NJ.

HILL, W. G. *et al.* (1986). 'DNA Fingerprint Analysis in Immigration Test-cases'. *N*, 322, 290.

HOEPPLI, R. (1959). *Parasites and Parasitic Infections in Early Medicine and Science.* University of Malaya Press, Singapore.

HOLBROOK, BRUCE (1981). *The Stone Monkey: An Alternative, Chinese-scientific, Reality.* Morrow, New York.

HSÜ, ELISABETH (1992). 'Transmission of Knowledge, Texts and Treatment in Chinese Medicine'. Ph.D. diss., Social Anthropology, University of Cambridge.

HSU, F. L. K. (Hsü Lang-kuang, 1952). *Religion, Science and Human Crises. A Study of China in Transition and its Implications for the West.* Routledge & Kegan Paul, London.

HU SHIU-YING (1980). *An Enumeration of Chinese Materia Medica.* Chinese University Press, Hong Kong. Standard source for English names.

HUANG, H. T. *et al.* (1988). 'Preliminary Experiments on the Identity of the Chiu shih 秋石 (Autumn mineral) in Medieval Chinese Pharmacopoeias'. Art. in *Abstracts*, Fifth International Conference on the History of Science in China, University of California, San Diego, CA, 5–10 August 1988, p. 182.

HUCKER, CHARLES O. (1985). *A Dictionary of Official Titles in Imperial China*. Stanford University Press, Stanford, CA.

HUGHES, THOMAS P. (1981). 'Convergent Themes in the History of Science, Medicine, and Technology'. *TCULT*, 22, 550.

HULSEWÉ, A. F. P. (1978). 'The Ch'in Documents Discovered in Hupei in 1975'. *TP*, 64, 4, 175, 338.

HULSEWÉ, A. F. P. (1985). *Remnants of Ch'in Law. An Annotated Translation of the Ch'in Legal and Administrative Rules of the Third Century B.C., Discovered in Yun-mêng Prefecture, Hupei Province, in 1975*. Sinica Leidensia, 17. Brill, Leiden.

HUMMEL, ARTHUR W. (ed.) (1943–4). *Eminent Chinese of the Ch'ing Period (1644–1912)*. 2 vols. US Government Printing Office, Washington, DC.

HUNNISETT, R. F. (1961). *The Mediaeval Coroner*. Cambridge University Press, Cambridge.

HYMES, ROBERT P. (1987). 'Not Quite Gentlemen? Doctors in Sung and Yüan', *CHIS*, 8, 9.

IDEMA, WILT (1977). 'Diseases and Doctors, Drugs and Cures. A Very Preliminary List of Passages of Medical Interest in a Number of Traditional Chinese Novels and Related Plays'. *CHIS*, 2, 37.

IMPERATO, P. J. (1968). 'The Practice of Variolation among the Songhai of Mali'. *TRSTMH*, 62, 6, 868.

IMPERATO, P. J. (1974). 'Observations on Variolation Practices in Mali'. *TGM*, 26, 429.

ISKANDAR, ALBERT Z. (1988). *Corpus medicorum graecorum supplementum orientale*. IV. *De optimo medico cognoscendo*. Akademie-Verlag, Berlin.

JAGGI, O. P. (1969–73). *History of Science and Technology in India*. 5 vols. Atma Ram, Delhi.

JEFFREYS, A. J., BROOKFIELD, J. F. Y. & SEMEONOFF, R. (1985). 'Positive Identification of an Immigration Test-case using Human DNA Fingerprints'. *N*, 317, 818.

JENNER, EDWARD (+1798). *An Inquiry into the Causes and Effects of the Variolae Vaccinae; a Disease discovered in some of the Western Counties of England, particularly Gloucestershire, and known by the Name of the Cow Pox*. Pr. for the author by Sampson Low, London.

JEWSON, NORMAN (1974). 'Medical Knowledge and the Patronage System in Eighteenth Century England'. *SOC*, 8, 369.

JOHNSTON, WILLIAM (1995). *The Modern Epidemic. A History of Tuberculosis in Japan*. Harvard East Asian Monographs, 162. Council on East Asian Studies, Harvard University, Cambridge, MA.

JURIN, JAMES (+1722). 'A Letter to the Learned Caleb Cotesworth . . . Containing a Comparison between the Mortality of the Natural Small Pox, and that Given by Inoculation'. *PTRS*, 32, 374, 213. Repr. W. & J. Innys, London, 1723.

JURIN, JAMES (+1724). *An Account of the Success of Inoculating the Small Pox in Great Britain. With a Comparison between the Miscarriages in that practice, and the Mortality of the Natural Small-pox*. J. Peele, London. 2nd ed. was also +1724.

KAHN, CHARLES (1963). 'History of Smallpox and its Prevention'. *AJDC*, 106, 597.

KANDEL, BARBARA (1979). '*Taiping Jing*. The Origin and Transmission of the "Scripture on General Welfare". The History of an Unofficial Text'. *MDGNVO*, 75, 1.

KARPLUS, H. (1973). *International Symposium on Society, Medicine and Law (Jerusalem, 1972)*. Elsevier, Amsterdam.

KATZ, PAUL R. (1995). *Demon Hordes and Burning Boats. The Cult of Marshal Wen in Late Imperial Chekiang*. State University of New York Press, Albany, NY.

KEEGAN, DAVID J. (1988). 'The "Huang-ti nei-ching": The Structure of the Compilation; the Significance of the Structure'. Ph.D. diss., History, University of California, Berkeley, CA.

KEELE, K. D. (1963). *The Evolution of Clinical Methods in Medicine*. Fitzpatrick Lectures, Royal College of Physicians, 1960–1. Pitman, London.

KELLAWAY, W. (1969). 'The Coroner in Mediaeval London'. Art. in *Studies in London History*, ed. W. Kellaway, p. 75. Hodder & Stoughton, London.

KENNEDY, PETER (+1715). *An Essay on External Remedies*. Andrew Bell, London.

KERR, D. J. A. (1956). *Forensic Medicine*. 6th ed. Black, London.

KINNEY, ANNE BEHNKE (ed.) (1995). *Chinese Views of Childhood*. University of Hawaii Press, Honolulu.

KIPLE, KENNETH (1993). *The Cambridge World History of Human Disease*. Cambridge University Press, Cambridge.

KIRKPATRICK, JAMES (+1743). *An Essay on Inoculation, Occasioned by the Small-pox being brought into South Carolina in the Year 1738. . . .* J. Huggonson, London.

KIRKPATRICK, JAMES (+1754). *The Analysis of Inoculation; Comprising the History, Theory, and Practice of it: with an Occasional Consideration of the Most Remarkable Appearances in the Small Pox*. J. Millan, London.

KITTREDGE, G. L. (1912). 'Some Lost Works of Cotton Mather'. *PMHS*, 45, 418.

VON KLAPROTH, HEINRICH JULIUS (1810). *Archiv für asiatische Litteratur, Geschichte und Sprachkunde*. Im Academischen Verlage, St Petersburg.

KLEBS, ARNOLD C. (1913a). 'The Historical Evolution of Variolation'. *JHHB*, 24, 69.

KLEBS, ARNOLD C. (1913b). 'A Bibliography of Variolation'. *JHHB*, 24, 83.

KLEBS, ARNOLD C. (1914). *Die Variolation im achtzehnten Jahrhundert. Ein historischer Beitrag zur Immunitätsforschung*. Zur historischen Biologie der Krankheitserreger. Materialen, Studien, und Abhandlungen, 7. Töpelmann, Giessen.

KLEBS, LUISE (1934). *Die Reliefs und Malereien des neuen Reiches (XVIII.–XX. Dynastie, ca. 1580–1100 v. Chr.). Material zur ägyptischen Kulturgeschichte*. Pt 1. *Szenen aus dem Leben des Volkes. AHAW/PH*, 9. 1.

KLEINMAN, ARTHUR (1980). *Patients and Healers in the Context of Culture. An Exploration of the Borderland between Anthropology, Medicine, and Psychiatry*. Comparative Studies of Health Systems and Medical Care, 3. University of California Press, Berkeley, CA.

KRISTELLER, PAUL OSKAR (1945). 'The School of Salerno. Its Development and its Contribution to the History of Learning'. *BIHM*, 17, 138.

KUHN, PHILIP A. (1990). *Soulstealers. The Chinese Sorcery Scare of 1768*. Harvard University Press, Cambridge, MA.

KURIYAMA, SHIGEHISA. (1995). 'Interpreting the History of Bloodletting'. *Journal of the History of Medicine and Allied Sciences*, 50, 1, 1.

LANGER, W. L. (1976). 'The Prevention of Smallpox before Jenner'. *SAM*, 234, 1, 112.

LAQUEUR, THOMAS W. (1990). *Making Sex: Body and Gender from the Greeks to Freud*. Harvard University Press, Cambridge, MA.

LARRE, CLAUDE & ROCHAT DE LA VALLÉE, ELISABETH (tr.) (1983). 'Plein ciel. Les authentiques de haute antiquité. Texte, présentation, traduction et commentaire du *Su Wen*, chap. I'. *Méridiens*, 61–2, 13.

LAU, D. C. (1982). *Chinese Classics. Tao Te Ching*. Chinese University Press, Hong Kong.

LAUFER, B. (1911). 'The Introduction of Vaccination into the Far East'. *OC*, 25, 525.

LAYTON, EDWIN T., Jr (1976). 'Technology and Science, or "Vive la petite différence"'. *PPSA*, 2, 139.

LAYTON, EDWIN T., Jr (1988). 'Science as a Form of Action: The Role of the Engineering Sciences'. *TCULT*, 29, 82.

LEAVITT, JUDITH WALZER (1990). 'Medicine in Context: A Review Essay of the History of Medicine'. *AHR*, 95, 5, 1471.

LEE, THOMAS H. C. (1985). *Government Education and Examinations in Sung China*. Chinese University Press, Hong Kong.

LEGGE, J. (1861). *The Chinese Classics. . . .* Vol. 1. *Confucian Analects, the Great Learning, and the Doctrine of the Mean*. Trübner, London.

LEGGE, J. (1872). *The Chinese Classics. . . .* Vol. 5, pts 1 and 2. *The Ch'un Ts'eu with the Tso Chuen*. Trübner, London. Translations of *Chhun chhiu, Tso chuan*.

LEGGE, J. (1885). *The Lĭ Kĭ*. SBE, 3. 2 vols. Oxford University Press, Oxford. Translation of *Li chi*.

LEGGE, J. (1891). *The Texts of Taoism*. 2 vols. Oxford University Press, Oxford. Contains *Tao tê ching, Chuang-tzu*, and four other titles.

LEPAGE, FRANÇOIS-ALBIN (1813). *Recherches historiques sur la médecine des Chinois*. Didot, Paris. Diss., Faculty of Medicine, Paris. Repr. Cercle sinologique de l'ouest, n.p. (1986?).

LEUNG, ANGELA KI CHE (1987). 'Organized Medicine in Ming-Qing China: State and Private Medical Institutions in the Lower Yangtze Region'. *LIC*, 8, 1, 134.

LI GUOHAO et al. (ed.) (1982). *Explorations in the History of Science and Technology in China. Compiled in Honour of the Eightieth Birthday of Dr. Joseph Needham, FRS, FBA*. Shang-hai Ku-chi, Shanghai.

LI YUN (1995). 'Aspects of the Doctor-Patient Relationship in Ancient China'. Art. in *Proceedings*, 14th International Symposium on the Comparative History of Medicine – East and West. Ishiyaku EuroAmerica, Tokyo, p. 55.

LINDBERG, DAVID C. (1992). *The Beginnings of Western Science. The European Scientific Tradition in Philosophical, Religious, and Institutional Context, 600 B.C. to A.D. 1450*. University of Chicago Press, Chicago. Textbook, mainly concerned with philosophy and mathematical sciences.

LITOLF, C. H. (1909–10). 'Le livre de la réparation des torts. Constations légales dans les cas de crimes contre des personnes en vue de la réparation du préjudice causé'. *RI*, 1909, 531, 676, 765, 881, 1017, 1107, 1217; 1910, 418. Separately pub., Imprimerie de l'Extrême-Orient, Hanoi, 1910.

LITTLE, A. G. & WITHINGTON, E. (1928). *Roger Bacon's 'De retardatione accidentium senectutis' cum aliis opusculis de rebus medicinalibus*. Pubs. Brit. Soc. Franciscan Studies, 14. Oxford University Press, Oxford.

LLOYD, G. E. R. (1987). *The Revolutions of Wisdom. Studies in the Claims and Practice of Ancient Greek Science*. Sather Classical Lectures, 52. University of California Press, Berkeley, CA.

LOCKHART, WILLIAM (1861). *The Medical Missionary in China. A Narrative of Twenty Years' Experience*. Hurst & Blackett, London.

LOEWE, MICHAEL (1968). *Everyday Life in Early Imperial China. During the Han Period 202 BC–AD 220*. Batsford, London.

LOEWE, MICHAEL (1993). *Early Chinese Texts. A Bibliographical Guide*. Society for the Study of Early China, Chicago.

LOEWE, MICHAEL (1997). 'The Physician Chunyu Yi and his Historical Background'. Art. in *En suivant la Voie Royale. Mélanges en hommage à Léon Vandermeersch*, ed. Jacques Gernet & Marc Kalinowski, p. 297. École Française d'Extrême-Orient, Paris. On Shun-yü I.

LONG, J. C. (1933). *Lord Jeffrey Amherst, Soldier of the King*. Macmillan, New York.

LOUDON, IRVINE (1986). *Medical Care and the General Practitioner, 1750–1850*. Clarendon Press, Oxford.

LU GWEI-DJEN & NEEDHAM, J. (1951). 'A Contribution to the History of Chinese Dietetics'. *ISIS*, 42, 1, 13. Submitted 1939.

LU GWEI-DJEN & NEEDHAM, J. (1963). 'China and the Origin of (Qualifying) Examinations in Medicine'. *PRSM*, 56, 63.

LU GWEI-DJEN & NEEDHAM, J. (1967). 'Records of Diseases in Ancient China'. Art. in *Diseases in Antiquity* ed. D. Brothwell & A. T. Sandison. Thomas, Springfield, IL, ch. 17.

LU GWEI-DJEN & NEEDHAM, J. (1980). *Celestial Lancets. A History and Rationale of Acupuncture and Moxa*. Cambridge University Press, Cambridge.

LU GWEI-DJEN & NEEDHAM, J. (1988). 'A History of Forensic Medicine in China'. *Medical History*, 32, 357.

LUCKIN, WILLIAM (1977). 'The Decline of Smallpox and the Demographic Revolution of the Eighteenth Century'. *SOCH*, 6, 793. Conference report.

MACDONALD, MICHAEL (1981). *Mystical Bedlam. Madness, Anxiety, and Healing in Seventeenth-century England*. Cambridge University Press, Cambridge. Study of the voluminous case records of a healer *ca.* +1600; revealing about the connections in therapy between learning and faith.

[MADDOX, ISAAC,] BISHOP OF WORCESTER (+1752). *A Sermon Preached before His Grace John, Duke of Marlborough, President, the Vice-Presidents and Governors of the Hospital for the Small-pox, and for Inoculation, at the Parish-Church of St. Andrew, Holborn, on Thursday March 5, 1732 . . .* H. Woodfall, London.

MAITLAND, CHARLES (+1723). *Mr. Maitland's Account of Inoculating the Small-pox*. J. Peele, London.

MAJOR, RALPH H. (1955). *Classic Descriptions of Disease. With Biographical Sketches of the Authors*. 4th ed. Thomas, Springfield, IL.

MALINOWSKI, BRONISLAW (1948). *Magic, Science and Religion, and Other Essays*, ed. Robert Redfield. Beacon Press, Boston, MA.

MANSON, PATRICK (1879). 'On Chinese Methods of Inoculation for Smallpox'. *CIMC/MR* (II, Special Series), 18, 59.

VON MARTELS, Z. R. W. M. (ed.) (1990). *Alchemy Revisited. Proceedings of the International Conference on the History of Alchemy at the University of Groningen. 17–19 April 1989*. E. J. Brill, Leiden.

MARTI-IBAÑEZ, FELIX (ed.) (1960). *Henry E. Sigerist on the History of Medicine*. MD Publications, New York.

MARTIN, ERNEST (1886). *Recueil des procédés au moyen desquels on lave quelqu'un d'une injure*. Leroux, Paris.

MASPERO, H. (1971). *Le Taoïsme et les religions chinoises*. Bibliothèque des histoires, 3. Gallimard, Paris. Collected posthumous papers; preface by M. Kaltenmark.

MASSEY, EDMUND (+1722). *A Sermon against the Dangerous and Sinful Practice of Inoculation, Preached at St. Andrew's, Holborn. . . .* W. Meadows, London.

MATHER, RICHARD B. (1976). *Shih-hsüo hsin-yü. A New Account of Tales of the World*. University of Minnesota Press, Minneapolis.

MATHER, RICHARD B. (1979). 'K'ou Ch'ien-chih and the Taoist Theocracy at the Northern Wei Court, 425–451'. Art. in Welch & Seidel (1979), p. 103.

MCGOWAN, D. J. (1884). 'The Introduction of Small-pox and Inoculation into China' in 'Report on the Health of Wenchow for the Half-year ended 31 March, 1884'. *CIMC/SS* 2, 27, 16.

MCKEOWN, THOMAS (1979). *The Role of Medicine: Dream, Mirage, or Nemesis?* 2nd ed. Princeton University Press, Princeton, NJ.

MCKNIGHT, BRIAN E. (1981). *The Washing Away of Wrongs*. Science, Medicine and Technology in East Asia, 1. Center for Chinese Studies, University of Michigan, Ann Arbor. Complete translation of *Hsi yüan chi lu*.

MCLEOD, KATRINA C. D. & YATES, ROBIN D. S. (1981). 'Forms of Chhin Law. An Annotated Translation of the *Feng chen shih*'. *HJAS*, 41, 111. On official forms for forensic examinations.

MCNEILL, WILLIAM H. (1977). *Plagues and Peoples*. Blackwell, Oxford.

MEAD, RICHARD (+1748). 'A Discourse on the Small Pox and Measles'. Repr. in *Medical Works of Dr. Richard Mead*. 2nd ed., 3 vols. A. Donaldson & J. Reid, Edinburgh, +1765. First publ. Leyden +1752?

MERTON, ROBERT K. (1938). 'Science, Technology and Society in Seventeenth Century England'. *OSIS*, 4, 360.

MIELI, ALDO (1938). *La science arabe, et son rôle dans l'évolution scientifique mondiale*. Brill, Leiden. Repr. Mouton, The Hague, 1966, with a bibliography and analytical index by A. Mazaheri.

MILLER, GENEVIÈVE (1956). 'Eighteenth-century Attempts to Attenuate Smallpox Virus – A Reappraisal'. *Actes*, VIIIe Congrès International d'Histoire des Sciences, Florence, 1956, vol. 2, p. 804.

MILLER, GENEVIÈVE (1957). *The Adoption of Inoculation for Smallpox in England and France*. University of Pennsylvania Press, Philadelphia.

MIYAJIMA, MIKINOSUKE (1923). 'The History of Vaccination in Japan'. *PRSM*, 16 (Hist. Med. Sect.), 23.

MIYASITA [MIYASHITA] SABURŌ (1976). 'A Historical Study of Chinese Drugs for the Treatment of Jaundice'. *AJCM*, 4, 3, 239.

MIYASITA SABURŌ (1977). 'A Historical Analysis of Chinese Formularies and Prescriptions. Three Examples'. *NIZ*, 23, 2, 283.

MIYASITA SABURŌ (1979). 'Malaria (*yao*) [i.e. *nüeh* 瘧] in Chinese Medicine during the Chin and Yüan Periods'. *ACTAS*, 36, 90.

MIYASITA SABURŌ (1980). 'An Historical Analysis of Chinese Drugs in the Treatment of Hormonal Diseases, Goitre, and Diabetes Mellitus'. *AJCM*, 8, 1, 17.

MONRO, ALEXANDER (+1765). *An Account of the Inoculation of Small Pox in Scotland*. Drummond & J. Balfour, Edinburgh. First of three physicians of this name.

MOORE, JAMES C. (1815). *The History of the Small Pox*. Longman, Hurst, Rees, Orme & Brown, London.

MORAN, BRUCE (ed.) (1991). *Patronage and Institutions. Science, Technology, and Medicine at the European Court, 1500–1750*. Boydell Press, Rochester, NY.

DE LA MOTTRAYE, AUBRY (+1727). *Voyages du sr. A. de La Motraye, en Europe, Asie & Afrique. Géographiques, historiques & politiques*. 2 vols. T. Johnson & J. van Duren, La Haye.

MOULE, A. C. (1921). 'The Wonder of the Capital'. *NCR*, 3, 12, 356. On two Sung books about Hangchow, *Tu chhêng chi shêng* 都城紀勝 and *Mêng Liang lu* 夢梁錄.

MÜNSTER, LADISLAO (1956). 'La Medicina legale a Bologna nel Quattrocento'. *Actes*, VIIIᵉ Congrès Internationale d'Histoire des Sciences. Florence, 1956, p. 687.

NAKAYAMA, SHIGERU & SIVIN, N. (eds.) (1973). *Chinese Science: Explorations of an Ancient Tradition*. MIT East Asian Science Series, 2. MIT Press, Cambridge, MA.

NAQUIN, SUSAN (1976). *Millenarian Rebellion in China. The Eight Trigrams Uprising of 1813*. Yale University Press, New Haven, CT.

NAQUIN, SUSAN (1981). *Shantung Rebellion. The Wang Lun Uprising of 1774*. Yale University Press, New Haven, CT.

NEEDHAM, J. (1964). *Time and Eastern Man*. The Henry Myers Lecture. *RAI/OP*, 1964. Repr. in Needham (1969), p. 218.

NEEDHAM, J. (1967). 'The Roles of Europe and China in the Evolution of Oecumenical Science'. *ADVS*, 24, 83.

NEEDHAM, J. (1969). *The Grand Titration. Science and Society in East and West*. George Allen & Unwin, London.

NEEDHAM, J. (1970). *Clerks and Craftsmen in China and the West. Lectures and Addresses on the History of Science and Technology*. Cambridge University Press, Cambridge.

NEEDHAM, J. (1980). 'China and the Origins of Immunology'. *EHOR*, 19, 1, 6. Abbreviated version of Huang Chan Lecture, University of Hong Kong; cf. Needham (1987).

NEEDHAM, J. (1983). 'Science, Technology, Progress, and the Breakthrough. China as a Case Study in Human History'. Lecture given at the Nobel Conference, Royal Swedish Academy, August.

NEEDHAM, J. (1987). 'China and the Origins of Immunology'. Art. in *Venezia e l'Oriente*, ed. L. Lanciotti. Olschki, Florence, p. 23.

NEEDHAM, J. & LIAO HUNG-YING (1948). 'The Ballad of Mêng Chiang-nü Weeping at the Great Wall'. *S*, 1, 194.

NEEDHAM, J. & LU GWEI-DJEN (1962). 'Hygiene and Preventive Medicine in Ancient China'. *JHMAS*, 17, 429.

NEEDHAM, J. & LU GWEI-DJEN (1969). 'Medicine and Culture in China'. Art. in Poynter (1969), p. 255, discussion, p. 285.

NEEDHAM, J. & LU GWEI-DJEN (1975). 'Manfred Porkert's Interpretations of Terms in Mediaeval Chinese Natural and Medical Philosophy'. *ANS*, 32, 491.

NEEDHAM, J., WANG LING, & PRICE, D. J. DE S. (1960). *Heavenly Clockwork. The Great Astronomical Clocks of Medieval China*. Cambridge University Press, Cambridge.

N[EEDHAM], M[ARCHAMONT] (+1665). *Medela medicinae. A Plea for the Free Profession, and a Renovation of the Art of Physick, out of the Noblest and Most Authentick Writers; shewing, the Publick Advantage of its Liberty; The Disadvantage that comes to the Publick by any sort of Physicians imposing upon the Studies and Practise of others; The Alteration of Diseases from their old State and Condition; the Causes of that Alteration; The Insufficiency and Uselessness of meer Scholastick Methods and Medicines, with a Necessity of New; Tending to the Rescue of Mankind from the Tyranny of Diseases, and of Physicians themselves, from the Pedantism of old Authors and present Dictators*. Lownds, London.

NEUGEBAUER, OTTO (1951). 'The Study of Wretched Subjects'. *ISIS*, 42, 111.

NGUYÊN TRÂN-HUÁN (1957). 'Biographie de Pien Tsio [Chhüeh]'. *BSEIC*, 32, 1, 59.

NIENHAUSER, WILLIAM H., JR (ed.) (1986). *The Indiana Companion to Traditional Chinese Literature*. Indiana University Press, Bloomington, IN.

NIIDA NOBORU (1950). 'The Industrial and Commercial Guilds of Peking and Religion and Fellow Countrymanship as Elements of their Coherence'. *FLS*, 9, 179.

NUTTON, VIVIAN (1988). 'Archiatri and the Medical Profession in Antiquity'. Art. in *From Democedes to Harvey. Studies in the History of Medicine*, Collected Studies Series, CS 277, Variorum, Aldershot, p. 191.

NUTTON, VIVIAN (1990). 'The Reception of Fracastoro's Theory of Contagion. The Seed that Fell among Thorns?' *OSIS*, 6, 196.

O'BOYLE, CORNELIUS (1994). 'Surgical Texts and Social Contexts: Physicians and Surgeons in Paris, *ca.* 1270 to 1430'. Art. in Garcia Ballester *et al.* (1994), p. 156.

OBRINGER, FRÉDÉRIC (1983). 'Les plantes toxiques du *Ben cao gang mu*'. Thèse pour la maîtrise de l'Institut National des Langues et Civilisations.

OBRINGER, FRÉDÉRIC (1997). *L'aconit et l'orpiment. Drogues et poisons en Chine ancienne et médiévale*. Penser la médecine, 4. Fayard, Paris.

OFFRAY DE LA METTRIE, J. (+1740). *Traité de la petite vérole, avec la manière de guérir cette maladie*. Huart, Briasson, Paris.

OVERMEYER, DANIEL L. (1976). *Folk Buddhist Religion. Dissenting Sects in Late Traditional China*. Harvard University Press, Cambridge, MA.

PANKHURST, R. (1965). 'The History and Traditional Treatment of Smallpox in Ethiopia'. *MH*, 9, 343.

PARÉ, AMBROISE (+1575). 'Traicté des rapports, et du moyen d'embaumer les corps morts'. Tractate in *Les oeuvres de M. Ambroise Paré, Conseiller et Premier Chirurgien du Roy. . . .* Buon, Paris. See Doe (1937).

PARISH, H. J. (1965). *A History of Immunisation*. Livingstone, Edinburgh.

PARISH, H. J. (1968). *Victory with Vaccines. The Story of Immunisation*. Livingstone, Edinburgh.

PEARSON, ALEXANDER (1805). See *Ying-chhi-li kuo hsin chhu chung tou chhi shu*.

PELLIOT, PAUL (1909). 'Notes de bibliographie chinoise. II. Le droit chinois'. *BEFEO*, 9, 123.

PHELPS, D. L. (1936). *A New Edition of the Omei Illustrated Guide Book [O shan t'u shuo or chih] by Huang Shou-Fu & T'an Chung-yo. . . .* West China Union University, Harvard-Yenching Institute Ser., 1. Jih-hsin Yin-shua Kung-yeh-shê, Chengtu.

DE POIROT, LOUIS. S. J. (tr.) (+1783). 'Instructions sublimes et familières de Cheng-Tzu-Quogen-Hoang-Ti'. *MCHSAMUC*, 9, 65. Introduction by the Yung-chêng emperor to the admonitions of his predecessor, Khang-hsi. Italian tr. from the Manchu by de Poirot, done into French by Mme. la Contesse de M**.

PORKERT, MANFRED (1974). *The Theoretical Foundations of Chinese Medicine. Systems of Correspondence*. MIT East Asian Science Series, 3. MIT Press, Cambridge, MA.

PORTER, ROY (ed.) (1985). *Patients and Practitioners. Lay Perceptions of Medicine in Pre-industrial Society*. Cambridge History of Medicine, 9. Cambridge University Press, Cambridge.

PORTER, ROY & PORTER, DOROTHY (1988). *In Sickness and in Health. The British Experience 1650–1850*. Fourth Estate, London.

PORTER, ROY & PORTER, DOROTHY (1989). *Patient's Progress. Doctors and Doctoring in Eighteenth-century England*. Polity Press, Cambridge.

POTHIER, R. J. & DE BRÉARD-NEUVILLE, M. (1818). *Pandectes de Justinien, mises dans un nouvel ordre. Avec les lois du Code et les nouvelles qui confirment, expliquent ou abrogent le Droit des Pandectes*. 25 vols. Dondy-Dupré, Paris.

POYNTER, F. N. L. (ed.) (1969). *Medicine and Culture. Proceedings of a Historical Symposium Organised Jointly by the Wellcome Institute of the History of Medicine, London, and the Wenner-Gren Foundation for Anthropological Research, New York*. Publications of the Wellcome Institute of the History of Medicine, n.s., 15. The Institute, London.

PRYNS, GWYN (1989). 'But What Was the Disease? The Present State of Health and Healing in African Studies'. *PP*, 124, 159.

PYLARINI, JACOB (+1715). *Nova et tuta variolas excitandi per transplantationem methodus; nuper inventa et in usum tracta; qua ritè peracta immunia in posterum praeservantur ab hujusmodi contagio corpora*. Jo. Gabrielem Hertz, Venice. At the time Pylarini was Venetian consul at Smyrna. Shorter version in *PTRS*, 1716, 29, 347, 393.

RAMSEY, MATTHEW (1987). *Professional and Popular Medicine in France, 1770–1830. The Social World of Medical Practice*. Cambridge History of Medicine. Cambridge University Press, Cambridge. Social context of important changes.

RANGER, TERENCE & SLACK, PAUL (ed.) (1996). *Epidemics and Ideas. Essays on the Historical Perception of Pestilence*. Cambridge University Press, Cambridge.

RAWSKI, EVELYN S. (1979). *Education and Popular Literacy in Ch'ing China*. University of Michigan Press, Ann Arbor, MI.

RAZZELL, PETER E. (1965a). 'Population Change in Eighteenth-century England. A Reappraisal'. *EHR*, 18, 131.

RAZZELL, PETER E. (1965b). 'Edward Jenner. The History of a Medical Myth'. *MH*, 9, 3, 216.

RAZZELL, PETER E. (1977a). *Edward Jenner's Cowpox Vaccine: The History of a Medical Myth*. Caliban, Firle, Sussex.

RAZZELL, PETER E. (1977b). *The Conquest of Smallpox. The Impact of Inoculation on Smallpox Mortality in Eighteenth-century Britain*. Caliban, Firle, Sussex.

READ, BERNARD E. (1936). *Chinese Medicinal Plants from the Pên Ts'ao Kang Mu* 本草綱目 *A.D. 1596 . . . a Botanical, Chemical and Pharmacological Reference List.* (Publication of the *Peking Natural History Bulletin*). Third ed., French Bookstore, Peiping. First ed., Peking Union Medical College, 1923. Compiled with Liu Ju-chhiang. Indexes and précis of botanical chapters of the *Pên tshao kang mu.*

RÉMUSAT, J. P. ABEL (1825–6). *Mélanges asiatiques. Ou, choix de morceaux de critique et de mémoires relatifs aux religions, aux sciences, aux coutumes, à l'histoire et à la géographie des nations orientales.* 2 vols. Dondey-Dupré, Paris.

RENOU, L. & FILLIOZAT, J. (1947–53). *L'Inde classique. Manuel des études indiennes.* Vol. 1, with the collaboration of P. Meile, A. M. Esnoul & L. Silburn, Payot, Paris. Vol. 2, with the collaboration of P. Demiéville, O. Lacombe & L. Silburn, Imprimerie Nationale, Paris.

RESTIVO, SAL P. (1979). 'Joseph Needham and the Comparative Sociology of Chinese and Modern Science'. *RSKSA*, 2, 25.

RIBEIRO SANCHES, ANTONIO NUNES (+1764). *De cura variolarum vaporarii ope apud Russos, omni memoria antiquioris usu recepti.* Ratione Medendi, Paris.

RIFAT OSMAN (1932). 'Sur l'inoculation antivariolique au 18ᵉ siècle'. *Comptes-rendus*, 9ᵉ Congrès International d'Histoire de la Médecine, Bucarest, 1932, p. 226.

ROSEN, GEORGE (1958). *A History of Public Health*. MD Monographs on Medical History, 1. MD Publications, New York.

ROSENBERG, CHARLES (1977). 'The Therapeutic Revolution: Medicine, Meaning and Social Change in Nineteenth-century America'. *PBM*, 20, 485.

ROSENBERG, CHARLES & GOLDEN, JANET (ed.) (1992). *Framing Disease. Studies in Cultural History*. Rutgers University Press, New Brunswick, NJ.

ROSENWALD, C. D. (1951). 'Variolation and Other Observations during the Smallpox Epidemic in the Southern Province of Tanganyika'. *MOFF*, March, 1.

DES ROTOURS, R. (1948). *Traité des fonctionnaires et traité de l'armée, traduites de la Nouvelle Histoire des Thang*. Bibl. de l'Inst. des Hautes Études Chinoises, 6. 2 vols. Brill, Leiden.

ROUSSEAU, GEORGE (ed.) (1990). *Languages of Psyche: Mind and Body in Enlightenment Thought*. University of California Press, Berkeley, CA.

RUFFER, M. A. & FERGUSON, A. R. (1911). 'Note on an Eruption Resembling That of Variola in the Skin of an Egyptian Mummy of the XXth Dynasty (−1200 to −1100)'. *JPB*, 15, 1. Repr. in Brothwell & Sandison (1967), p. 346.

RUSSELL, PATRICK (+1768). 'An Account of Inoculation in Arabia. In a Letter from Dr. Patrick Russell, Physician, at Aleppo, to Alexander Russel, M.D. F.R.S'. *PTRS*, 58, 140.

SABBAN, FRANÇOISE (1986). 'Court Cuisine in Fourteenth-century Imperial China. Some Culinary Aspects of Hu Sihui's Yinshan zhengyao'. *Food and Foodways*, 1, 161.

SAMBURSKY, S. (1956). *The Physical World of the Greeks*. Routledge & Paul, London.

SAMBURSKY, S. (1959). *The Physics of the Stoics*. Routledge & Paul, London.

SAMOGGIA, L. (1965). 'I medici della famiglia Zancari all'inizio del secolo XIV in Bologna'. *BSM*, 137, 99.

SARASOHN, LISA T. (1993). 'Nicolas-Claude Fabri de Peiresc and the Patronage of the New Science in the Seventeenth Century'. *ISIS*, 84, 1, 70.

SARTON, GEORGE (1927–48). *Introduction to the History of Science*. 3 vols. Williams & Wilkins, Baltimore, MD.

SCHAFER, EDWARD H. (1956). 'The Development of Bathing Customs in Ancient and Mediaeval China, and the History of the Floreate Clear Palace'. *JAOS*, 76, 57.

SCHAFER, EDWARD H. (1967). *The Vermilion Bird. T'ang Images of the South*. University of California Press, Berkeley, CA.

SCHIPPER, K. M. (1975). *Concordance du Tao-tsang. Titres des ouvrages*. Publications de l'École Française d'Extrême-orient, 102. École Française d'Extrême-orient, Paris.

VON SCHRÖTTER, E. (1919). 'Inoculation for Smallpox in India. A Literary Reference in the *Sacteya* of Dhanwantari'. *ZAF*, 16, 244.

SCHULTZ, SIMON (+1677). 'De modo emtionis variolarum ab infectis'. *MCMP* (Breslau ed. of 1678), 8, 22.

SCOTT, S. P. (1932). *The Civil Law. Including the Twelve Tables, the Institutes of Gaius, the Rules of Ulpian, the Opinions of Paulus, the Enactments of Justinian and the Constitutions of Leo. Translated from the Original Latin, Edited, and Compared with all Accessible Systems of Jurisprudence Ancient and Modern*. 17 vols. Central Trust Co., Cincinnati, OH.

SEIFFERT, G. & TU CHÊNG-HSING (1937). 'Zur Geschichte der Pocken und Pockenimpfung'. *AGMN*, 30, 26.

SHEPPARD, H. J. (1962). 'The Origin of the Gnostic-Alchemical Relationship'. *Scientia*, 97, 146.

SHEPPARD, H. J. (1981). 'Alchemy'. Art. in Bynum *et al.* (1981), p. 9.

SHERRINGTON, SIR CHARLES (1948). 'Sir Charles Sherrington and Diphtheria Antitoxin'. *N*, 161, 266.

SHRYOCK, RICHARD HARRISON (1936). *The Development of Modern Medicine. An Interpretation of the Social and Scientific Factors Involved.* University of Pennsylvania Press, Philadelphia. Repr., University of Wisconsin Press, Madison, WI, 1979.

SIGERIST, HENRY E. See Marti-Ibañez (1960).

SIMILI, ALESSANDRO (1969). 'Tre caratteristiche inquisizioni a Bologna nel secolo 14'. *EPI*, 3, 2, 115.

SIMILI, ALESSANDRO (1973). 'The Beginnings of Forensic Medicine in Bologna'. Art. in Karplus (1973), p. 91.

SIMILI, ALESSANDRO (1974). *Storia della Medicina Legale. EPI*, Monograph Series.

SIMON, J. (1857). *Papers Relating to the History and Practice of Vaccination.* N.p., London.

SIMPSON, D. (+1789). *A Discourse on Inoculation for the Smallpox.* N.p., Birmingham.

SINGER, C. (1913). *The Development of the Doctrine of Contagium vivum, +1500 to +1700.* Privately pr., London.

SINGER, C. (1957). *A Short History of Anatomy and Physiology from the Greeks to Harvey.* Dover, New York. Revised from *The Evolution of Anatomy.* Kegan Paul, Trench & Trubner, London, 1925.

SINGER, C. & SINGER, D. W. (1913). 'The Development of the Doctrine of Contagium vivum (1500 to 1700)'. *Proceedings*, XVIIth International Congress of Medicine, London, Sect. 23, p. 187.

SINGER, C. & SINGER, D. W. (1917). 'The Scientific Position of Girolamo Fracastoro with Special Reference to the Sources, Character, and Influence of his Theory of Infection'. *AMH*, 1, 1.

SINGER, C. *et al.* (1954–8). *A History of Technology.* 5 vols. Oxford University Press, Oxford.

SIVIN, N. (1968). *Chinese Alchemy: Preliminary Studies.* Harvard Monographs in the History of Science, 1. Harvard University Press, Cambridge, MA.

SIVIN, N. (1976). 'Chinese Alchemy and the Manipulation of Time'. *ISIS*, 67, 513. Repr. in Sivin (1977), p. 108.

SIVIN, N. (ed.) (1977). *Science and Technology in East Asia. Selections from Isis.* Science History Publications, New York.

SIVIN, N. (1978). 'On the Word Taoism as a Source of Perplexity. With Special Reference to the Relations of Science and Religion in Traditional China'. *HOR*, 17, 303.

SIVIN, N. (1979). 'Report on the Third International Conference on Taoist Studies'. *Bulletin*, Society for the Study of Chinese Religions, Fall, 7, 1.

SIVIN, N. (1982). 'Why the Scientific Revolution Did Not Take Place in China – Or Didn't It? The Edward H. Hume Lecture, Yale University, 1981'. *CHIS*, 5, 45.

SIVIN, N. (1987). *Traditional Medicine in Contemporary China. A Partial Translation of* Revised Outline of Chinese Medicine *(1972) with an Introductory Study on Change in Present-day and Early Medicine.* Center for Chinese Studies, University of Michigan, Ann Arbor, MI.

SIVIN, N. (1988). 'Science and Medicine in Imperial China – The State of the Field'. *JAS*, 47, 1, 41.

SIVIN, N. (1989). 'A Cornucopia of Reference Works for the History of Chinese Medicine'. *CHIS*, 9, 29.

SIVIN, N. (1990). 'Research on the History of Chinese Alchemy'. Art. in von Martels (1990), p. 3. Repr. in Sivin (1995b), ch. VIII.

SIVIN, N. (1991). 'Over the Borders: Technical History, Philosophy, and the Social Sciences'. *CHIS*, 10, 69, repr. in Sivin (1995a), ch. VIII.

SIVIN, N. (1993). 'Huang ti nei ching'. Art. in Loewe (1993), p. 196.

SIVIN, N. (1995a). *Science in Ancient China. Researches and Reflections.* Collected Studies Series, CS 506. Variorum, Aldershot.

SIVIN, N. (1995b). *Medicine, Philosophy, and Religion in Ancient China. Researches and Reflections.* Collected Studies Series, CS 512. Variorum, Aldershot.

SIVIN, N. (1995c). 'Text and Experience in Classical Chinese Medicine'. Art. in Bates (1995), p. 177.

SIVIN, N. (1995d). 'Taoism and Science'. Art. in Sivin (1995b), ch. VII.

SIVIN, N. (1995e). 'The Myth of the Naturalists'. Art. in Sivin (1995b), ch. IV.

SIVIN, N. (1995f). 'State, Cosmos, and Body in the Last Three Centuries B.C.'. *HJAS*, 55, 5.

SIVIN, N. (1995g). 'Emotional Counter-therapy'. Art. in Sivin (1995b), ch. II.

SIVIN, N. (1998). 'On the Dates of Yang Shang-shan and the *Huang-ti nei ching t'ai su*'. *CHIS*, 15, 29.

SKINNER, G. WILLIAM (ed.) (1977). *The City in Late Imperial China.* Studies in Chinese Society. Stanford University Press, Stanford, CA.

SKINNER, G. WILLIAM (1985). 'Presidential Address: The Structure of Chinese History'. *JAS*, 44, 2, 271. For debate see *JAS*, 45, 4, 721 and 48, 1, 90.

SKINNER, G. WILLIAM (1987). 'Sichuan's Population in the Nineteenth Century: Lessons from Disaggregated Data'. *LIC*, 8, 1, 1.

SLEESWYK, ANDRÉ WEGENER & SIVIN, N. (1983). 'Dragons and Toads. The Chinese Seismoscope of A.D. 132'. *CHIS*, 6, 1.

SLOANE, HANS, SIR (+1707–25). *A voyage to the islands Madera, Barbados, Nieves, S. Christophers and Jamaica, with the natural history . . . of the last of those islands; to which is prefix'd an introduction, wherein is an account of the inhabitants, air, waters, diseases, trade, &c. . . . Illustrated with the figures of the things describ'd.* 2 vols. B. M., London.

SMITH, F. PORTER (1871). *Contribution towards the Materia Medica and Natural History of China. For the Use of Medical Missionaries and Native Medical Students.* American Presbyterian Mission Press, Shanghai.

SMITH, KIDDER, JR *et al.* (1990). *Sung Dynasty Uses of the I Ching.* Princeton University Press, Princeton, NJ.

SMITH, SYDNEY (1951). 'The History and Development of Forensic Medicine'. *BMJ*, 1, 599.

SMITH, WESLEY D. (1979). *The Hippocratic Tradition.* Cornell University Press, Ithaca, NY.

SPENCE, JONATHAN (1975). *Emperor of China. Self-portrait of K'ang-hsi.* Knopf, New York.

VAN DER SPRENKEL, SYBILLE (1962). *Legal Institutions in Manchu China.* London School of Economics Monographs on Social Anthropology, 24. Athlone, London. 2nd ed., 1977.

STEARNS, R. P. & PASTI, G. (1950). 'Remarks upon the Introduction of Inoculation for Smallpox into England'. *BIHM*, 24, 107.

STEWARD, ALBERT N. (1930). *The Polygonaceae of Eastern Asia.* Contributions, 88. Gray Herbarium, Harvard University.

STRICKMANN, MICHEL (1977). 'The Mao Shan Revelations. Taoism and the Aristocracy'. *T'oung Pao*, 63, 1.

STRICKMANN, MICHEL (1981). *Le Taoïsme du Mao Chan. Chronique d'une révélation.* Mémoires de l'Inst. des Hautes Études Chinoises, 17. The Institute, Paris.

STUART, G. A. (1911). *Chinese Materia Medica. Vegetable Kingdom. Extensively revised from Dr. F. Porter Smith's Work.* American Presbyterian Mission Press, Shanghai. Expansion of Smith (1871).

SUN, E-TU ZEN (1961). *Ch'ing Administrative Terms. A Translation of The Terminology of the Six Boards with Explanatory Notes.* Harvard East Asian Studies, 7. Harvard University Press, Cambridge, MA.

TAMBIAH, S. J. (1968). 'The Magical Power of Words'. *Man*, n.s., 3, 2, 175.

TARRY, EDWARD (+1721). 'Letter to Sir Hans Sloane on Smallpox Inoculation at Aleppo during the Epidemic of 1706' (written 1 August +1721). British Library, Sloane MSS, 4061, fol. 164. Pr. in Sloane (+1707–25), pp. 517ff., and Moore (1815), pp. 230–2.

TAYLOR, F. S. (1957). *A History of Industrial Chemistry.* Heinemann, London.

TAYLOR, F. S. & SINGER, C. (1956). 'Pre-scientific Industrial Chemistry'. Art. in Singer *et al.* (1954–8), vol. 2, p. 347.

TENG SSU-YÜ (1943). 'Chinese Influence on the Western Examination System'. *HJAS*, 7, 267.

TENG SSU-YÜ (tr.) (1968). *Family Instructions for the Yen Clan. 'Yen shih chia hsün' by Yen Chih-t'ui (531–591).* T'oung Pao Monographs, 4. Brill, Leiden.

THOMAS, J. A. C. (1975). *The Institutes of Justinian. Text, Translation and Commentary.* North Holland, Amsterdam.

THOMAS, J. A. C. (1976). *Textbook of Roman Law.* North Holland, Amsterdam.

THOMPSON, LAURENCE G. & SEAMAN, GARY (1993). *Chinese Religions: Publications in Western Languages 1981 through 1990.* Association for Asian Studies, Ann Arbor, MI.

THORNDIKE, LYNN (1923–58). *A History of Magic and Experimental Science.* 8 vols. Columbia University Press, New York.

THORNDIKE, LYNN (1955). 'The True Place of Astrology in the History of Science'. *ISIS*, 46, 273.

TIMONE (OR TIMONI), EMANUEL[E] (+1712). 'Historia variolarum quae per insitionem excitantur'. Pr. in de la Mottraye (+1727) as Appendix I, vol. 2.

TIMONE, EMANUELE (+1714). 'Historia variolarum, quae per insitionem excitantur'. *AER* (Leipzig), 33, 382. Repr. *ACLCNCE* (Nuremberg), 1717, 5, obs. II. Also in 'An Account, or History, of the Procuring the Small Pox by Incision, or Inoculation, As it has for Some Time Been Practised at Constantinople' (letter of December 1713). *PTRS*, 29, 339, 72. MSS of Timone's account survive in Sweden, France and Germany. See Miller (1957), pp. 55ff.

TISSOT, P. A. (tr.) (1806). *Code et novelles de Justinien. Novelles de l'empereur Léon, fragments de Gaius, d'Ulpian et de Paul. . . .* Behmer, Metz.

TOMLINSON, GARY (1993). *Music in Renaissance Magic. Toward a Historiography of Others.* University of Chicago Press, Chicago.

TOPLEY, MARJORIE (1974). 'Cosmic Antagonisms: A Mother–child Syndrome'. Art. in Wolf (1974), p. 233.

TREASE, GEORGE EDWARD & EVANS, WILLIAM CHARLES (1989). *Trease and Evans' Pharmacognosy.* 13th ed. Baillière Tindall, London. 1st ed. 1902.

TWITCHETT, DENIS (ed.) (1979). *The Cambridge History of China.* Vol. 3. *Sui and T'ang China, 589–906, pt 1.* Cambridge University Press, Cambridge.

TWITCHETT, DENIS & LOEWE, MICHAEL (eds.) (1986). *The Cambridge History of China.* Vol. 1. *The Ch'in and Han Empires, 221 B.C.–A.D. 220.* Cambridge University Press, Cambridge.

TYLER, VARRO E. (1988). *Pharmacognosy.* 9th ed. Lea & Febiger, Philadelphia.

ULLMANN, M. (1978). *Islamic Medicine.* Islamic Surveys, 2. Edinburgh University Press, Edinburgh.

ULRICH, LAUREL THATCHER (1990). *A Midwife's Tale. The Life of Martha Ballard, Based on her Diary, 1785–1812.* Knopf, New York.

UNSCHULD, PAUL ULRICH (1978). *Medical Ethics in Imperial China. A Study in Historical Anthropology.* University of California Press, Berkeley, CA.

UNSCHULD, PAUL ULRICH (1985). *Medicine in China. A History of Ideas.* University of California Press, Berkeley, CA.

UNSCHULD, PAUL ULRICH (1986a). *Medicine in China. A History of Pharmaceutics.* University of California Press, Berkeley, CA.

UNSCHULD, PAUL ULRICH (1986b). *Medicine in China. Nan-Ching. The Classic of Difficult Issues.* University of California Press, Berkeley, CA.

UNSCHULD, PAUL ULRICH (ed.) (1988). *Approaches to Traditional Chinese Medical Literature. Proceedings of an International Symposium on Translation Methodologies and Terminologies.* Kluwer Academic Publishers, Boston, MA.

UNSCHULD, PAUL ULRICH (1989). *Forgotten Traditions in Ancient Chinese Medicine. The I-hsüeh Yüan Liu Lun of 1757 by Hsü Ta-Ch'un.* Paradigm Publications, Brookline, MA.

VEITH, ILZA (1954). 'Plague and Politics'. *BIHM*, 28, 408.

VERELLEN, FRANCISCUS (1989). *Du Guangting (850–933). Taoïste de cour à la fin de la Chine médiévale.* Mémoires, 30. Collège de France, Institut des Hautes Études Chinoises, Paris.

VESALIUS, ANDREAS (+1543). *De humani corporis fabrica libri septem.* Johannes Oporinus, Basel.

VOLK, PETER & WARLO, H. J. (1973). 'The Role of Medical Experts in Court Proceedings in the Mediaeval Town'. Art. in Karplus (1973), p. 101.

VOLLMER, H. (1938). 'Studies on the Biological Effects of Coloured Light'. *APTH*, 19, 197.

WALEY, ARTHUR (1938). *The Analects of Confucius.* George Allen & Unwin, London.

WALLS, H. J. (1974). *Forensic Science. An Introduction to Scientific Crime Detection.* 2nd ed. Sweet & Maxwell, London.

WALTER, REV. RICHARD (+1750). *Lord Anson's Voyage around the World, 1740 to 1744.* Society for Promoting Christian Knowledge, London.

WANG CHI-MIN & WU LIEN-TÊ (1936). *History of Chinese Medicine. Being a Chronicle of Medical Happenings in China from Ancient Times to the Present Period.* 2nd ed. National Quarantine Service, Shanghai. First published 1932.

WANG YI-T'UNG (tr.) (1984). *Record of Buddhist monasteries in Lo-yang.* Princeton University Press, Princeton, NJ.

WARD, W. (+1795). *A View of the History, Literature, and Religion of the Hindus.* 4 vols. n.p., London.

WARE, JAMES R. (1966). *Alchemy, Medicine, and Religion in the China of A.D. 320: The Nei P'ien of Ko Hung (Pao-p'u tzu).* MIT Press, Cambridge, MA.

WARRING STATES WORKING GROUP, UNIVERSITY OF MASSACHUSETTS, AMHERST, MA (1993–). 'Notes' and 'Queries'. Informally circulated.

WATERHOUSE, BENJAMIN (1800–2). *A Prospect of Exterminating the Smallpox.* William Hilliard, Cambridge, MA.

WEBER, MAX (1922/1964). *The Religion of China. Confucianism and Taoism,* tr. Hans H. Gerth. The Macmillan Company, New York, 1964. Translation from the German (Tübingen, 1922).

WEINSTEIN, STANLEY J. (1987). *Buddhism under the T'ang.* Cambridge University Press, Cambridge.

WELCH, HOLMES H. (1957). *The Parting of the Way. Lao Tzu and the Taoist Movement.* Beacon Press, Boston, MA.

WELCH, HOLMES H. & SEIDEL, ANNA (eds.) (1979). *Facets of Taoism. Essays in Chinese Religion.* Yale University Press, New Haven, CT.

WERNER, E. T. CHALMERS (1922). *Myths and Legends of China.* Harrap, London.

WERNER, E. T. CHALMERS (1932). *A Dictionary of Chinese Mythology.* Kelly & Walsh, Shanghai.

WESTMAN, ROBERT (1977). 'Magical Reform and Astronomical Reform: The Yates Thesis Reconsidered'. Art. in *Hermeticism and the Scientific Revolution,* ed. Robert Westman & J. E. McGuire, p. 1. William Andrews Clark Memorial Library, University of California, Los Angeles, CA.

WHITE, B. (1924). *Smallpox and Vaccination in the U.S.A.* Harvard University Press, Cambridge, MA.

WIDMER, ELLEN (1996). 'The Huanduzhai of Hangzhou and Suzhou: A Study in Seventeenth-century Publishing'. *HJAS*, 56, 1, 77.

WIEDEMANN, E. (1915). 'Beiträge zur Geschichte der Naturwissenschaften, XLV. Zahnärztliches bei den Muslimen'. *SPMSE*, 47, 127. Repr. in Wiedemann (1970), vol. 2, p. 181.

WIEDEMANN, E. (1970). *Aufsätze zur arabischen Wissenschaftsgeschichte.* 2 vols. Olm, Hildesheim. Repr. of 79 articles from *SPMSE*.

WILHELM, HELLMUT (1984). 'Notes on Some Sung Shih-hua'. Art. in *Sung Studies, In Memoriam Etienne Balasz.* Ser. II, Civilisation, 3. Hautes Études, Paris, p. 267.

WILHELM, RICHARD (1924/1950). *I ging. Das Buch der Wandlungen.* 2 vols. Diederichs, Jena. Eng. tr. C. F. Baynes. Bollingen Series, 19. 2 vols. Pantheon, New York, 1950.

WILHELM, RICHARD (1928). *Frühling und Herbst des Lü Bu We.* Diederichs, Jena. Translation of *Lü shih chhun chhiu.*

WILHELM, RICHARD (1930). 'Li Gi', das Buch der Sitte des älteren und jungeren Dai (i.e. both Li chi and Ta Tai li chi). Diederichs, Jena.

WILKINSON, ENDYMION P. (1972). 'Chinese Merchant Manuals and Route Books'. CSWT, 2, 3, 1.

WILLIAMS, PERROT (+1723). 'Part of Two Letters concerning a Method of Procuring the Small Pox, Used in South Wales. From Perrot Williams, M. D., Physician, Haverford West, to Dr. Samuel Brady, Physician to the Garrison at Portsmouth'; 'Part of a Letter from the Same Learned and Ingenious Gentleman, upon the Same Subject, to Dr. Jurin, Royal Society Secretary'. PTRS, 32, 262.

WILSON, C. ANN. (1984). 'Philosophers, Iosis, and Water of Life'. PLPLS/LH, 29, 5.

WISE, THOMAS ALEXANDER (1867). Review of the History of Medicine [among the Asiatic Nations]. 2 vols. J. Churchill, London.

WOLF, ARTHUR P. (ed.) (1974). Religion and Ritual in Chinese Society. Stanford University Press, Stanford, CA.

WONG, K. CHIMIN & WU LIEN-TEH (1936). See Wang Chi-min & Wu Lien-tê (1936).

WOODVILLE, WILLIAM (+1796). The History of Inoculation of the Smallpox in Great Britain. Comprehending a Review of All the Publications on the Subject. With an Experimental Inquiry into the Relative Advantages of Every Measure which has been deemed Necessary in the Process of Inoculation. Vol. 1. James Phillips, London. No further vols. published.

WRIGHT, ARTHUR F. (1979). 'The Sui Dynasty (581–617)'. Art. in Twitchett (1979), p. 48.

WRIGHT, RICHARD (+1723). 'A Letter on the same Subject ["buying the smallpox"] from Mr. R. W., Surgeon at Haverford West [to Mr. Sylvanus Bevan, Apothecary in London]'. PTRS, 32, 375, 267.

WRIGHT, WILMER C. (1930). Hieronymi Fracastorii De contagione et contagiosis morbis et eorum curatione, libri III. Translation and notes. History of Medicine Series, 2. G. P. Putnam, London.

WU LIEN-Tê (1931). 'The Early Days of Western Medicine in China'. JRAS/NCB, 9.

WU, PEI-YI (1990). The Confucian's Progress. Autobiographical Writings in Traditional China. Princeton University Press, Princeton, NJ.

WU YIYI (1993). 'A Medical Line of Many Masters: A Prosopographical Study of Liu Wansu and his Disciples from the Jin to the Early Ming'. CHIS, 11, 36.

WYLIE, A. (1867). Notes on Chinese Literature. Shanghai. Repr. of Shanghai 1922 ed., Vetch, Peiping, 1939.

YAMADA KEIJI (1991). 'Anatometrics in Ancient China'. CHIS, 10, 39.

YANG, LIEN-SHÊNG (1961). 'Schedules of Work and Rest in Imperial China'. Art. in Studies in Chinese Institutional History. Harvard-Yenching Institute Studies, 20. Harvard University Press, Cambridge, MA, p. 18.

YATES, FRANCES A. (1964). Giordano Bruno and the Hermetic Tradition. Routledge & Kegan Paul, London.

YATES, FRANCES A. (1979). The Occult Philosophy in the Elizabethan Age. Routledge & Kegan Paul, London.

YATES, FRANCES A. (1982). Lull and Bruno. Collected essays, 1. Routledge & Kegan Paul, London.

YATES, FRANCES A. (1983). Renaissance and Reform: The Italian Contribution. Collected Essays, 2. Routledge & Kegan Paul, London.

YATES, FRANCES A. (1984). Ideas and Ideals in the North European Renaissance. Collected Essays, 3. Routledge & Kegan Paul, London.

ZACCHIA, PAOLO (+1628). Quaestiones medico-legales. Guglielmo Facciotti, Rome.

VON ZAREMBA, R. W. (1904). 'Die Heilkunst in China. Eine geschichtliche Skizze'. JAN, 9, 103, 158, 201, 257.

ZHAO HONGJUN (CHAO HUNG-CHÜN) (1991). 'Chinese versus Western Medicine: A History of their Relations in the Twentieth Century'. CHIS, 10, 21.

ZITO, ANGELA & BARLOW, TANI E. (eds.) (1994). Body, Subject, and Power in China. University of Chicago Press, Chicago.

TABLE OF CHINESE DYNASTIES

	夏	HSIA kingdom (legendary?)		c. −2000 to c. −1520
	商	SHANG (YIN) kingdom		c. −1520 to c. −1030
	周	CHOU dynasty (Feudal Age)	Early Chou period	c. −1030 to −722
			春秋 CHHUN CHHIU period	−722 to −480
			戰國 Warring States (CHAN KUO) period	−480 to −221
First Unification	秦	CHHIN dynasty		−221 to −207
	漢	HAN dynasty	CHHIEN HAN (Earlier or Western)	−202 to +9
			HSIN interregnum	+9 to +23
			HOU HAN (Later or Eastern)	+25 to +220
First Partition	三國	SAN KUO (Three Kingdoms period)		+221 to +265
			蜀 SHU (HAN)	+221 to +264
			魏 WEI	+220 to +265
			吳 WU	+222 to +280
Second Unification	晉	CHIN dynasty	Western Chin	+265 to +317
			Eastern Chin	+317 to +420
	劉宋	(Liu) SUNG dynasty		+420 to +479
Second Partition		Northern and Southern Dynasties (NAN PEI CHHAO)	齊 CHHI dynasty	+479 to +502
			梁 LIANG dynasty	+502 to +557
			陳 CHHÊN dynasty	+557 to +589
			魏 { Northern (Thopa) WEI dynasty	+386 to +535
			Western (Thopa) WEI dynasty	+535 to +556
			Eastern (Thopa) WEI dynasty	+534 to +550
			北齊 Northern CHHI dynasty	+550 to +577
			北周 Northern CHOU (Hsienpi) dynasty	+557 to +581
Third Unification	隋	SUI dynasty		+581 to +618
	唐	THANG dynasty		+618 to +906
Third Partition	五代	WU TAI (Five Dynasty period)	(Later Liang, Later Thang (Turkic), Later Chin (Turkic), Later Han (Turkic) and Later Chou)	+907 to +960
			遼 LIAO (Chhitan Tartar) dynasty	+907 to +1124
			West LIAO dynasty (Qarā-Khiṭāi)	+1124 to +1211
			西夏 HSI HSIA (Tangut Tibetan) state	+986 to +1227
Fourth Unification	宋	SUNG dynasty		+960 to +1279
			宋 Northern SUNG dynasty	+960 to +1126
			宋 Southern SUNG dynasty	+1127 to +1279
			金 CHIN (Jurchen Tartar) dynasty	+1115 to +1234
	元	YÜAN (Mongol) dynasty		+1260 to +1368
	明	MING dynasty		+1368 to +1644
	清	CHHING (Manchu) dynasty		+1644 to 1911
	民國	Republic		1912
	人民共和國	People's Republic		1949

N.B. When no modifying term in brackets is given, the ruling house was Han. During the Eastern Chin period there were no less than eighteen independent States (Hunnish, Tibetan, Hsienpi, Turkic, etc.) in the north. The term 'Liu chhao' (Six Dynasties) is often used by historians of literature. It refers to a succession of southern dynasties with their capital at Nanking from the beginning of the +3rd to the end of the +6th centuries, including (San Kuo) Wu, Chin, (Liu) Sung, Chhi, Liang and Chhen.

ROMANISATION CONVERSION TABLES

BY ROBIN BRILLIANT

PINYIN/MODIFIED WADE-GILES

Pinyin	Modified Wade-Giles	Pinyin	Modified Wade-Giles	Pinyin	Modified Wade-Giles
a	a	chi	chhih	dui	tui
ai	ai	chong	chhung	dun	tun
an	an	chou	chhou	duo	to
ang	ang	chu	chhu	e	ê, o
ao	ao	chuai	chhuai	en	ên
ba	pa	chuan	chhuan	eng	êng
bai	pai	chuang	chhuang	er	êrh
ban	pan	chui	chhui	fa	fa
bang	pang	chun	chhun	fan	fan
bao	pao	chuo	chho	fang	fang
bei	pei	ci	tzhu	fei	fei
ben	pên	cong	tshung	fen	fên
beng	pêng	cou	tshou	feng	fêng
bi	pi	cu	tshu	fo	fo
bian	pien	cuan	tshuan	fou	fou
biao	piao	cui	tshui	fu	fu
bie	pieh	cun	tshun	ga	ka
bin	pin	cuo	tsho	gai	kai
bing	ping	da	ta	gan	kan
bo	po	dai	tai	gang	kang
bu	pu	dan	tan	gao	kao
ca	tsha	dang	tang	ge	ko
cai	tshai	dao	tao	gei	kei
can	tshan	de	tê	gen	kên
cang	tshang	dei	tei	geng	kêng
cao	tsho	den	tên	gong	kung
ce	tshê	deng	têng	gou	kou
cen	tshên	di	ti	gu	ku
ceng	tshêng	dian	tien	gua	kua
cha	chha	diao	tiao	guai	kuai
chai	chhai	die	dieh	guan	kuan
chan	chhan	ding	ting	guang	kuang
chang	chhang	diu	tiu	gui	kuei
chao	chhao	dong	tung	gun	kun
che	chhê	dou	tou	ha	ha
chen	chhên	du	tu	hai	hai
cheng	chhêng	duan	tuan	han	han

Pinyin	Modified Wade-Giles	Pinyin	Modified Wade-Giles	Pinyin	Modified Wade-Giles
hang	hang	kui	khuei	mu	mu
hao	hao	kun	khun	na	na
he	ho	kuo	khuo	nai	nai
hei	hei	la	la	nan	nan
hen	hên	lai	lai	nang	nang
heng	hêng	lan	lan	nao	nao
hong	hung	lang	lang	nei	nei
hou	hou	lao	lao	nen	nên
hu	hu	le	lê	neng	nêng
hua	hua	lei	lei	ng	ng
huai	huai	leng	lêng	ni	ni
huan	huan	li	li	nian	nien
huang	huang	lia	lia	niang	niang
hui	hui	lian	lien	niao	niao
hun	hun	liang	liang	nie	nieh
huo	huo	liao	liao	nin	nin
ji	chi	lie	lieh	ning	ning
jia	chia	lin	lin	niu	niu
jian	chien	ling	ling	nong	nung
jiang	chiang	liu	liu	nou	nou
jiao	chiao	lo	lo	nu	nu
jie	chieh	long	lung	nü	nü
jin	chin	lou	lou	nuan	nuan
jing	ching	lu	lu	nüe	nio
jiong	chiung	lü	lü	nuo	no
jiu	chiu	luan	luan	o	o, ê
ju	chü	lüe	lüeh	ou	ou
juan	chüan	lun	lun	pa	pha
jue	chüeh, chio	luo	lo	pai	phai
jun	chün	ma	ma	pan	phan
ka	kha	mai	mai	pang	phang
kai	khai	man	man	pao	phao
kan	khan	mang	mang	pei	phei
kang	khang	mao	mao	pen	phên
kao	khao	mei	mei	peng	phêng
ke	kho	men	mên	pi	phi
kei	khei	meng	mêng	pian	phien
ken	khên	mi	mi	piao	phiao
keng	khêng	mian	mien	pie	phieh
kong	khung	miao	miao	pin	phin
kou	khou	mie	mieh	ping	phing
ku	khu	min	min	po	pho
kua	khua	ming	ming	pou	phou
kuai	khuai	miu	miu	pu	phu
kuan	khuan	mo	mo	qi	chhi
kuang	khuang	mou	mou	qia	chhia

Pinyin	Modified Wade-Giles	Pinyin	Modified Wade-Giles	Pinyin	Modified Wade-Giles
qian	chhien	shu	shu	xian	hsien
qiang	chhiang	shua	shua	xiang	hsiang
qiao	chhiao	shuai	shuai	xiao	hsiao
qie	chhieh	shuan	shuan	xie	hsieh
qin	chhin	shuang	shuang	xin	hsin
qing	chhing	shui	shui	xing	hsing
qiong	chhiung	shun	shun	xiong	hsiung
qiu	chhiu	shuo	shuo	xiu	hsiu
qu	chhü	si	ssu	xu	hsü
quan	chhüan	song	sung	xuan	hsüan
que	chhüeh, chhio	sou	sou	xue	hsüeh, hsio
qun	chhün	su	su	xun	hsün
ran	jan	suan	suan	ya	ya
rang	jang	sui	sui	yan	yen
rao	jao	sun	sun	yang	yang
re	jê	suo	so	yao	yao
ren	jên	ta	tha	ye	yeh
reng	jêng	tai	thai	yi	i
ri	jih	tan	than	yin	yin
rong	jung	tang	thang	ying	ying
rou	jou	tao	thao	yo	yo
ru	ju	te	thê	yong	yung
rua	jua	teng	thêng	you	yu
ruan	juan	ti	thi	yu	yü
rui	jui	tian	thien	yuan	yüan
run	jun	tiao	thiao	yue	yüeh, yo
ruo	jo	tie	thieh	yun	yün
sa	sa	ting	thing	za	tsa
sai	sai	tong	thung	zai	tsai
san	san	tou	thou	zan	tsan
sang	sang	tu	thu	zang	tsang
sao	sao	tuan	thuan	zao	tsao
se	sê	tui	thui	ze	tsê
sen	sên	tun	thun	zei	tsei
seng	sêng	tuo	tho	zen	tsên
sha	sha	wa	wa	zeng	tsêng
shai	shai	wai	wai	zha	cha
shan	shan	wan	wan	zhai	chai
shang	shang	wang	wang	zhan	chan
shao	shao	wei	wei	zhang	chang
she	shê	wen	wên	zhao	chao
shei	shei	weng	wêng	zhe	chê
shen	shen	wo	wo	zhei	chei
sheng	shêng, sêng	wu	wu	zhen	chên
shi	shih	xi	hsi	zheng	chêng
shou	shou	xia	hsia	zhi	chih

Pinyin	Modified Wade-Giles	Pinyin	Modified Wade-Giles	Pinyin	Modified Wade-Giles
zhong	chung	zhuang	chuang	zou	tsou
zhou	chou	zhui	chui	zu	tsu
zhu	chu	zhun	chun	zuan	tsuan
zhua	chua	zhuo	cho	zui	tsui
zhuai	chuai	zi	tzu	zun	tsun
zhuan	chuan	zong	tsung	zuo	tso

MODIFIED WADE-GILES/PINYIN

Modified Wade-Giles	Pinyin	Modified Wade-Giles	Pinyin	Modified Wade-Giles	Pinyin
a	a	chhio	que	chua	zhua
ai	ai	chhiu	qiu	chuai	zhuai
an	an	chhiung	qiong	chuan	zhuan
ang	ang	chho	chuo	chuang	zhuang
ao	ao	chhou	chou	chui	zhui
cha	zha	chhu	chu	chun	zhun
chai	zhai	chhuai	chuai	chung	zhong
chan	zhan	chhuan	chuan	chü	ju
chang	zhang	chhuang	chuang	chüan	juan
chao	zhao	chhui	chui	chüeh	jue
chê	zhe	chhun	chun	chün	jun
chei	zhei	chhung	chong	ê	e, o
chên	zhen	chhü	qu	ên	en
chêng	zheng	chhüan	quan	êng	eng
chha	cha	chhüeh	que	êrh	er
chhai	chai	chhün	qun	fa	fa
chhan	chan	chi	ji	fan	fan
chhang	chang	chia	jia	fang	fang
chhao	chao	chiang	jiang	fei	fei
chhê	che	chiao	jiao	fên	fen
chhên	chen	chieh	jie	fêng	feng
chhêng	cheng	chien	jian	fo	fo
chhi	qi	chih	zhi	fou	fou
chhia	qia	chin	jin	fu	fu
chhiang	qiang	ching	jing	ha	ha
chhiao	qiao	chio	jue	hai	hai
chhieh	qie	chiu	jiu	han	han
chhien	qian	chiung	jiong	hang	hang
chhih	chi	cho	zhuo	hao	hao
chhin	qin	chou	zhou	hên	hen
chhing	qing	chu	zhu	hêng	heng

Modified Wade-Giles	Pinyin	Modified Wade-Giles	Pinyin	Modified Wade-Giles	Pinyin
ho	he	kao	gao	lieh	lie
hou	hou	kei	gei	lien	lian
hsi	xi	kên	gen	lin	lin
hsia	xia	kêng	geng	ling	ling
hsiang	xiang	kha	ka	liu	liu
hsiao	xiao	khai	kai	lo	luo, lo
hsieh	xie	khan	kan	lou	lou
hsien	xian	khang	kang	lu	lu
hsin	xin	khao	kao	luan	luan
hsing	xing	khei	kei	lun	lun
hsio	xue	khên	ken	lung	long
hsiu	xiu	khêng	keng	lü	lü
hsiung	xiong	kho	ke	lüeh	lüe
hsü	xu	khou	kou	ma	ma
hsüan	xuan	khu	ku	mai	mai
hsüeh	xue	khua	kua	man	man
hsün	xun	khuai	kuai	mang	mang
hu	hu	khuan	kuan	mao	mao
hua	hua	khuang	kuang	mei	mei
huai	huai	khuei	kui	mên	men
huan	huan	khun	kun	mêng	meng
huang	huang	khung	kong	mi	mi
hui	hui	khuo	kuo	miao	miao
hun	hun	ko	ge	mieh	mie
hung	hong	kou	gou	mien	mian
huo	huo	ku	gu	min	min
i	yi	kua	gua	ming	ming
jan	ran	kuai	guai	miu	miu
jang	rang	kuan	guan	mo	mo
jao	rao	kuang	guang	mou	mou
jê	re	kuei	gui	mu	mu
jên	ren	kun	gun	na	na
jêng	reng	kung	gong	nai	nai
jih	ri	kuo	guo	nan	nan
jo	ruo	la	la	nang	nang
jou	rou	lai	lai	nao	nao
ju	ru	lan	lan	nei	nei
jua	rua	lang	lang	nên	nen
juan	ruan	lao	lao	nêng	neng
jui	rui	lê	le	ni	ni
jun	run	lei	lei	niang	niang
jung	rong	lêng	leng	niao	niao
ka	ga	li	li	nieh	nie
kai	gai	lia	lia	nien	nian
kan	gan	liang	liang	nin	nin
kang	gang	liao	liao	ning	ning

Modified Wade-Giles	Pinyin	Modified Wade-Giles	Pinyin	Modified Wade-Giles	Pinyin
niu	niu	sang	sang	thê	te
no	nuo	sao	sao	thêng	teng
nou	nou	sê	se	thi	ti
nu	nu	sên	sen	thiao	tiao
nuan	nuan	sêng	seng, sheng	thieh	tie
nung	nong	sha	sha	thein	tian
nü	nü	shai	shai	thing	ting
o	e, o	shan	shan	tho	tuo
ong	weng	shang	shang	thou	tou
ou	ou	shao	shao	thu	tu
pa	ba	shê	she	thuan	tuan
pai	bai	shei	shei	thui	tui
pan	ban	shên	shen	thun	tun
pang	bang	shêng	sheng	thung	tong
pao	bao	shih	shi	ti	di
pei	bei	shou	shou	tiao	diao
pên	ben	shu	shu	tieh	die
pêng	beng	shua	shua	tien	dian
pha	pa	shuai	shuai	ting	ding
phai	pai	shuan	shuan	tiu	diu
phan	pan	shuang	shuang	to	duo
phang	pang	shui	shui	tou	dou
phao	pao	shun	shun	tsa	za
phei	pei	shuo	shuo	tsai	zai
phên	pen	so	suo	tsan	zan
phêng	peng	sou	sou	tsang	zang
phi	pi	ssu	si	tsao	zao
phiao	piao	su	su	tsê	ze
phieh	pie	suan	suan	tsei	zei
phien	pian	sui	sui	tsên	zen
phin	pin	sun	sun	tsêng	zeng
phing	ping	sung	song	tsha	ca
pho	po	ta	da	tshai	cai
phou	pou	tai	dai	tshan	can
phu	pu	tan	dan	tshang	cang
pi	bi	tang	dang	tshao	cao
piao	biao	tao	dao	tshê	ce
pieh	bie	tê	de	tshên	cen
pien	bian	tei	dei	tshêng	ceng
pin	bin	tên	den	tsho	cuo
ping	bing	têng	deng	tshou	cou
po	bo	tha	ta	tshu	cu
pu	bu	thai	tai	tshuan	cuan
sa	sa	than	tan	tshui	cui
sai	sai	thang	tang	tshun	cun
san	san	thao	tao	tshung	cong

Modified Wade-Giles	Pinyin	Modified Wade-Giles	Pinyin	Modified Wade-Giles	Pinyin
tso	zuo	tzhu	ci	yao	yao
tsou	zou	tzu	zi	yeh	ye
tsu	zu	wa	wa	yen	yan
tsuan	zuan	wai	wai	yin	yin
tsui	zui	wan	wan	ying	ying
tsun	zun	wang	wang	yo	yue, yo
tsung	zong	wei	wei	yu	you
tu	du	wên	wen	yung	yong
tuan	duan	wo	wo	yü	yu
tui	dui	wu	wu	yüan	yuan
tun	dun	ya	ya	yüeh	yue
tung	dong	yang	yang	yün	yun

INDEX

(1) The various parts of hypenated words are treated as separate words in the alphabetical sequence. It should be remembered that, in accordance with the convention adopted, some Chinese proper names are written as separate syllables, while others are written as one word.

(2) In the arrangement of Chinese words, Chh- and Hs- follow normal alphabetical sequence, and *ü* is treated as equivalent to *u*.

(3) References to footnotes are only given where the subject is not covered otherwise on the page. They are indicated by n suffix to the page number.

(4) Page numbers in *italics* refer to illustrations.